Annals of Theoretical Psychology

Volume 3

A Continuation Order Plan is available for this series. A continuation order will bring delivery of each new volume immediately upon publication. Volumes are billed only upon actual shipment. For further information please contact the publisher.

Annals of Theoretical Psychology

Volume 3

Edited by
K. B. MADSEN
Laboratory of General Psychology
The Royal Danish School of Educational Studies
Copenhagen, Denmark

and
LEENDERT P. MOS
Center for Advanced Study in Theoretical Psychology
University of Alberta
Edmonton, Alberta, Canada

Plenum Press • New York and London

The Library of Congress has catalogued this serial title as follows:

Annals of theoretical psychology. — Vol. 1- — New York, N.Y.: Plenum Press, 1984-
v. : ill. ; 23 cm.
Annual.
ISSN 0747-5241 = Annals of theoretical psychology.

1. Psychology—Philosophy—Periodicals.
BF38.A53 150′.5—dc19 84-644088
AACR 2 MARC-S

Library of Congress [8501]

ISBN-13: 978-1-4612-9507-5 e-ISBN-13: 978-1-4613-2487-4
DOI: 10.1007/978-1-4613-2487-4

Softcover reprint of the hardcover 1st edition 1985

A Division of Plenum Publishing Corporation
233 Spring Street, New York, N.Y. 10013

Contributors

Mario Bunge, Foundations and Philosophy of Science Unit, McGill University, Montreal, Quebec, Canada

M. C. Corballis, Department of Psychology, University of Auckland, Auckland, New Zealand

P. C. Dodwell, Department of Psychology, Queen's University, Kingston, Ontario, Canada

Hubert C. J. Duijker, Late of the Department of Psychology, University of Amsterdam, Amsterdam, The Netherlands

Edward Erwin, Department of Philosophy, University of Miami, Coral Gables, Florida

H. J. Eysenck, Institute of Psychiatry, University of London, De Crespigny Park, London, England

Michael E. Hyland, Department of Psychology, Plymouth Polytechnic, Plymouth, Devon, England

Richard F. Kitchener, Department of Philosophy, Colorado State University, Fort Collins, Colorado

Bo Larsson, The Swedish Psychoanalytical Society, Västerlånggatan 60, Stockholm, Sweden

Carl Lesche, Department of Psychology, University of Stockholm, Stockholm, Sweden

Stig Lindholm, Department of Education, Stockholm University, Stockholm, Sweden

Lars B. Lofgren, School of Medicine, University of California at Los Angeles, Los Angeles, California

K. B. Madsen, Laboratory of General Psychology, Royal Danish School of Educational Studies, Copenhagen NV-2400, Denmark

Willis F. Overton, Department of Psychology, Temple University, Philadelphia, Pennsylvania

Gerard Radnitzky, Department of Philosophy of Science, University of Trier, Trier D-5500, West Germany

H. V. Rappard, Department of Theoretical Psychology, Free University, Amsterdam, The Netherlands

Alexander Rosenberg, Departments of Philosophy and Social Science, Syracuse University, Syracuse, New York

Joseph R. Royce, Center for Advanced Study in Theoretical Psychology, University of Alberta, Edmonton, Alberta, Canada

C. Sanders, Department of Theoretical Psychology, Free University, Amsterdam, The Netherlands

Stephen P. Stich, Committee on the History and Philosophy of Science, University of Maryland, College Park, Maryland

Preface

For readers acquainted with the first two volumes it should come as no surprise that the manuscripts in Volume 3 have the inimitable style of their authors. While several of the manuscripts and commentaries by our nonnative English speaking authors required editorial adjustment, every effort was made to honor their lingual sensitivities. Indeed, it is one of the objectives of the *Annals* to preserve the distinctiveness of the contributor's text (with due respect for citation and reference requirements).

Plans for this volume began in late 1982 when Joseph Royce invited K. B. Madsen, whose work has been largely devoted to the comparison of theories, to edit a volume of the *Annals* devoted to metatheory. It was left to Professor Madsen to decide upon the major contributors and commentators. Professor Eysenck's contribution to the present volume complements a paper he contributed to Volume 1 of the *Annals* entitled "The place of individual differences in a scientific psychology." Together, these two papers constitute a succinct statement of his views on the discipline. Professor Madsen himself contributed to Volume 2 of the *Annals* a paper entitled "The hypothesis quotient: A quantitative estimation of the testability of a theory," which is an exemplary application of his "systematology" referred to in his *Introduction* to the present volume. Professor Bunge's contribution to the *Annals* is the first by a philosopher (fortunately several philosophers have served as commentators). We hope others will follow. (It was Professor Bunge who submitted the late Professor Dalbir Bindra's Volume 1 contribution, "Cognition: Its origin and future in psychology," to K. B. Madsen for inclusion in the present volume. Although the decision to place Professor Bindra's paper in our inaugural volume was a good one, it did deprive K. B. Madsen of a valuable addition to this collection on metatheory.) The contribution by our colleagues from the Netherlands is especially appropriate in view of their extensive writings on the topic available only in the Dutch language. Professor Sanders, who heads an institute expressly devoted to theoretical studies, sought contact with our Center at the University of Alberta some ten years ago. We are certain that, together with a recent visit here by Professor J. V. Rappard, their contribution will further serve to advance cooperation between our two centers. The difficult and penetrating work of Dr. Carl Lesche is perhaps not well known to North American scholars and his contribution here will surely begin to rectify this deficiency. In any case, the persistent reader will find here, together with the commentaries, a very rich perspective, not only on psychoanalysis but on psychology. Finally, Joseph Royce's pursuit of pluralism reflects a lifelong concern with this topic both in his theoretical and experimental work.

Again, there has been some delay in the publication of this volume, but less so than in the case of the two previous volumes. Most of the major manuscripts were available in either final or first draft form by July, 1983. Commentary and final drafts of the major manuscripts were completed about a year later. Illness prevented Dr. Lesche from translating his reply to commentators and we are grateful to Professor Herman Tennessen of our Center for doing so. H. C. J. Duijker died in 1983 just after completing his commentary on the Sanders and Rappard paper. Professor Sanders kindly prepared the biographical note that appears with Professor Duijker's commentary.

The University of Alberta extended to me the opportunity to be editorially active. Fortunately, I continue to enjoy the excellent assistance of Mrs. E. Murison and Mrs. F. Rowe, Center secretaries. Also gratefully acknowledged is the assistance of Mr. David Sharplin, University of Alberta librarian, in searching for references and citations. Professor Madsen's contribution is noted by his editorship and yet I must mention that his extensive correspondence in preparation of this volume was all accomplished in his own script!

LEENDERT P. MOS

Call for Papers and Commentary

Although space for the first five volumes is committed, readers are invited to submit papers and comments for subsequent volumes in this annual series.

Papers (up to 15,000 words) concerned with substantive theory, metatheory, or a mixture thereof are eligible. We are also soliciting commentary (up to 2,500 words) on previously published papers and commentaries.

Send the original and four carbon copies, following APA guidelines, to Leendert P. Mos or Joseph R. Royce, Editors, Center for Advanced Study in Theoretical Psychology, University of Alberta, Edmonton, Alberta, Canada, T6G 2E9.

Contents

1

Psychological Metatheory

An Introduction to Volume 3

K. B. Madsen

This volume contains papers and commentaries about the metatheoretical problems of psychology. Hence, we shall begin it with a definition of metatheory and a formulation of our conception of its relation to theoretical psychology in general. In order to do this, we must start with a definition of *theoretical psychology*. We shall summarize the various published definitions of theoretical psychology in this way: *Theoretical psychology can be defined as the metascientific study of psychological theories and theory-problems*. Thus, we have defined theoretical psychology as a sub-discipline or, rather, as an application of *metascience*. We must therefore define metascience and also relate metatheory and theoretical psychology to the more general concept, metascience. That is the subject of the first section of this introductory paper; the second section will present a systematic summary of the papers in this volume.

1. Metascience

1.1. A Concept of Science

1.1.1. Introduction

Before we define metascience it would be convenient to have a definition of *science*. This is a major problem and some people would prefer to postpone the formulation of such a definition until after a long metascientific exposition. But we prefer to start with a preliminary definition, which may later be revised.

Our conception of science contains the following three components.

K. B. Madsen • Laboratory of General Psychology, Royal Danish School of Educational Studies, Copenhagen NV-2400, Denmark.

1.1.2. Empirical Research

For most scientists and philosophers the word *science* means first and foremost empirical research, the results of which are descriptions of observations. Some philosophers of science believe that the concept of science should be exclusively identified with empirical research and its descriptions. This was especially the case with the nineteenth-century philosophers of science including August Comte and the continental positivists, as well as John Stuart Mill and the English empiricists.

After the First World War, however, this restrictive conception of science was enlarged by the logical empiricists (neopositivists) such as Bertrand Russell, Ludwig Wittgenstein, Rudolf Carnap, and other members of the Vienna circle.

1.1.3. Theoretical Thinking

The logical empiricists added theoretical thinking to empirical research as one of the components of science. The role of theoretical thinking should be to produce theories, which were conceived of as sets of testable hypotheses along with explanatory models. Therefore, according to this conception, there are two scientific processes: empirical research and theoretical thinking, which produce two kinds of scientific expositions: descriptions and theories.

This conception of science was dominant in the Western world until after the Second World War, when a new enlarged conception of science was developed by Karl Popper and his many—more or less critical—followers, such as Mario Bunge, Norwood Russell Hanson, Thomas S. Kuhn, Imre Lakatos, Michael Polanyi, Stephen Toulmin, Håkan Törnbohm, Gerard Radnitzky, and others.

1.1.4. Philosophical Thinking

These new philosophers of science added philosophical thinking to theoretical thinking and empirical research. The role of philosophical thinking is to produce a philosophical background or frame of reference, called *paradigm* by Kuhn and *metaphysical research program* by Popper. The author has used the term *metalevel* including sets of *metatheses* for this philosophical part of a scientific text. The metalevel may be divided into two subparts (see Figure 1):

1. *Philosophy of the world* includes the ontological (metaphysical) *world-hypotheses* and the overall *metamodel*. These philosophical world-hypotheses must be distinguished from the scientific hypotheses.

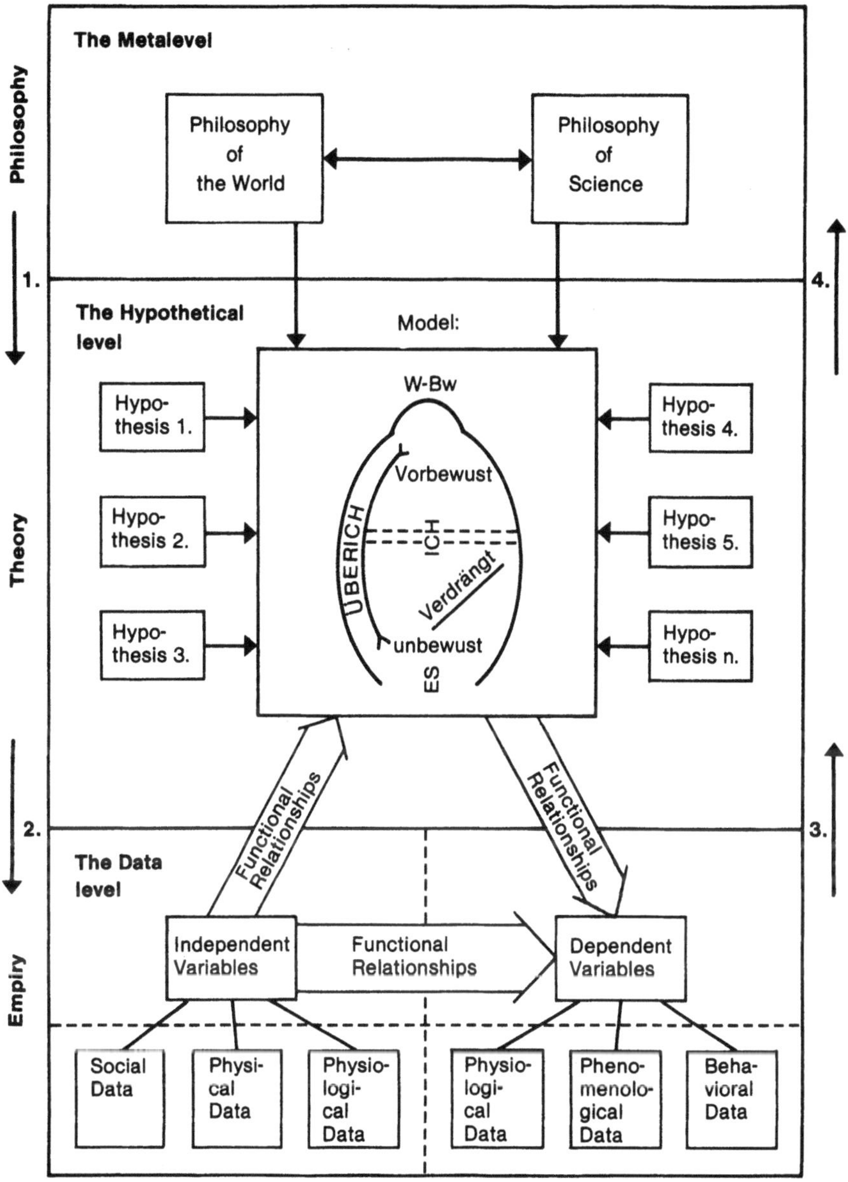

Figure 1. Our metatheory, called "systematology," conceives of a scientific text as consisting of three levels of abstraction: the *metalevel*, containing metatheses (i.e., propositions about the philosophy of the world and the philosophy of science); the *hypothetical level*, containing hypotheses and explanatory models; and the *datalevel*, containing datatheses (i.e., general functional relationships and specific descriptive propositions).

Scientific hypotheses, which belong to the hypothetical level of the scientific text, are testable, whereas philosophical world-hypotheses are not. However, in conjunction with a metamodel the world-hypotheses constitute the overall, generic background for empirical research and descriptions (the datalevel), as well as theoretical explanations and interpretations of the world (the hypothetical level).

2. *Philosophy of science* includes metatheses about epistemological, metatheoretical, and methodological problems. These metatheses are often formulated in a prescriptive language as *rules, norms,* or *ideals* for scientific research and the construction of scientific theories. This aspect of the metalevel has a guiding or directing role, related to the hypothetical level and the datalevel.

1.1.5. Definition of *Science*

We conclude this brief exposition of the development of the conception of science by a summarizing definition. Science can be defined as the social-cultural system of individuals who are engaged in empirical research, theoretical and philosophical thinking. It produces scientific texts, which in their complete versions include three levels of abstraction: the philosophical metalevel, the theoretical hypothetical level, and the empirical datalevel. We are now in a position to formulate our definition of *metascience.*

1.2. Definition of *Metascience*

1.2.1. Introduction

We will start with a preliminary and very brief definition of metascience as the general term for all studies pertaining to science. In accordance with our previous definition of science, we can also classify the metascientific disciplines on three levels of abstraction: the philosophical level, the theoretical level, and the empirical level. Each level contains one or more metascientific disciplines.

1.2.2. The Philosophy of Science

On the most abstract, philosophical metalevel of metascience we have the philosophy of science. This metascientific discipline derives from the philosophy of knowledge in general, that is, from epistemology. Philosophy of science therefore developed *after* the emergence of modern science in the Renaissance. It was especially Immanuel Kant,

August Comte, and John Stuart Mill who founded this branch of philosophy. However, the philosophy of science was not organized as an independent discipline with its own journals and professional societies until this century. Modern philosophy of science often deals with epistemological problems (of knowledge and truth in general) as well as with more specific problems concerning scientific theories and methods. Philosophical thinking about these problems often results in the formulation of prescriptions (rules, norms, or ideals) for how to construct scientific theories and use scientific methods in such a way that the theories and methods are accepted as genuinely scientific. It is by this prescriptive thinking that the philosophy of science can be distinguished from the next level, the hypothetical level of metascience, which also deals with scientific theories and methods, but in a more hypothetical or explanatory manner.

1.2.3. Metatheory (*Wissenschaftstheorie*)

In the years after the Second World War different kinds of theories about science developed. These theories were not exclusively based on the philosophy of science, but more on the empirical studies of science (such as the history of science and other disciplines which we shall deal with in the next paragraph). As examples of these new kinds of theories about science, we may mention Kuhn's well-known theory (1962), but also the theories of Lakatos (1970) and Hanson (1957). What these and other modern theories about science have in common is that they are based more upon the history of science than upon the philosophy of science, although the authors are also well versed in the philosophy of science. But the principal difference between these kinds of theories about science and the philosophy of science is that these modern theories about science are formulated as testable (scientific) theories, whereas the philosophy of science is often prescriptive in nature.

Therefore these modern theories about science belong to the theoretical or hypothetical level of the general category metascience. This level has relationships both to the philosophical level, which may have been the inspiration for their formulation, and to the empirical level, from which the data for testing the theories are drawn (Figure 2).

Before we turn our attention to the empirical studies of science, we must deal with another problem of terminology: What is a proper name for this modern "theory about science"? In Anglo-American literature the term *philosophy of science* is often used in such an encompassing manner that it includes both the philosophy of science in the narrow sense (belonging to the philosophical level) and theory about science

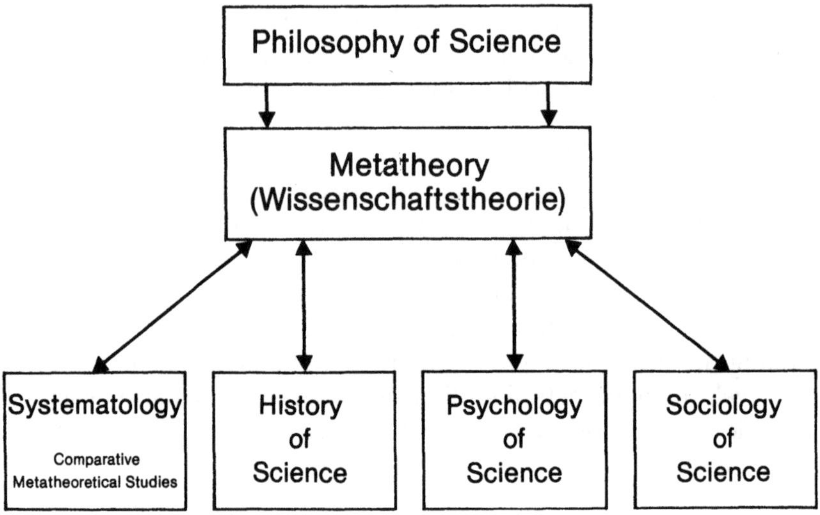

Figure 2. Three levels of metascience: 1. the philosophical level (philosophy of science), 2. the hypothetical level (metatheory), and 3. the empirical level (containing systematology, history of science, psychology of science, and sociology of science).

(belonging to the theoretical or hypothetical level). In German the term *Wissenschaftstheorie* is often used as equivalent to this modern theory about science (there are similar words in the Scandinavian languages.) The English term "theory *about* science" may be confused with "theory *of* science," which may be understood as "a scientific theory." Therefore, the author suggests that the term *metatheory* be used in English and in such a broad sense that it will include methodology. This broad meaning is equivalent to *Wissenschaftstheorie*. We may therefore conclude this paragraph with a definition: *Metatheory* can be defined as theory about scientific theories and methods, which may be inspired by philosophy of science and can be tested by empirical studies of science. We turn now to these empirical studies of science.

1.2.4. Empirical Sciences of Science

The empirical or datalevel of metascience includes several disciplines which are more or less well established as independent scientific disciplines (see Figure 2).

The oldest and most well established discipline is the history of science, which has been in existence for many years and has its own journals and professional societies. Even a subdiscipline of this discipline, the history of psychology, flourishes in textbooks, journals and societies.

Many modern metatheories (in the sense defined above) are based especially upon historical studies of science. This is the case with the theories of Hanson, Kuhn, and Lakatos.

Another empirical metascientific discipline is the sociology of science, established between the two world wars. Often the name "science of science" is used exclusively to refer to this discipline. But this is, of course, too narrow a use of this term since the (empirical) science of science includes several disciplines. In addition to the two already mentioned—the history and sociology of science—we shall mention some less well established disciplines.

The psychology of science is a metascientific discipline not yet organized into its own journals and professional societies. However, several empirical studies exist in this area, both of the psychology of scientific knowledge (for example, Maslow, 1966, and Royce, 1973) and the psychology of scientists' personality (Roe, 1953; Coan, 1979).

In Figure 2 we placed the psychology of science between the history of science and the sociology of science because two additional disciplines exist on its borders. Thus, between sociology of science and psychology of science we have a discipline called the "social psychology of science," which studies scientific teams. A well-known example is the work of Pelz (1958, 1964). Another metascientific discipline may be found between the history of science and the psychology of science, namely, the "psychobiography of science." A representative example is Gruber's study of Charles Darwin (1974).

The last metascientific discipline to be mentioned here is the comparative study of scientific theories. As the classic work in this discipline we may refer to the six volumes edited by a pioneer in theoretical psychology, namely, Sigmund Koch, entitled: *Psychology: A Study of a Science* (see Koch, 1959). This major endeavor in theoretical psychology analyzes about 80 psychological theories. The analyses are done either by the authors themselves or by another expert of the theory (in cases in which the author died). These analyses follow a common metatheoretical outline established by the editor. Unfortunately, the planned comparative volume seven was never published.

Inspired by Koch and other theoretical psychologists and philosophers of science, I made a comparative study of about 50 psychological theories (Madsen, 1959, 1974, 1975). I suggested the term *systematology* for these comparative studies of theories, the purpose of which was to contribute to a general metatheory. Since the term *theory* is used with very different meanings, I have suggested instead the use of the term *scientific text* in accordance with the following definition: A scientific text is a text which contains one, two, or three of the following levels of abstraction: the descriptive level, the hypothetical level and the philosophical level. With this definition of a scientific text, we can define systematology as the comparative, metatheoretical study of scientific texts.

This new metascientific discipline is closely related to the history of science, because scientific texts are also important materials for historical studies. The major difference between the history of science and systematology is that in historical studies other empirical materials are used in addition to scientific texts with the purpose of describing (and perhaps explaining) historical development or evolution, whereas the purpose of systematology is to contribute to a general metatheory.

I am at present working on a combined historical and systematological study of psychological theories in which the main texts in psychology are studied with systematological methods and the results are then organized into a historical frame of reference (inspired by Kuhn; see Madsen, in press).

1.3. Concluding Definitions

We presented a preliminary definition of metascience in the introduction and may now conclude with a more extensive definition which summarizes the presentation of the various metascientific disciplines.

Metascience can be defined as a common term for all those studies that have science as their object and which are organized on one or more of the following three levels of abstraction:

1. the philosophical (often prescriptive) level, containing the philosophy of science
2. the theoretical (hypothetical) level, containing metatheory (in a broad sense, including methodology)
3. the empirical level, containing the history of science, the psychology of science, the sociology of science, and systematology

A note about the term systematology: The author introduced the term *psychological systematology* in *Theories of Motivation* (Madsen, 1959),wherein the term was defined as "the study of psychological systems (theories)." Later, the term *systematology* was used in Madsen (1974) and as the title of a book (in Danish, subtitled *A Comparative Metatheory for Psychologists*, Madsen, 1975).

Still later, the author discovered that the term had been used by other authors. The Austrian philosopher Franz Körner used it in *Die Anarchie der philosophischen Systeme* in 1929. According to a symposium on systematology published in *Metaphilosophy*, Vol. 13, No. 3 and 4, July/

October 1982, Körner used *systematology* as a name for the comparative metaphilosophical studies of philosophical systems.

This is parallel to our use of systematology as a name for the comparative metatheoretical studies of scientific theories. However, the term was used even earlier by the German philosopher J. Heinrich Lambert in his posthumously published *Philosophische Schriften* (Band I, 1782, Band II, 1787). Volume II contains a "Fragment einer Systematologie" (pp. 385–413). This is a rough outline of what today would be called a general systems theory, because it deals with both physical systems, political systems, and intellectual systems (including scientific theories). Thus the term *systematology* has been used for at least two hundred years with different but overlapping meanings: essentially, the comparative, metatheoretical study of scientific texts (theories).

We defined theoretical psychology in the introduction as the metascientific study of psychological theories and theory-problems. This definition does not need extensive reformulation; rather, we may present a shorter version. Thus, the "metascientific study of psychological theories and theory-problems" could be referred to by the phrase *psychological metatheory*. On this account, the definition of theoretical psychology is the same as psychological metatheory. According to this definition the history of psychology and other empirical studies of psychology (except systematology) do not belong to theoretical psychology unless their purpose is (as is that of systematology) to contribute to psychological metatheory.

Using these definitions, we shall try in the next section to organize the five contributions to this volume.

2. The Content of This Volume

2.1. Introduction

This section will present the content of the five papers in this volume utilizing our three-level classification of the texts into the philosophical, theoretical, and empirical levels.

2.2. The Philosophical Level

2.2.1. C. Sanders and H. V. Rappard

The major part of this paper is on the philosophical level. In the first section on scientific knowledge the authors present a structural epistemology, a theory about the acquisition of knowledge. In the second section on science and reality they present an ontology derived from the work of the British philosopher P. F. Strawson. This ontology is the basis for a classification of the science into natural and humanistic sciences. In addition, a classification of sciences and theories according to "structural levels" is presented, but this belongs more to our next level of abstraction.

2.2.2. Mario Bunge

This paper starts with a short presentation of Bunge's philosophy of science. It includes:

1. *a general outlook*, including a so-called naturalistic ontology (a materialism or neutral monism), a realistic epistemology, and a research ethics
2. *a problematics* in terms of which all cognitive problems may be regarded as scientific problems
3. *a set of aims*, namely, the description, explanation, and prediction on the basis of data and laws
4. *a methodics*, referring to a set of scientific methods and justifiable techniques or tactics

This philosophy of science is used to evaluate three schools of psychology: behaviorism, mentalism, and neuropsychology, but this evaluation belongs to the next level of abstraction.

2.2.3. Carl Lesche

This paper presupposes a philosophy of science created by the Frankfurt school, especially Karl-Otto Apel's version. The main feature of this philosophy is the distinction among:

1. *the natural sciences*, the aims of which are explanation, prediction, and technical application
2. *the hermeneutic-humanistic sciences*, the aims of which are interpretation and understanding (especially self-understanding)
3. *the critical social sciences*, the aims of which are critical analysis and emancipation

This philosophy of science is applied to psychoanalysis, but again that belongs to the next level of abstraction.

2.2.4. H. J. Eysenck and Joseph R. Royce

These two papers belong mainly to the next level of abstraction, to which we now turn.

2.3. The Theoretical Level

2.3.1. C. Sanders and H. V. Rappard

On the theoretical (that is, metatheoretical and methodological) level, these two authors present several classifications of the sciences in addition to those already mentioned in connection with the analysis at the philosophical level. They specify:

1. the normative versus the pure sciences
2. systematic versus field psychology

Furthermore, within systematic psychology they distinguish among three structural levels (and three kinds of theories):

1. the level of basic psychological functions, represented by mechanistic theories (e.g., Hull and Freud)
2. the level of persons represented by organismic theories (e.g., Bertalanffy and Piaget)
3. the level of social behavior represented by humanistic and personalistic theories

These three types of theories have different aims (explanation versus interpretative understanding) and different methods of testing (experiments versus clinical observation). Many other metatheoretical and methodological problems are considered in the Sanders and Rappard paper, but we refer the reader to the paper itself.

2.3.2. Mario Bunge

On the theoretical level, Bunge applies his philosophy of science as outlined above. This philosophy of science is used for an analysis of behaviorism, mentalism, and neuropsychology.

Behaviorism, which is scientific in a very restricted sense, is characterized as follows:

1. The general outlook: a restricted naturalism and primitive realism
2. Problematics: restricted to behavior
3. Aims: only descriptions and predictions, no deeper explanations
4. Methodics: restricted to behavioral observations

Mentalism, which is nonscientific, is characterized as follows:

1. The general outlook: mentalism occurs in two slightly different versions: (a) substantialism, which presupposes an immaterial substance interacting with the brain (e.g., John Eccles' theory) and (b) functionalism, in which not substance but form (organization) is important. The mind is regarded as a program for a complicated information-processing machine. Functionalist mentalism is the most popular version, according to Bunge. Both versions are alike with regard to the following:
2. Problematics: the domain of all psychological problems
3. Aims: understandings or explanations of the function of the mind, but without general laws
4. Methodics: introspective methods and intuition

Neuropsychology (or psychobiology) is characterized as follows:

1. The general outlook: a naturalistic ontology and a realistic epistemology
2. Problematics: all problems addressed by both behaviorism and mentalism and also the problems of development
3. Aims: descriptions, predictions, and explanations on the basis of general laws
4. Methodics: to start with problems and then formulate hypotheses and models tested by experiments. In addition, Bunge argues for a systems orientation, favoring systemic integration from below.

Had Bunge's analysis included more concrete examples from the history of psychology, we would have placed it on the empirical (metascientific) level, but as it is restricted to three idealized types of psychologies, it properly belongs to the theoretical (metatheoretical) level.

2.3.3. Carl Lesche

This paper belongs mainly to the theoretical (that is, metatheoretical) level. Lesche presupposes the philosophy of science put forward by the Frankfurt school and applies the metatheory proposed by the Swedish metascientist Håkon Törnebohm.

The results of Lesche's analysis state that psychoanalysis is both a therapeutic technique and a form of scientific research. As scientific research, it is a hermeneutic humanistic science and not a natural science.

The theories of psychoanalysis may be divided into clinical theories and metapsychological theories. *Metapsychology* may be regarded as the metatheory of the clinical theories.

We have placed the main part of Lesche's paper on the metatheoretical level, because it is conducted in a rather abstract way. However, Lesche has (see Lesche & Stjernholm Madsen, 1976) conducted a more empirical (metascientific) analysis of Freud's *Introductory Lectures* since as a practicing psychoanalyst he is very familiar with psychoanalysis as therapeutic technique.

2.3.4. H. J. Eysenck

This paper is largely based upon historical studies of the evolution of psychology and the sciences in general. Therefore, much of the paper belongs to the empirical (metascientific) level. However, a metatheory (belonging to the theoretical level) is also presented. The main thesis of this metatheory is that there is an evolution in science from "weak" to "strong" theories, and this evolution is accompanied by *an evolution in the philosophy of science from positivism through logical empiricism to Popper and Kuhn.*

Strong theories are characterized as (1) based upon many observations, (2) integrating laws from different areas, (3) having simple mathematical laws, and (4) making precise predictions. Such strong theories may be tested by negative testing (falsification), whereas weak theories may be tested positively (by verification).

In addition to the classification of theories along this weak-strong continuum, Eysenck also introduces a classification of theories in the form of generality. The most general theories are kinds of weltanschauung, which may contain components of political ideology. Such an ideological weltanschauung may strongly influence the formulation of more specific theories and empirical research. For example, he points to the concepts of equality and environmentalism which inhibited both Russian and American psychology. We shall return to other historical examples in Eysenck's paper below.

2.3.5. Joseph R. Royce

The major thesis of Royce's paper is "that theoretical pluralism is a characteristic of all sciences, both mature and immature, but that it is particularly characteristic of immature sciences such as psychology." Royce admits that theoretical pluralism creates problems in psychology and hence that we "need a method for appraising theories in order to determine their degree of theoretical power." His proposed method of appraisal is called *dialectical analysis* and may be used in the following way.

The first step is to distinguish between complementary and competitive theories. Competitive theories may be especially difficult to compare (they may even be incommensurables, as they may presuppose different philosophies of science—or paradigms in Kuhn's sense—with different sets of standards for evaluation); however, Royce claims that dialectical analysis (the comparison of theories) can be conducted by appealing to a metatheory (and philosophy of science) which may be the basis for arbitration. He demonstrates how the dialectical analysis can proceed, but this demonstration belongs to the next level of abstraction.

2.4. The Empirical Level

2.4.1. Introduction

It may be convenient for the reader to be reminded that the words "empirical level" in our terminology, applied to metascience, refer to *the empirical level of metascience,* including data from the history of science, the psychology of science, and other empirical, metascientific disciplines (see Figure 2).

The most thoroughly developed of these empirical, metascientific disciplines is the history of science. It is also the only one represented in the papers included in this volume. Only two of our papers, those by Eysenck and Royce, present material belonging to the empirical level of metascience. The remaining papers are mainly on the theoretical level of metascience (that is, the level of metatheory and methodology) and on the philosophical level of metascience (that is, the level of the philosophy of science).

2.4.2. H. J. Eysenck

Eysenck claims that the two disciplines of psychology, experimental general psychology and correlational differential psychology, have developed without much communication, to the great disadvantage of both. He further avers that it is possible to integrate the two disciplines so that psychology may produce general laws with "built-in" consideration of individual differences. As a concrete example of this integration, he discusses the possibility of using measurements of reaction time and evoked potentials as indicators of genetically determined differences in intelligence. Eysenck further demonstrates how American and Soviet egalitarianism and ideological environmentalism have had an inhibiting influence on the formulation of theories of intelligence and empirical research in this area. Thus, Eysenck's metatheory is supported by data from the history of psychology.

2.4.3. Joseph R. Royce

Royce supports his metatheory empirically by historical data from the evolution of factor analytical theories of personality and intelligence. Thus, theories by Spearman, Thurstone, Thomson, Cattell, Eysenck, and Guilford are metatheoretically analyzed and compared dialectically for theoretical power. In the case of factor analytical theories of intelligence, he finds that these theories are competitive, whereas the factor analytical theories of personality are complementary. After further analysis, Guilford's theory is the only theory of intelligence which is retained. Among the theories of personality, those of Cattell, Eysenck, and Guilford are retained as complementary and ready for a synthesis. Thus, Royce both supports his metatheory by historical examples and demonstrates the usefulness of his method of dialectical analysis.

Table 1. Analytical Classification of the Content of the Papers in this Volume

Authors	Concentrated restatement of main theses
	Philosophical level (i.e., philosophy of science)
1. Bunge	Science = naturalistic ontology, realistic epistemology; aims; descriptions, predictions, and explanations; testable methods.
2. Lesche	Sciences are naturalistic, hermeneutic, or critical-social.
3. Sanders and Rappard	Science = structured knowledge acquisition.
	Theoretical level (i.e., metatheory and methodology)
1. Bunge	Neuropsychology is the only scientific psychology.
2. Eysenck	Scientific evolution goes from weak to strong theories.
3. Lesche	Psychoanalysis is a hermeneutic science.
4. Royce	Scientific evolution is multitheoretical competition.
5. Sanders and Rappard	Psychological theories belong to different structural levels.
	Empirical level (i.e., empirical, metascientific disciplines)
1. Eysenck	The evolution of theories about intelligence and personality supports the evolutionary metatheory.
2. Royce	Dialectical analysis of factor analytical theories supports the pluralistic metatheory.

2.4.4. Conclusion

Although we have presented summaries of all five major papers in this volume, we have not referred to the highly valuable commentaries. Instead, we refer the reader to the commentaries themselves and conclude here with a "summary of the summaries" in the classification scheme (presented in Table 1).

3. References

Coan, R. W. (1979). *Psychologists: Personal and theoretical pathways.* New York: Irvington.

Gruber, H. E. (1974). *Darwin on man: A psychological study of scientific creativity.* New York: Dutton.

Hanson, N. R. (1957). *Patterns of discovery.* Cambridge: Cambridge University Press.

Koch, S. (Ed.). (1959–1963). *Psychology: A study of a science* (Vols. 1–6). New York: McGraw-Hill.

Kuhn, T. S. (1962). *The structure of scientific revolutions.* Chicago: University of Chicago Press.

Lakatos, I. (1970). Falsification and the methodology of scientific research-programs. In I. Lakatos & A. Musgrave (Eds.), *Criticism and the growth of knowledge* (pp. 91–196). Cambridge: Cambridge University Press.

Lambert, J. H. (1967, 1969). *Philosophische Schriften.* Berlin: Vol. 1, 1782, Vol. 2, 1787. Photographic Reprint. Hildesheim: Georg Olms Verlagsbuch-handel.

Lesche, C., & Stjernholm Madsen, E. (1976). *Psykoanalysens videnskabsteorie.* Copenhagen: Munksgaard.

Madsen, K. B. (1959). *Theories of motivation.* Copenhagen: Munksgaard. (4th ed., 1968).

Madsen, K. B. (1974). *Modern theories of motivation.* Copenhagen: Munksgaard and New York: Wiley.

Madsen, K. B. (1975). *Systematology: Sammenlignende Videnskabteori.* Copenhagen: Munksgaard.

Madsen, K. B. (in press). *A history of psychology in metascientific perspective.* Copenhagen: The National Research Foundation.

Madsen, K. B. (in press). *Comparative metatheory for psychologists.* New York: Ablex Publishing.

Maslow, A. H. (1966). *The psychology of science.* New York: Harper & Row.

Pelz, D. (1958). Social factors in the motivation of engineers and scientists. *School, Science, and Mathematics, 58,* 417–429.

Pelz, D. (1964). Freedom in research. *International Science and Technique, 31* 54–66.

Roe, A. (1953). *The making of a scientist.* New York: Dodd, Mead.

Royce, J. R. (1973). The present situation in theoretical psychology. In B. B. Wolman (Ed.), *Handbook of general psychology* (pp. 8–21). Englewood Cliffs: Prentice-Hall.

2

The Place of Theory in a World of Facts

H. J. Eysenck

Abstract. When it is frequently said, following Kuhn, that social science in general, and psychology in particular, is in a preparadigmatic phase, this may be interpreted to mean that there are no widely accepted general theories covering important areas. Psychology and other social sciences appear to suffer from the added disadvantage that not only are such theories and paradigms lacking, but professional members of these groups often regard this lack of theory as a virtue and proclaim a lack of interest in theories in general, adopting a low-level sort of empirical pragmatism.

This paper argues for the vital importance of theory in psychology and gives a number of examples to demonstrate the empirical value of such theories in gaining new and better knowledge. To quote Lewin's famous saying: "There is nothing more practical than a good theory." This is extended to empirical research as well as to practical application, and it is suggested that psychology should be more interested in theories, generate theories more readily, and come to grips with the general importance of theories in scientific work. Only in this way, it is suggested, can psychology join the ranks of the properly accredited sciences and take its rightful place.

1. The Developmental Concept of Scientific Theories

Psychology as a whole has not been very hospitable to theory, and relatively little thought has gone into the problem of just what the function of theory might be in science generally and in psychology in particular, how best such theories might serve the purposes of the psychologist, and whether it is desirable or even possible to dispense with theory (Bergman, 1951; Spence, 1944). Lewin (Marrow, 1969) is indeed credited with the statement, "There is nothing as practical as a good theory," but even he did not discuss in any detail just what made

H. J. Eysenck • Institute of Psychiatry, University of London, De Crespigny Park, London SE5 8AF, England.

a theory "good," or indeed what kind of theories might be particularly useful in psychology.[1]

Philosophers, of course, have debated such issues as these endlessly although seldom with psychology in mind (Suppe, 1974). As the contributors to this volume make clear, logical positivism and the "received view" are now decisively rejected by philosophers of science, and it is curious to think that it is precisely these rejected views which are perhaps the only ones widely known and accepted amongst psychologists (Bergman & Spence, 1941). What is this received view? Shortly after the First World War, philosophers of science began to construe scientific theories as axiomatic calculae which were given a partial observation interpretation by means of correspondence rules. For the most sophisticated presentation of this view, see Carnap (1962, 1966) and Hempel (1952, 1965, 1966). See also Tarski (1941, 1956). It was Putnam (1962) who first referred to this type of analysis as the received view on theories, and, as Suppe points, out virtually every significant result obtained in the philosophy of science between the 1920s and 1950s either employed or tacitly assumed the received view (Bergman, 1954, 1957; Harré, 1975; Quine, 1962). Then we have the period during which a number of attacks were mounted challenging the very conception of theories in scientific knowledge, beginning perhaps with Toulmin (1953).

Some of these attacks were directed at specific features of the received view, such as the notion of partial interpretation, and the observational-theoretical distinction (Achinstein, 1965; and Putnam, 1962). Other critics advanced alternative philosophies of science which rejected the received view out of hand, and proceeded to argue for some other conception of theories in scientific knowledge (Hanson, 1958; Toulmin, 1953). As Suppe (1974) points out:

> These attacks were so successful that by the late 1960s the general consensus had been reached among philosophers of science that the Received View was inadequate as an analysis of scientific theories; derivatively, the analyses of other aspects of the scientific enterprise (for example, explanation) erected upon the Received View became suspect and today are subject to much criticism. At the same time the various proposed alternatives to the Received View have been subjected to strong critical attack, and none of them has gained general acceptance among philosophers of science. (p. 4)

The received view, of course, was the product of logical positivism and it is odd that it survived long after logical positivism itself had been rejected. An explanation might be that positivism had tried to force all

[1]I am indebted to Imre Lakatos for enlightening discussions on this topic; but for his untimely death I would have benefitted even more from his incisive comments.

empirical knowledge into a scientific mold, and it was possible to reject positivism as a general epistemology on the grounds that not all empirical knowledge was like scientific knowledge. Rejecting logical positivism as the general epistemology was compatible with the willingness to concede that positivism was adequate as an analysis of scientific knowledge, and logical positivism thus became a philosophy of science and continued to be acceptable as the philosophy dealing with a restricted range of empirical knowledge, namely, scientific knowledge. However that may be, there is no doubt that at the moment philosophy of science is in a state of turmoil; having rejected both logical positivism and the received view, it is desperately searching for a new unifying theory of scientific theories (Feyerabend, 1975).

This search is only of marginal interest to psychologists, and instead of following Suppe and others in scrutinizing the ramifications of alternative theories, we might better examine not theories of science but theories within science. However, it is difficult to do this without having some general point of view which coordinates one's thinking, and this is difficult to achieve without at the same time having some idea of just what constitutes a science, as opposed to nonscience, pseudoscience, and so forth. Thus it is difficult not to devote a few words to what Popper (1959, 1974a, 1974b) calls the "demarcation" dispute, that is, the problem of what is the distinguishing mark of science, as opposed to the various disciplines which claim to be scientific (Popper singles out astrology, psychoanalysis, and Marxism) but are not. Popper's views are too well known to repeat them here; so are those of Lakatos (1968), Kuhn (1962), and many others. Here I will rather depart from present-day controversies (Lakatos & Musgrave, 1970). Most of the controversies deal with wholly developed sciences, particularly physics and astronomy; I think from the point of view of the psychologist it is more useful to think of the different metatheories of science as taking a developmental course, rather than as being opposed to each other in a fundamental sort of way. Figure 1 illustrates the view I have put forward before (Eysenck, 1976b).

The view there expressed runs counter to the usual assumption that scientific theories, and theories about the nature of scientific theories, have a universal application, but this is almost certainly not so. Scientific concepts develop in the course of history, and different methods of investigation may be appropriate at different stages. As Figure 1 illustrates, usually development starts with ordinary observation and induction; on the basis of these, the investigator develops a hunch that certain features of the observations may be invariant, that is, the sun might rise again tomorrow because in the past it has always risen again after setting.

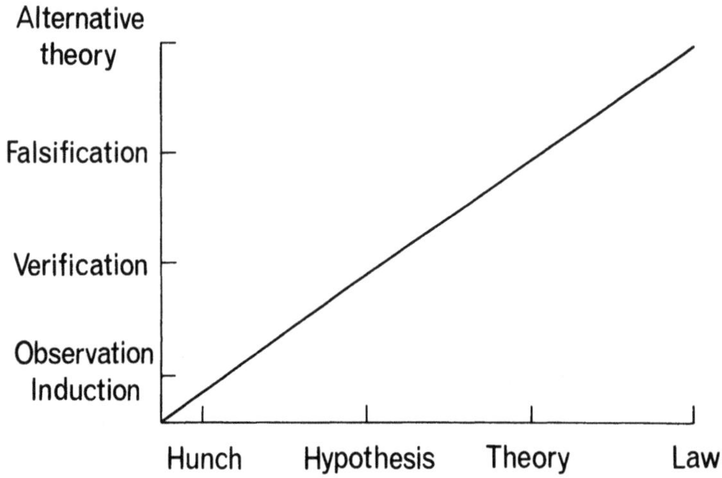

Figure 1. Changes in the nature of scientific theories corresponding to the level of development of a particular science. From Eysenck, 1976b.

Gradually limited hypotheses are formed, for instance, that the sun is moving around the earth, or vice versa. At this early stage, verification is sought of such hypotheses, and falsification is not very important, there are so many areas of ignorance that apparent falsification may not be as destructive to the hypothesis as it might be at a later stage. (A failure to observe stellar parallax did not render Copernicus's heliocentric hypothesis nugatory.) Gradually hypotheses become more firmly established, and related ones are seen to have certain features in common; out of these related hypotheses a theory is formed, such as Newton's theory of gravitation. Such a theory is highly specific in its predictions, and consequently falsification becomes important, although even at this stage simple falsification is not enough to overthrow a theory, as Lakatos has shown. Gradually theory develops into law; we tend to refer to theories which have become well established as natural or scientific *laws*. Falsification of laws is almost anathema; the anomalies in the precession of the perihelion of Mercury were known for centuries, but they were not admitted as disproof of Newton's laws. What is required is a Kuhnian revolution, in the form of an alternative theory; it needed Einstein's theory of relativity to overthrow Newton's theory. Falsification in the simple factual sense was not enough.

We can see that this developmental concept of scientific theory, and the nature of scientific thought and conceptualization, embraces all the various attempts to demarcate science as opposed to nonscience. The earliest observation–induction phase clearly corresponds very largely to the stress Bacon laid on these features of the scientific approach, and this stage may thus justifiably be called Baconian. The second stage, of hypothesis formation and verification with the stress on the latter, might be labeled the logical positivism stage, applicable to a higher level of organization and observation at which specific hypotheses are put forward and the attempt is to verify them. Once we get to the stage of wide-ranging theories, Popperian methods of falsification assume greater importance; it is taken for granted at this stage that many deductions have been and can be verified, and then it becomes more important to seek for falsification.[2] Finally, once these wide-ranging theories have become laws they become a paradigm in the Kuhnian sense, and replacement of one paradigm by another implies a revolution and a marked discontinuity in the theoretical concepts used (Krige, 1980).

Much of the debate between philosophers of science has centered on which of these different approaches is the *correct* one; from the point of view of the practicing scientist, however, there cannot be a *correct* answer to his quest for guidance in this choice of theories. This choice, and the conceptualization of science as a developing enterprise, requires rather that attention be paid to the stage of development of a given science; this will determine the kind of hunches, hypotheses, and theories which are appropriate to that science at that time. All sciences begin at a level which would be rejected by logical positivists and Popperians alike; there is a simple scrabbling for facts of the most elementary kind, uncoordinated and ill-defined. Gradually science pulls itself up by its bootstraps, a process which may not be pretty but which has received rather less attention from philosophers of science than have the later, rather more coordinated stages. Nevertheless, from the point of view of a very young science (like psychology) it is precisely the earlier stages that are of much interest, and it is doubtful whether the psychologist can receive much help from the philosophical conceptualizations of the more advanced sciences. There are, of course, links between all these stages, but there are also marked differences, and it behooves us to pay attention to these differences as well as to the similarities.

It might be thought that the stages in these developments illustrated in Figure 1 relate to an increase in rigor, in the sense that theories and confirmations which might pass muster at an early stage of theory development might not do so at a later stage. This certainly is the view of Popper, who regarded psychoanalysis as outside the scientific pale, not

[2]Actually, as Grünbaum (1976) has pointed out, Popper has misinterpreted Bacon, who was as much a falsificationist as Popper himself. I have here used the names of Bacon and others simply to denote in a rather simplified manner positions in the scheme of Figure 1, regardless of the fact that actually their thoughts were more complex than would fit easily into such a simple scheme.

because it failed inductively but rather because it did not produce theories which could be falsified and hence failed to accord with his demarcation principle. As Grünbaum (1976, 1977, 1979, 1981) has pointed out, the boot is on the other foot. Freud's theory would pass by Popper's criterion but fail by Bacon's. Popper suggested that psychoanalytic theory was a prime illustration of his thesis that inductively countenanced confirmations can easily be found for nearly every theory, as we look for them. But as Grünbaum (1981) points out, Popper ignores that the inductivist legacy of Bacon and Mill gives no methodological sanction to the ubiquitous "confirmation" claimed by some of these Freudians and Adlerians whom he had encountered in his early years. Popper regarded inductivism as probatively promiscuous because he believed the ubiquitous confirmations claimed for psychoanalysis to be sanctioned by the Bacon/Mill tenets. Grünbaum (1981) comments:

> It is ironic that Popper should have pointed to psychoanalytic theory as a prime illustration of his thesis that inductively countenanced confirmations can easily be found for nearly every theory, if we look for them. Being replete with a host of etiological and other causal hyotheses, Freud's theory is challenged by neo-Baconian inductivism to furnish a collation of positive instances from *both* experimental and control groups, if they are to be inductively *supported* instances. But . . . if such instances do exist, the psychoanalytic method is quite unable to furnish them. Moreover, to this day, analysts have not furnished the kinds of instances from controlled enquiries that are *inductively* required to lend genuine support to Freud's specific etiologies of the neuroses. Hence it is precisely Freud's theory which furnishes poignant evidence that Popper has caricatured the inductivist tradition by his thesis of easy inductive confirmability of nearly every theory! (p. 103)

Grünbaum is undoubtedly correct in his view that inductivism may pose a more serious threat to theories such as the Freudian than does the Popperian view. Indeed, Popper seems to be torn between two views which are antithetical to each other, namely, that the theories of Freud and Marx are unscientific because no testable deductions can be derived from them and the contrary view that both theories have been disproved![3] Eysenck and Wilson (1973), in their discussion of the experimental study of Freudian theories suggest that certain quite definite deductions can be made from the Freudian hypothesis and have indeed been made by Freud himself; it follows that the theory is scientific by Popper's standards (as being falsifiable), but it is not scientific by neo-Baconian standards (because these predictions are in fact either unsupported or falsified). Thus, to take a very simple case, the purely

[3]Popper (1974a) claims that Marxism was falsifiable and was indeed falsified; it was then rendered unfalsifiable. On the other hand, Freud's theory was unfalsifiable from the start.

symptomatic treatments of neurotic disorders, and even simple spontaneous remission, result in cures which do not give rise to symptom substitution or relapse; this should not happen on Freudian principles and consequently constitutes a falsification of the Freudian theory. As the deduction from Freudian principles was made by Freud himself and accepted by all his disciples, the disconfirmation is especially noteworthy.

The differences in these various forms of conceptions of scientific theories and the role of induction, verification, and falsification is not, as a consequence, a simple question of increasing rigor; it is more a question of increasing complexity, interrelation of different hypotheses, and extension of the factual realm over which the theory extends. But above all, it is a question of the general strength or weakness of the theory which is involved, and this dictates to a large extent the kind of research problem that is most fruitful at any particular stage of development. We must next turn to a consideration of the nature of strength and weakness of scientific theories.

2. "Strong" and "Weak" Theories in Science

Psychologists hold quite widely differing views about the usefulness and the general role of theory in psychology. Some stress simple inductive methods exclusively, maintaining that the time is not yet—or perhaps may never be!—when more ambitious generalizations of a theoretical kind would be appropriate. Others emphasize the similarities between all sciences and conclude that what is good for physics must be good for psychology. In all the discussion that has taken place, sight is often lost of a very simple fact which, to my mind, is crucial. Scientific theories differ among themselves to such an extent that any discussion about the place of theory in science should really be restyled: a discussion about the place of *theories* in science. In particular, there is a continuum ranging all the way from *weak* to *strong* theories, and failure to pay attention to the distinction between these two kinds of theories renders much discussion, and many criticisms, quite meaningless. This continuum, of course, is very similar in nature to that shown in Figure 1.

The theory which is always quoted as the perfect example of the application of the hypothetico-deductive method is of course Newton's theory of gravitation. This is a good example of a strong theory in science, and it may be worthwhile to look at those features of it which make it so. In the first place, it is based on a very large number of accurate observations, made over many years by large groups of people. In the second place, it brings together a number of subfields in which quantitative laws—such as those of Kepler and Galileo—had already been discovered and verified. In the third place, the phenomena in question were relatively clear-cut and unambiguous; in particular, they were not embedded in or entangled with groups of other phenomena. In the fourth place, the mathematical relations in question were not of a very complex order compared, say, with modern atomic theory. In the fifth place, and largely as a consequence of the preceding points, predictions were uncommonly straightforward and precise; verification and confirmation of deductions did not give rise to special problems.

Although for all these reasons Newton's theory is a good example for the beginner, it is seriously misleading for the practical scientist. Very few scientific theories are in fact strong theories of this kind; most of them lie toward the opposite end of a continuum going from strong to weak. The typical weak theory in science shows all the opposite characteristics to those mentioned in connection with Newton's. Only few observations, and these of doubtful accuracy, are available. Few quantitative or even qualitative laws, universally established, are available in subfields. The nature of the phenomena in question is by no means clear-cut or well understood. Mathematical relations are often very complex, and predictions are neither straightforward nor precise. What, it may be asked, is the use of such a weak theory?

The answer may be given in the words of the famous physicist J. J. Thomson: "A theory in science is a policy rather than a creed." I quoted these words on the front page of my book *Dynamics of Anxiety and Hysteria* (Eysenck, 1957) to indicate the heuristic nature of the theory there developed, and I believe that they contain the key to a proper understanding of the function of theory in psychology—where nearly all theories are weak theories almost by definition. The value of a weak theory, to put it briefly, lies in the fact that *it directs attention to those problems which most repay study from a systematic point of view*; in Thomson's words, it defines a policy of action and research. It is by giving rise to worthwhile research, rather than by necessarily being right, that a weak theory makes its greatest contribution to science.

An example or two may make clearer what I have in mind. On the basis of the theory developed in *Dynamics of Anxiety and Hysteria,* I predicted that extraverted people would have greater reminiscence effects after massed practice than would introverted people. Using the pursuit rotor, several investigators have tested this deduction. All those using a practice period of five minutes have verified the deduction; none of those using a practice period of 90 seconds have done so. This illustrates two points. In the first place, our knowledge of the subfield of reminiscence was too restricted to allow of a more precise specification of differences between extraverts and introverts; clearly, length of practice should have been specified as an important parameter but could not be so specified because of lack of knowledge in this respect. In the second place, the partial verification of the deduction is of considerable interest; it suggests relationships between personality and learning theory which are worthy of closer study. At the same time, the direction which such study should take is emphasized by the findings; we clearly must concentrate on time relations in the practice period and presumably also in the rest period. In other words, the theory leads us to a more precise study of the growth and decay of reactive inhibition, which is supposed to underlie reminiscence; the fact that positive relations with extraversion have been firmly established *for certain time intervals* suggests that a policy of research concentrating on this aspect will not be a waste of time (Eysenck & Frith, 1977).

These considerations suggest that it is relatively absurd, particularly at an early stage of development of a scientific theory, to construct a score board, with each success and each failure of prediction written in, to give a kind of batting average. There are usually many reasons for the (apparent) failure of a prediction, but far fewer for its success; consequently, failures are of much less interest than successes in evaluating a theory at an early stage. In particular, failures may arise not because the theory is in error but because the deduction made in a particular subfield makes use of a theoretical model in that subfield which is incorrect; this does not in any way invalidate the general theory. As an example, we may take the prediction that extraverts would show a more pronounced bowing in the serial position curve effect in nonsense syllable learning. According to the Hull–Lepley theory, this effect is due to inhibition of delay, and the hypothesis that extraverts are more prone than introverts to generate inhibitory potential mediates a clear-cut prediction. This prediction could not be verified, however, and neither could another one linking depressant drugs with an increased bowing effect. A special experiment was therefore carried out to test the Hull–Lepley theory, by comparing bowing effects with and without intervals between successive presentations of the nonsense syllable series; the differences which should have appeared according to the Hull–Lepley theory failed to materialize, and consequently it was concluded that the theory was itself in error (Eysenck, 1959). Failure to verify the deduction from the general personality theory was due, therefore, not to an error in the theory itself, but to an error in that part of learning theory used to mediate a particular prediction.

Even when there is no such error, predictions may not be verifiable for a variety of reasons although the theory is in fact correct. Two well-known examples are the failure to observe parallax in stellar positions, which was one of the most direct predictions made from Copernicus's heliocentric theory of the planetary system, and the failure to discover the capillaries which according to Harvey's theory of the circulation of the blood should intervene between arteries and veins. The (correct) explanations given by Copernicus and Harvey, namely, that the stars were too far away, and the capillaries too small, to make observation of the predicted effect possible with available instruments, were not at the time susceptible to proof. Scientists tended to accept the theories in question because they unified a large number of facts, although these apparently crucial deductions remained unverified for a long time.

Altogether, the notion of an *experimentum crucis* to decide the correctness of a theory, or to decide between alternative theories, is one which appears more frequently in the pages of popular expositions of scientific method than in actual practice. It will be remembered that two members of the Thomson family, father and son, were both awarded the Nobel price in physics, one for showing conclusively that light was of the nature of a *particle*, the other that it was of the nature of a *wave*! Having performed crucial experiments to prove both of these alternative theories regarding the nature of light, physicists are still left with something which sometimes behaves like a wave, sometimes like a particle; they have also learned rather painfully that crucial experiments are seldom as crucial as they are supposed to be—even in connection with a strong theory!

We may put the whole matter slightly differently, by following a discussion given by Cohen and Nagel (1936). They take as their example Foucault's famous experiment in which he showed that light travels faster in air than in water. This was considered a crucial experiment to decide between two hypotheses: H_1, the hypothesis that light consists of very small particles travelling with enormous speeds, and H_2, the hypothesis that light is a form of wave motion. H_1 implies the proposition P_1 that the velocity of light in water is *greater* than in air, whereas H_2 implies the proposition P_2 that the velocity of light in water is *less* than in air. "According to the doctrine of crucial experiments, the corpuscular hypothesis of light should have been banished to limbo once and for all." However, as is well known, contemporary physics has revived the corpuscular theory in order to explain certain optical effects which cannot be explained by the wave theory. What went wrong?

As Cohen and Nagel (1936) point out,

> In order to deduce the proposition P_1 from H_1, and in order that we may be able to perform the experiment of Foucault, many *other* assumptions, K, must be made about the nature of light and the instruments we employ in measuring its velocity. Consequently, it is not the hypothesis H_1 alone which is being put to the test by the experiment—it is H_1 and K. (p. 115)

The logic of the crucial experiment may therefore be put in this fashion: If H_1 and K, then P_1; if now experiment shows P_1 to be false, then either H_1 is false or K (in part or complete) is false.

> If we have good grounds for believing that K is not false, H_1 is refuted by the experiment. Nevertheless the experiment really tests both H_1 *and* K. If in the interest of the coherence of our knowledge it is found necessary to revise the assumptions contained in K, the crucial experiment must be reinterpreted, and it need not then decide against H_1. (p. 115)

We may now indicate the relevance of this discussion to our distinction between weak and strong theories. Strong theories are elaborated on the basis of a large, well-founded, and experimentally based set of assumptions, K, so that the results of new experiments are interpreted almost exclusively in terms of the light they throw on H_1, H_2 . . . H_n. Weak theories lack such a basis, and results of new experiments may be interpreted with almost equal ease as disproving H as disproving K (see the example given above of serial position learning effects). The relative importance of K can of course vary continuously, giving rise to a continuum; the use of the terms *strong* and *weak* is merely intended to refer to the extremes of this continuum, not to suggest the existence of two quite separate types of theories. In psychology, K is infinitely less strong than it is in physics, and consequently theories in psychology inevitably lie towards the weaker pole.

Weak theories in science, then, generate research the main function of which is to investigate certain problems which, but for the theory in question, would not have arisen in that particular form; their main purpose is not to generate predictions the main use of which is the direct verification or infirmation of the theory. This is not to say that such theories are not weakened if the majority of predictions made are infirmed; obviously there comes a point when investigators turn to more promising theories after consistent failure with a given hypothesis, however interesting it may be. My intention is merely to draw attention to the fact, which will surely be obvious to most scientifically trained people, that both proof and failure of deductions from a scientific hypothesis are more complex than may appear at first sight and that the simple-minded application of precepts derived from strong theories to a field like psychology may be extremely misleading. Ultimately, as Conant has emphasized, scientific theories of any kind are not discarded because of failures of prediction,but only because a better theory has been advanced; this may account for the longevity of Hull's system in spite of the many onslaughts on it in recent years.

There is a further characteristic of weak theories, as contrasted with strong, which deserves mention. In strong theories the different postulates are *interdependent*; it is not possible to change one without changing the rest, indeed without throwing overboard the whole theory. In weak theories, such interdependence is much less marked, and changes in one part of the theory are quite permissible without the necessity of altering other parts as well. Thus Hull's theory of inhibition is peripheral and "work"-oriented; I have preferred (for various experimental reasons) to work with a central theory of inhibition rather more akin to Pavlov's. This substitution, as well as many others, can be made without extending the framework of Hull's theory unduly; nothing of this kind would have been possible with Newton's theory of gravitation. Weak theories are very flexible; that is why they are such good guides for research; strong theories have an air of "take it or leave it" which makes them superior as guides to action but also less likely to lead to important new discoveries. In a similar manner, and also dealing with reminiscence, is the substitution of a consolidation theory for an inhibition one (Eysenck & Frith, 1977). This is a totally different concept, yet it could explain all the known phenomena just as well as does the inhibition hypothesis, with the crucial addition that it also explains certain phenomena which previously could not be explained.

There is, of course, one point which all scientific theories have in common and which decisively sets them off from nonscientific theories. However weak a theory may be, it must generate predictions which admit of experimental or observational investigation. In other words, the theory must be *reality-oriented* and the manipulations of reality implicit in its testing must be capable of being made explicit without ambiguity. The hypothesis that planetary motions can be explained in terms of angels pushing the planets around on their courses, preordained by God, is not a scientific theory because it suggests no direct experimental investigation. It is an interesting question whether such concepts as the Oedipus complex, or the superego, or the archetype, have characteristics which make them suitable for use in scientific theories, however weak, or whether they are outside science altogether; it would take us too far to discuss this point in detail (Eysenck & Wilson, 1973).

One of the main consequences of having to deal with weak rather than with strong theories is that attention should shift from consideration of right and wrong to considerations of the fruitful or the useless. It is difficult enough to disprove a strong theory; it is almost impossible to disprove a weak one. The point has been well made by Roley (1959):

> Ultimately, theory must answer to the facts, but this is not the only requirement placed on a theoretical system. Logical consistency, economy of assumptions, and even a degree of elegance are by no means secondary factors in determining the overall staying power of a theory. These patrician qualities are quite unlikely to mature, however, if the demand for direct descriptive capability is too insistent. It is not . . . contended that theory construction should be totally unresponsible to the general body of knowledge about behaviour. Rather it is held that point-by-point testing of isolated facets of a theory against specific behavioural phenomena or experimental findings is at odds with the whole purpose of theoretical abstraction. Suggestive hypotheses should not be put directly to drudgery but should be entertained for a while, as rare and welcome guests. It might be thought that all this is as much applicable to a strong as to a weak theory, as indeed implied in Lakatos' model of "hard core" plus protective belt as the structure of a theoretical research programme. Perhaps the main difference would lie in the relative prominence of the "hard core" and the "protective belt" respectively! (p. 130)

Weak theories usually imply the absence of precise, trustworthy data, and it is interesting that in the history of science most strong theories started out, in fact, as weak theories; their very existence stimulated the accumulation of precise data which later transformed these theories into strong ones. The heliocentric theory of Copernicus was based on wretchedly poor and inaccurate data; indeed, available observations were so erroneous and unreliable that they could not be used to arbitrate between the Copernican and the Ptolemaic theories. It was a weak theory in every respect, but by its very existence it encouraged astronomers to seek for ever more accurate observations, until Tycho Brahe finally gathered data reliable enough for Kepler to verify his three laws of planetary motion; these in turn mediated Newton's great synthesis. This would appear to be the answer to those who feel that the publication of weak theories is premature and should await more precise measurement and other advances in knowledge; such precise measurement, and such other advances, are usually the consequence of the interest aroused by a challenging new theory, however weak. Without the existence of the theory, there would be little motivation for the more precise measurements to be made. Every strong theory started out as a weak one (but of course not every weak theory necessarily becomes a strong one); premature demands on such a theory for accuracy and rigor appropriate to strong theories are less likely to lead to advances than a realistic insight into the limitations of weak theories.

Roley (1959) goes on to point out that absolutist notions of validity in scientific theory find no counterpart in physics. He distinguishes three major types of validity:

> First, a theory may be valid in a *subjunctive* sense. That is, it may describe the behaviour of entities under conditions that never obtain in fact. The Galilean law of falling bodies, and the ideal gas laws are examples. Second, a theory may be *locally* true, that is, may hold over certain ranges of the relevant variables. Hooke's law of elasticity is a clear-cut example, and the Newtonian laws of motion (holding for "middle-sized" phenomena) are now accepted in this sense. Finally, a law may hold *statistically*, being supported by large numbers of observations although there are local exceptions. The standard example is the law of increasing entropy. In each case, the postulated "law" can be defined as the potential limit of observational approximations as certain stipulations are satisfied. (p. 131)

Roley goes on to point out:

> It may well be that behavioural laws must invoke all three of these forms of licence in relating predictions to the world of observation. In other words, we may be reduced to predicting relationships among parameters of statistical distributions which can be empirically sampled within certain limits only, and which are subject to extraneous disturbing factors. (p. 131)

In summary of this discussion, it might perhaps be fair to repeat that when we are dealing with a strong theory, successful predictions are commonplace and do not do much to enhance the validity of the theory, whereas failures to predict correctly are very serious and may be disastrous for the theory in question.[4] When we are dealing with a weak theory, however, the reverse is true. Successful predictions, particularly if unlikely on commonsense grounds, are important and valuable and do much to support the right of the theory to be considered seriously. Failures, however, are not unequivocal enough to be taken too seriously; they certainly do not damage the theory in the same way as do failures in the case of a strong theory.

We may put this point of view in terms of information theory. If we may take K as certain (or near certain), then the confirmation or information of H gives us one bit of information (provided both outcomes

[4]Successful predictions from strong theory may of course enhance the validity of the theory if they are novel and not of the same sort as previous predictions. Eddington's (1920) observations of the gravitational bending of light predicted by the general relativity theory very much enhanced the credibility of that theory.

are equiprobable.) If we take K as quite uncertain, then the failure of H to be verified may have many causes; its success, however, is unlikely to have a cause other than the one specified. Consequently, success gives us more information than failure. It gives us information not only about H but also about K. Success implies that both H and K are essentially correct; failure implies that either H or any part of K may be in error, and this is too vague to be very useful. Failure may of course be used to suggest more complex relationships, improvements in technique, or limitations in choice of parameters, but it is not in itself very informative. Information gained by successful tests of a strong theory is largely redundant, but failure is highly informative. The reverse is true of weak theories.

An example from physics may take this statement clearer. There are certain stars, the so-called "white dwarfs," to which Sir Arthur Eddington's (1920) mass–luminosity principle does not apply; they are supposed to be much more massive than might be expected from their luminosity (this is because in them matter reaches an extraordinary high degree of density at which the ordinary gas laws cease to apply). Use has been made of this condition to furnish another proof for relativity theory. This theory predicts that the apparent frequency of a periodic phenomenon, such as atomic vibration, is changed when the source of the vibration is situated in a strong gravitational field. The slowing down of the vibration shows itself in a slight displacement of spectral lines toward the red; it was observed in 1925 by Adams in the spectrum of the Companion of Sirius. Vaucouleurs (1957) points out: "As the observed effect was in good agreement with the theoretical prediction, both the theory and the existence of extremely dense matter in white dwarfs were confirmed." Here the relativity theory (H_1) is being tested in relation to a body of knowledge (K) not itself too firmly established; confirmation of the hypothesis is likely only if both H_1 and K are correct, and consequently contributes much to our knowledge of both relativity theory and of the nature of white dwarfs. Failure of the hypothesis could have been due to errors in the relativity theory or to lack of true knowledge of the compositon of white dwarfs; it would therefore have been ambiguous at best and not very informative.

What has been said in this section should not, of course, be taken to mean that the psychologist has unbridled licence to theorize to his heart's content, regardless of negative results. For a theory to be regarded seriously it must show some evidence, both internally and externally, that it can mediate experimental predictions. But once this evidence is available, early failures of *some* predictions should not be taken too seriously; they should be regarded as a challenge to discover the causes of the failure, rather than as necessarily disproving the theory. This, of course, implies a tolerance of ambiguity, an ability to suspend judgment, which is often difficult to maintain. Nevertheless, the whole history of science argues powerfully against rash decisions on the basis of inadequate data; time and time again we have seen buried theories rise afresh from the ashes to which they had been prematurely consigned.

Nor should expectations be too high of close relationships between experimental variables. Much refinement will have to go into the raw scores obtained from psychological tests before we can begin to claim to measure one single variable rather than a mixture of the most diversified traits, abilities, and attitudes. The best that can be expected is a set of low correlations usually in the expected direction, but occasionally directly opposed to prediction; on such a foundation we can then begin to erect the infinitely complex set of laws and functional relationships, concepts, and definitions which will ultimately, shorn of ambivalence and ambiguity, constitute that proper science of behavior and personality which so obviously does not exist at the present time, except possibly as a foundation for a palimpsest.

3. The Two Disciplines of Scientific Psychology

Before turning to more detailed theories within psychology, let me discuss at some length certain overall views which masquerade as theories, which are seldom verbalized, and which have determined very strongly (and adversely) the whole structure of scientific psychology. Attention was drawn to these "meta theories" by Cronbach (1957) in his presidential address to the APA, which had the same title as this section. Broadly speaking, he pointed out that there are two different traditions within psychology

> two historic streams of method, thought and affiliation which run through the last century of our science. One stream is *experimental psychology*; the other, *correlational psychology*. Dashiell (1938) optimistically forecast a confluence of these two streams, but that confluence is still in the making. Psychology continues to this day to be limited by the dedication of its investigators to one or the other method of enquiry rather than to scientific psychology as a whole. (p. 671)

The experimental method is characterized by the simple fact that the experimentalist studies the effects of certain manipulations of the environment, which may take place mostly in the laboratory but may also take place in everyday life situations. He is concerned with general laws, the effects of his manipulation on the average person, and the elaboration of laws which would adequately represent the observed changes.

By contrast, the correlational psychologist is concerned with individual differences; he observes the different reactions of people under identical conditions and searches for regularities by looking for correlations between people in different situations. In a manner of speaking the experimentalist is concerned with the mean, the correlationist with the standard deviation; one looks for generality, the other specificity. Both clearly are concerned with important aspects of human behavior, and one would have expected them to collaborate in the solution of the very complex and profound problem presented by human behavior. As Cronbach points out, however, this is not so. The two groups are hardly on speaking terms; they know little of each other's facts and theories; they read different journals. This is clearly a tragedy for psychology, and it is the result of a rather unthinking acceptance of traditions which are based on unacceptable theories which are seldom formulated but whose hold is very tenacious indeed.

Let us first of all look upon the sins of the experimentalist and the implicit theories (metatheories?) governing his conduct. He is following a general theoretical formulation of the problem of psychology which may be called functional; in this formulation, he looks upon the dependent variable as a function of the independent variable, and by manipulating the latter and studying the variation in the former he expects to be able to delineate general laws which will become the basis of a true science. His general formulation: $a = f(b)$, that is, a is the function of b, is pervasive, and, within limits, of course perfectly acceptable. It is clearly advantageous to see to what extent varying b causes changes in a, and insofar as that is the program of the experimentalist it can hardly be faulted. In exactly the same way can we say that physics or chemistry or astronomy studies functional relationships of a similar kind. This notion of functional relationships has of course given rise to the old-fashioned behaviorist stimulus–response method of analysis; the stimulus is the independent variable, the response the dependent variable, and what is to be studied is the functional relationship between the two.

The simple comparison with physics, however, leaves out of account a very fundamental aspect of the functional relationships recognized by physicists. Let us consider Hooke's law of elasticity: Stress $= k \times$ strain, where k is a constant (the modulus of elasticity) that depends on the nature of material and type of stress used to produce the strain. This constant k, that is, the stress–strain ratio, is called Young's modulus and is illustrated (with certain simplifications) in Figure 2a. A and B are two metals differing in elasticity; they are stressed by increasing loads, and

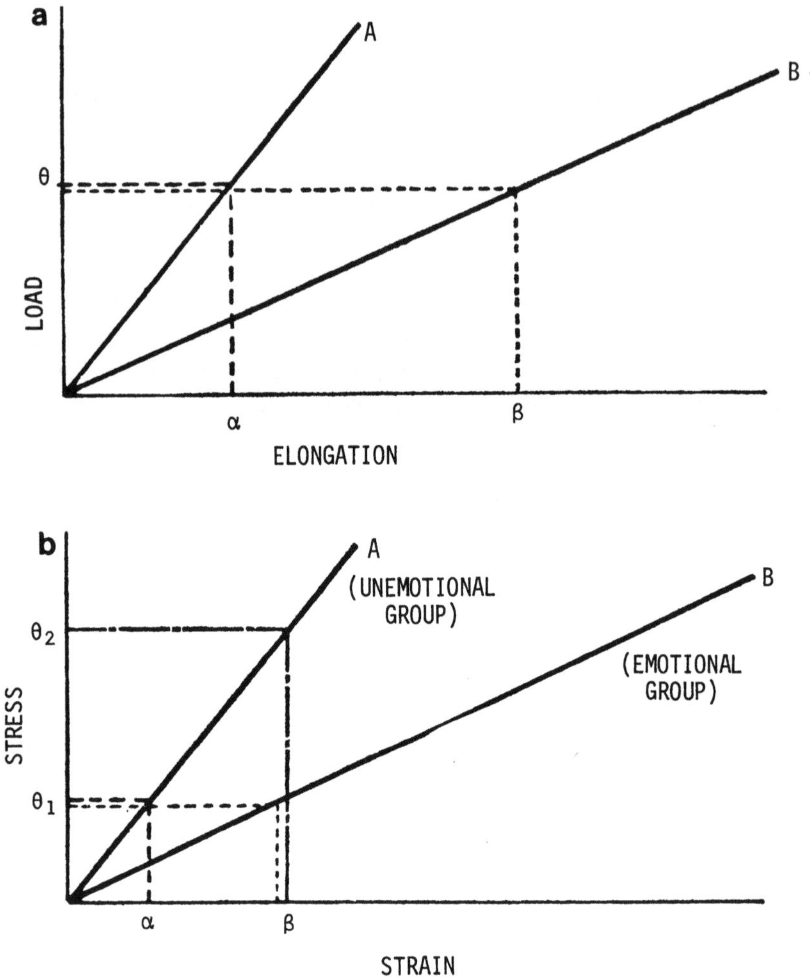

Figure 2. The concepts of stress and strain in physics (top diagram) and psychology (bottom diagram). From Eysenck, 1975b.

the elongation corresponding to each load is plotted on the abscissa. It will be seen that identical loads give rise to quite divergent elongations, α and β. Thus the physicist does not have the simple stimulus (independent variable)—response (dependent variable) relationship; it also incorporates the term *k*, which corresponds to the notion of *organism* in psychology. The more recent formulation: S—O—R attempts to do justice to this situation, but such a formulation certainly carries with it the implication that we ought to change our functional law to read: $a = f(b,O)$, in other words, *a* is a function not only of *b* but also of the particular nature of the organism which is being stimulated, just as the elongation of the wire in the experiment detailed in Figure 2 is a function partly of the nature of the wire, partly of the weight used to stretch it. As far as physics is concerned, strain would correspond to the experimental method, k to the individual differences or correlational method, and no physicist would ever doubt that both are essential for a united science (Eysenck, 1975b).

Let us use the same argument in relation to psychology. Figure 2b illustrates an analysis of human behavior (physiological, verbal, expressive) in an experimental situation productive of emotion. Again the stress (independent variable) is plotted on the ordinate, and the strain (dependent variable) on the abscissa; A and B represent an emotionally stable and an emotionally unstable individual or group of individuals respectively. Identical stress θ_1 gives rise to quite different strains α and β. It would require stress θ_2 to make the strain in A individuals equal to that produced by θ_1 in B individuals. Differences between θ_1 and θ_2 are the kinds of differences traditionally studied by experimental psychologists; differences between A and B are the kinds of differences traditionally studied by personality psychologists, believers in the importance of constitutional factors, and clinical psychologists. In order to understand what is going on, in order to formulate theories, and to make predictions, we must incorporate both types of factors in our model.

This model may be called compensatory or substitutional in the sense that one type of variable can compensate for or be substituted for the other (Savage & Eysenck, 1964). To produce a certain elongation greater than x we can either increase the weight suspended from the wire or choose a wire the k of which indicates greater pliability. On the human side, consider a series of experiments carried out by Rosenbaum (1953, 1956). He found that threat of a strong shock led to greater generalization of a voluntary response than did threat of a weak shock; this would be a typical experimental approach. He also discovered that anxious subjects showed greater generalization to *identical* stimuli than did nonanxious subjects; this would be the typical correlational or individual differences approach.[5] To obtain a given degree of generalization, we can thus either change the strength of the threatened shock or choose more or less anxious subjects; the two can be traded against each other. Savage and Eysenck (1964) have carried out a series of experiments with rats illustrating the same point. Their chapter is headed "Definition and Measurement of Emotionality," indicating another important function

[5]It is interesting to note that Rosenbaum published one paper in the *Journal of Experimental Psychology*, the other in the *Journal of Abnormal and Social Psychology*, thus ensuring that very few psychologists would read *both* papers!

of this whole theoretical concept of experimental and individual variables as being to some degree interchangeable.

We may ask ourselves whether a given test or measure which is used to indicate differences in emotionality or anxiety, whether in humans or in animals, is in fact a good measure of that variable, and this whole question of validity of tests has of course always been a very difficult one. It can be solved, however, along the lines of the above argument by looking at experiments which are generally accepted as increasing an organism's degree of emotionality. We can then argue that if experimental modification *x* could produce an increment in emotionality, which can be indexed by performance *y*, then if our personality test *a* is a good measure of emotion or anxiety, people having high scores on this test should behave like persons subjected to a strong dose of *x*, that is, show strong *y*, whereas people having low scores on *a* should have low scores on *y*. Thus this conception of the relationship between experimental and individual differences variables gives us a powerful tool for demonstrating the validity of personality measures and adding substantially to the range of application of the experimental design.

Another important consequence of these theoretical considerations is that if the suggestions made above are true, then it follows that we cannot make verifiable predictions from general laws without incorporating specifically a variable *k* which refers to the constitution of the individual for whom the prediction is being made. Consider an experiment reported by Jensen (1962). He studied the number of errors in serial learning as a function of the rate of stimulus presentation; there were two rates, one of 2 seconds and one of 4 seconds. The traditional experimental psychologist would regard this problem as meaningful and soluble; either differences in rate of stimulus presentation cause a difference in the number of errors to criterion, or they do not. Jensen argued that this test imposes a stress on the subject and that the resulting strain would be indexed in terms of an increased number of errors when the shorter rate of stimulus presentation was employed, as compared with the longer rate. (There is an obvious pressure with the quick rate of presentation which might well increase the general stress of the test situation.)

This stress would impose differential strain on the individuals respectively high and low on trait anxiety, and Jensen measured this trait anxiety with the neuroticism scale of the Maudsley Personality Inventory. Contrasting subjects scoring high and low respectively on this scale, he found that for low scorers (i.e., nonemotional individuals) the added stress of shortening the rate of stimulus presentation produced no effect at all; they made 63 errors on the average for the long rate and 64 errors for the short rate. But for the high scorers there was a tremendous difference; in the long rate of presentation condition they only made 46 errors, far fewer than the unemotional subjects, whereas in the short rate of presentation conditions they made 90 errors! In other words, the results of the experiment can only be understood in terms of the person × condition interaction; leaving out the differential personality effect makes complete nonsense out of any simple averaging of the results. Such averaging would tell us that there was a mild and nonsignificant effect of shortening the rate of presentation, when in reality there was no effect for nonemotional subjects and a very strong effect for emotional ones.

I feel very strongly that just as it would be meaningless in physics to leave out the constant *k* in dealing with predictions of the kind considered, so it is meaningless in psychology to try and frame general laws without taking into account individual differences. In fact, this is even more so in psychology than in physics because of certain very relevant differences between these two sciences. In physics we can usually proceed along very analytic lines by experimentally excluding certain variables and by completely dissecting the objects of our interest until we are dealing with simple elements or alloys of known composition. In psychology, however, we cannot do this. By definition, we are dealing with organisms and their behavior; as a consequence, we are not allowed to cut up the organism in such a way as to isolate certain aspects. The integrity of the oranism must be maintained, and that means inevitably that the personality, the intelligence, and other important functions of the organism will play an important part in whatever measurement we may be concerned with. To relegate these individual differences to the error term, as experimentalists are wont to do, simply means that the error term will become enormously exaggerated in size and the main effects (unless in trivial and obvious experiments) will be much smaller than is acceptable in a scientific discipline. We can rescue these variables from the error term by looking at interactions, and many experiments have shown that these interaction terms can frequently be much more important than the so-called main effects (Eysenck, 1967, 1981). This general rule has certain important consequences for the design of experiments meant to test perfectly general hypotheses in experimental psychology which have been spelled out and illustrated elsewhere (Eysenck, 1976b). The many different experiments there quoted to illustrate this point suggest that the use of personality parameters in experimental psychology is not only permissive but *mandatory* (Eysenck, 1981). An experiment in which personality variables are theoretically likely to play a part should never be planned without either including these personality variables in the analysis of variance design or at least measuring them in order to use them as moderator variables in the final analysis.

Anything less is an abuse of the scientific method and is likely to lead to the absurd state in which much of experimental psychology finds itself, namely a failure to replicate, an inability to formulate wide-ranging theoretical conceptions, and an error term which far outweighs the main effects in the experiment. A combination of the experimental and the individual differences approaches is essential if psychology is to become a science at last.

It might be thought that the studies mentioned do not deserve the title of *experimental* if this term is interpreted in the traditional way of changing the independent variable and observing the resulting changes in the dependent variable in a laboratory setting. I believe that this interpretation is too narrow. Physicists and astronomers usually consider Eddington's observation of the gravitational effect of the sun on light coming from a star situated apparently close to the eclipsed orb as being an experiment, although the eclipse was not of course physically produced by Sir Arthur. In a similar way MZ and DZ twins are not actually produced by the experimenter; like Sir Arthur, he is making use of experiments conducted by nature, going to considerable pains to ensure that conditions of observation are such as to make the experiment as well controlled as if it had been conducted under laboratory conditions. Such an empirical trial of a given hypothesis is far removed from the usual correlational or factor analytic study of the psychometrist and deserves to be considered closer to the rubric of experiment considered in its narrowest aspect.

4. The Contribution of Experimental Psychology to Research in Individual Differences

In the last section we have discussed the need for experimental psychology (and also social, educational, industrial, clinical, and other areas of applied psychology) to take into account and incorporate in experimental designs the major relevant dimensions of personality. It should not be imagined, however, that trading between experimental and individual differences in psychology is entirely one way; nothing could be further from the truth. In this section we will consider the importance of introducing experimental methods and theories into the field of individual differences, using as our example the study of intelligence. Readers with a philosophical background will note that I have concentrated here as elsewhere on giving only a brief discussion of the general theoretical viewpoint and much more space to a discussion of specific application of that theoretical point of view. In addition, it will be noted that I have concentrated on a historical treatment. The reason for adopting these methods of presentation is twofold. In the first place, I am not a professional philosopher, although I have received some training in that discipline; neither are most of my readers likely to be philosophers. Consequently an *ad oculos* demonstration of the points made, in their historical development, will be more easily intelligible and more compelling than would a purely theoretical discussion. In addition, I have always believed that philosophy of science is crucially dependent on an accurate history of science; the history of science supplies the empirical foundations for any pronouncements that the philosopher of science may make in the hope that it might be helpful or enlightening for the working scientist. Hence my stress on historical examples and developments.

The developments here discussed have been documented extensively in *A Model for Intelligence* (Eysenck, 1983); hence no extensive references will be made to the literature. The topic to be discussed is the development of the notion of intelligence and the theories concerning it which have been developed by psychologists and others during the past hundred years or so. In looking at this development, we note from the beginning an antithesis which became quite clear in the views and methods of measurement advocated by Sir Francis Galton, on the one hand, and Alfred Binet, on the other. For Galton, intelligence was a single determinant of cognitive behavior, entering into all types of activity which could be called cognitive rather than emotional or conative. This activity he conceived to be determined largely by genetic causes, and as a consequence he recommended that it be measured by as direct an index of neural functioning as could be devised. In particular, he recommended the study of reaction time; the development of electrophysiological measurement, such as the EEG, would undoubtedly have attracted his interest and would have led him to advocate the use of these more recondite procedures.

Binet, on the other hand, believed that there were a number of *different* abilities determining cognitive functioning, such as verbal ability, numerical ability, suggestibility, memory, and so forth. His view was that these should all be measured and that intelligence was a rather artificial concept based on the sum of all these different abilities. Furthermore, he seemed to believe that an individual's abilities were determined to a large extent, if not entirely, by environment, socioeconomic factors, cultural influences, and education. Thus we see right from the beginning a difference in conception which has persisted ever since.

The Galtonian tradition was continued by Charles Spearman, who introduced factor analysis into the debate and believed that he had demonstrated the existence of a general factor of intelligence by means of his method of tetrad differences. Thurstone and later Guilford also used factor analysis and believed that they succeeded in demonstrating the existence of a number of primary abilities, to some extent resembling those suggested by Binet. Guilford at the moment claims in his model of intellect that already he has evidence for something like 80 or more of the 120 different abilities he postulates. It will be seen that factor analysis has been unable to settle the question, and this illustrates very clearly my own belief that problems of this kind cannot be settled by the individual differences approach only, using correlational or factor analytic methods, but that experimental tests of specific hypotheses are needed.

In my very first contribution to this field (Eysenck, 1939) I suggested that a reanalysis of Thurstone's original data indicated a compromise between the two extreme positions, namely, the existence of a strong general factor accompanied by a number of rather weaker and independent primary abilities. This view was later accepted by Thurstone and Spearman and is probably more widely accepted now than either extreme. However, it cannot be claimed that on the basis of correlational studies such a view can find any direct and incontrovertible *proof*; there are many psychometric and empirical reasons for preferring it to the extreme views originally adopted by Spearman and now by Guilford (in opposite directions!), but reasonable expectations and compromise solutions are a long way removed from the convincing proof that scientists rightly demand for the adoption of a particular hypothesis.

When we come to the question of tests used and the determination of genetic influences, we find that until quite recently the balance was very much in favor of Binet. He introduced tests of an educational character, depending to some or even to a large degree on acquired knowledge, skills, and habits, and consequently inevitably subject to a strong determination of scores indicative of individual differences by environmental factors. Genetic studies summarized elsewhere (Eysenck, 1979) indicate that even so the proportion of the total variance on such tests accounted for by genetic factors is 80%, that accounted for by environmental factors only 20%, but in practical terms this 20% is still a quite large proportion of the total variance and makes it impossible to draw any conclusions regarding individual persons. Heritability estimates are always population estimates and cannot be applied to individuals; hence tests of this type are inherently incapable of telling us very much about the specific genetic endowment of a given person.

It is interesting to consider historically why Binet won out and Galton lost in the type of measurement of intelligence mostly used by psychologists. An early study of the ability of reaction time experiments to discriminate between more and less intelligent individuals was carried out early this century by Clark Wissler at Columbia University with negative results, and this study so much influenced psychologists (and found its way into most textbooks) that until quite recently very little work had been done along these lines. Acceptance of Wissler's results can only be explained in terms of the zeitgeist, because his experiment must be considered to be one of the worst ever done in experimental psychology. It is well known that to measure reaction time properly a large number of determinations have to be made, as there is considerable variance around the mean; yet Wissler only tested each individual a very small number of times. It is well known that to obtain a reasonable correlation of any given test with intelligence the range of ability must be large and should be roughly equivalent to that obtaining in the whole population; yet Wissler chose advanced university students at Columbia University, showing a very small range of ability. And finally it is well known that as a measure of intelligence we must use a proper IQ test; Wissler chose achievement tests, which are known to correlate very little if at all with IQ tests in the kind of population he used. In these circumstances a low correlation was inevitable and proved nothing. Yet his research has been cited again and again to indicate the lack of value of Galton's suggestion!

Recent work has demonstrated conclusively that reaction time in fact correlates very highly with Binet-type measures of IQ such as the Wechsler test. Admittedly tests of simple reaction time only correlate less than .5 with IQ in children or adults showing the normal spread of IQs, but other measures of reaction time have been found much more useful. Hicks's law has been used to look at complex reaction times; this law simply states that for any individual reaction times will increase in a linear fashion as the logarithm to the base 2 of the number of stimuli presented. The slope thus defined differs from person to person, with people having a high IQ showing a smaller increase in reaction time with increase in the number of stimuli than do individuals having a lower IQ. Using this slope or simply correlating IQ with reaction times to some eight differential stimuli gives us correlations in the neighborhood of .50, uncorrected for attenuation, which means that the true correlation must be somewhere in the neighborhod of .70 or thereabouts.

Another variable which has been found to be highly correlated with IQ (even more so than reaction time) has been the *variability* in reaction times shown by a given individual. People with a high IQ tend to show low variability, people with a low IQ high variability. This feature thus can be added to our determination of IQ by reaction time, making the correlation even higher than before.

Other similar developments have been related to the measurement of what has come to be called *inspection time*. Here the subject is shown on the tachistoscope two lines, one very obviously much longer than the other; he has to say which is longer. (Backward masking is used to avoid the use of after-images, etc.). The lines are presented for very short periods of time a number of times, with the position of the long and short lines varying on a random basis, of course, and the experimenter determines the shortest duration of exposure at which the subject gives an accurate estimate 97.5% of the time. It has been found that high-IQ individuals have shorter inspection time than low-IQ individuals, and a similar finding has been made when the test was translated into the auditory field. Inspection time techniques work better with individuals of below- and above-average IQ, and the correlations with below-average IQ subjects again extend into the 70s. We thus have here another correlate of Binet-type IQ which follows the theories of Galton rather than those of Binet.

More important than these developments, however, have been recent applications of theories and methods of electrophysiological measurement of so-called evoked potentials. When the EEG of a given individual is taken and a sudden visual or auditory stimulus is introduced, a series of rapid waves can be seen which constitute the so-called evoked potential. These waves extend over a period of 500 milliseconds at most, but for measurement purposes only the first 250 milliseconds are considered. It has been known for some dozen years that the latencies and amplitudes of these waves show some slight correlation with IQ, not usually above .3; but as these correlations were difficult to replicate, and as they were not high enough to be of any great importance, little has been made of these findings. In addition, there seemed to be little theoretical basis for these findings, and purely pragmatic correlations of this kind are of little interest in science.

In recent years there has been a considerable change in this position, due largely to the theoretical work of Alan Hendrickson and the measurement of IQ by means of evoked potential, following this theoretical work, of Elaine Hendrickson (Eysenck, 1983). Essentially what these two have done is this: Alan Hendrickson developed a theory of information processing through the cortex, which goes into considerable physiological and biochemical detail. It would be going too far to reproduce this theory here, but two aspects of it are essential for an understanding of the new type of measurement undertaken by Elaine Hendrickson. The first point of this theoretical development is that the evoked potential, in all its details, is *homologous* with the detailed information concerning the stimulus which is being propagated through the nervous system.

The second aspect to be considered is the fact that in the propagation of the message errors will occur. What the theory now says is simply that people in whom many errors occur will have a low IQ, whereas people in whom few errors occur will have a high IQ. We thus have here a very simple causal hypothesis about the reasons why certain individual differences occur, and this theory is eminently testable.

How does this theory influence our measurement? Clearly, latencies and amplitudes are largely irrelevant to the measurement of IQ following this particular hypothesis. What is important is the following: As is well known, evoked potentials have a low signal-to-noise ratio and hence have to be averaged over a number of time-locked presentations; in our own work we have usually used 90 such presentations. Now if the resulting averaged curve is homologous with the message in all its details, under ideal conditions of no error, then the presence of error should reduce the *complexity* of the curve, until with individuals in whom a great deal of error is present only the broadest outline should be shown, that is, simple waves without any of the numerous small ups and downs so characteristic of evoked potentials. In other words, what we should be measuring is the *complexity* of the waveform, and this is what in effect has been done by the Hendricksons.

The theory predicts that this complexity measure should correlate highly with IQ, and in one of their studies Elaine Hendrickson found a correlation of .83 with the WAIS over a population of some 200 school children, with a mean IQ and a standard deviation closely resembling the normal population. (In other studies she found similar correlations for adult populations, also using other tests of IQ such as the Matrices.) These results thus show a very high correlation between evoked potential and IQ, a correlation which would be higher still if corrected for attenuation.

A factor analysis of the correlations between the 11 subtests of the Wechsler and the evoked potential complexity measure showed a very prominent general factor on which the evoked potential had a loading of .77; the highest loading observed for any of the Wechsler Tests was .82 (for the Similarities test). We must now turn to a theoretical consideration of the consequences which may be derived for a better understanding of intelligence from all these data (Eysenck, 1985).

Following Binet, most American psychologists have tended to look for an environmental interpretation of the individual differences found on ordinary IQ tests. Most would probably admit that heredity plays a very important part, but interest has largely centered on the possibility of changing a person's IQ by educational, cultural, and other modification of the environment. Head Start is but one example of the interest shown in this possibility. Such an interest is understandable in view of the mixed nature of the typical IQ test, giving as it does scores which are determined partly by genetic, partly by environmental factors. It is obvious that results from such tests will be difficult to interpret and will give rise to endless controversies, such as those attending Jensen's suggestion of genetic factors' being responsible for racial differences in IQ. In the nature of things it is almost impossible to answer such questions in a direct manner, and we are reduced to complex statistical argumentation which is unlikely to prove totally convincing to the adherents of either side. Clearly it would be very advantageous if, by going back to Galton's approach, we could derive a measurement technique which would as far as possible approach the genotype without the inclusion of cultural, socioeconomic, and educational factors. The study summarized above suggests that this possibility is a very real one; measures of reaction time, inspection time, and evoked potential all correlate between .70 and .80 with traditional IQ tests, and a look at the detailed statistics suggests that such a test as evoked potential must be very near to being an almost pure measure of innate ability, without the addition of culturally determined factors (Eysenck & Barrett, 1985).

Such a hypothesis can be tested and has been tested in the following manner. Twenty-five children from low socioeconomic status homes and 25 children from high socioeconomic status homes were given the Wechsler, with a difference of 23 points between them, which amounts to 1.64 standard deviations using the standard deviation found in the total group from which these were extracted. If we accept the estimate of 80% of the total variance on the Wechsler being contributed by heredity, 20% by environment as correct, and if the evoked potential measure is a true measure of genetic differences only, then we would predict that on this evoked potential measure the difference between the children will be reduced by 20% (i.e., the 20% contributed to the Wechsler by environmental factors), and this is actually what we do find (Eysenck, 1983). It is not suggested that this single small-scale experiment *proves* the accuracy of the evoked potential as a measure of the genotype; it is merely quoted as an example of the type of deduction which can be made from hypotheses of this kind. Obviously many more such experiments are needed and much larger numbers of subjects will be required before we can rest content with demonstration as actively proving the point.

The implications of these findings for a theory of intelligence are considerable. On Galton's hypothesis, the high correlations between reaction time and evoked potential on the one hand and the different tests constituting the Wechsler on the other are to be expected; indeed, one would have expected that Wechsler subtests having high loadings on the general factor of intelligence would have high correlations with these external criteria and tests having low loadings would have low correlations. This application of *criterion analysis* (Eysenck, 1950) has indeed been shown to work along these lines, and it is difficult to see how such proportionality could arise if we disallow the existence of a strong and general factor of intelligence running through all the subtests of the Wechsler. (The very high loading of the EEG measure on a factor analysis of the Wechsler subtests is another way of stating the same set of interrelations.) Thus introduction of an external, experimentally defined criterion makes it possible for us to go beyond the vicious circle of purely relational and factor analytic investigations and to make a more formal test of the Galtonian hypothesis (Eysenck & Barrett, 1985). Clearly, the finding does not deny the existence of additional "primary abilities" such as are proposed in the compromise solution, but the data completely rule out solutions such as those suggested by Guilford, that is, the absence of a general factor of intelligence and the distribution of the total variance over a large number of special abilities.

Criterion analysis is a quite general and very powerful test of certain types of psychological hypotheses, linking psychometric research on individual differences with direct experimental analysis. Two illustrations may be given, both deriving from work reported by Jensen (1981a). The first relates to inbreeding depression. This is a genetic phenomenon manifested in the offspring of parents who are genetically related (father-daughter, cousins, etc.). Such inbred offspring show a depression or diminution in those characteristics which are in some degree genetically influenced by directional dominance, such as intelligence. This is due to the fact that when recessive alleles detract from the positive expression of a trait, inbreeding increases the chances that recessive alleles from each parent will be paired at the same loci on the chromosomes, thereby diminishing the phenotypic expression of the trait. The degree of inbreeding depression for any given trait is normally assessed by comparing measurements of the trait in the inbred offspring of genetically related parents with measurements in the offspring of unrelated parents.

Now it is known that intelligence is inherited in a mode which includes dominant and recessive genes (Eysenck, 1979), and consequently one would expect that tests which correlate highly with intelligence (have high *g* loadings) would show a *greater* degree of inbreeding depression than tests having a low correlation with intelligence (having low *g* loadings). Jensen plotted the *g* factor loadings of 11 WISC subtests as a function of the percent of inbreeding depression on subtest scores. (The tests are *I*nformation, *C*omprehension, *A*rithmetic, *S*imilarities,

*V*ocabulary, *P*icture Completion, *P*icture *A*rrangement, *B*lock *D*esign, *O*bject *A*ssembly, *D*igit *S*ymbol, and *M*azes.) It will be seen (Figure 3) that there is a very high correlation of 0.76 between the *g* loading and the percentage of inbreeding depression, verifying this particular deduction from the theory. (The correlation is unexpectedly high because the range of *g* loadings is of course very restricted, going from .5 to .8. Had the range been greater, clearly the correlation would have been very much higher.)

The second example, also taken from Jensen (1981a), presents a test of what he calls the Spearman hypothesis of white–black differences. Spearman (1927, p. 379) hypothesized that the varying magnitudes of the mean differences between whites and blacks in standardized scores on a variety of mental tests were directly related to the size of the tests' loading on *g*. Jensen re-analyzed nine independent studies giving data making such an analysis possible and found strong confirmation for Spearman's hypothesis. The largest set of relevant data was based on the GATB, a widely used and well-standardized battery of tests of cognitive abilities; the comparison used populations of white and black subjects of well over 1,000. Figure 4 shows the *g* loadings, the white–black mean difference (in standard score units), and the average correlation of each aptitude scale with 12 well-known standard tests of IQ or general intelligence, for each of the nine aptitudes measured by the GATB. There is clearly a close similarity between the profiles for these three variables, and correlations calculated by Jensen among the three

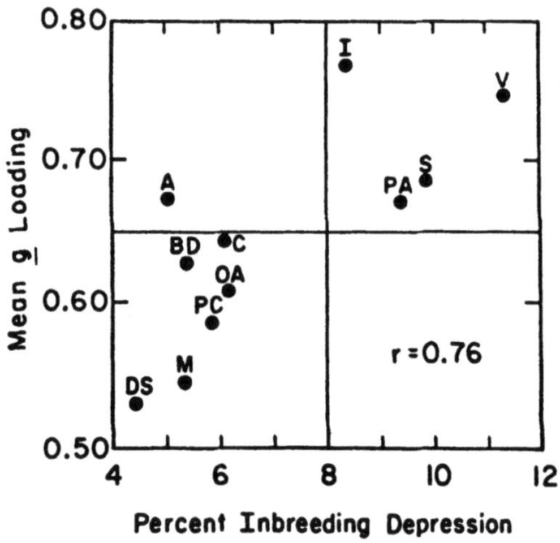

Figure 3. Mean *g* loadings and percentage inbreeding depression, as shown in 11 Wechsler subscales. From Jensen, 1981a, with permission.

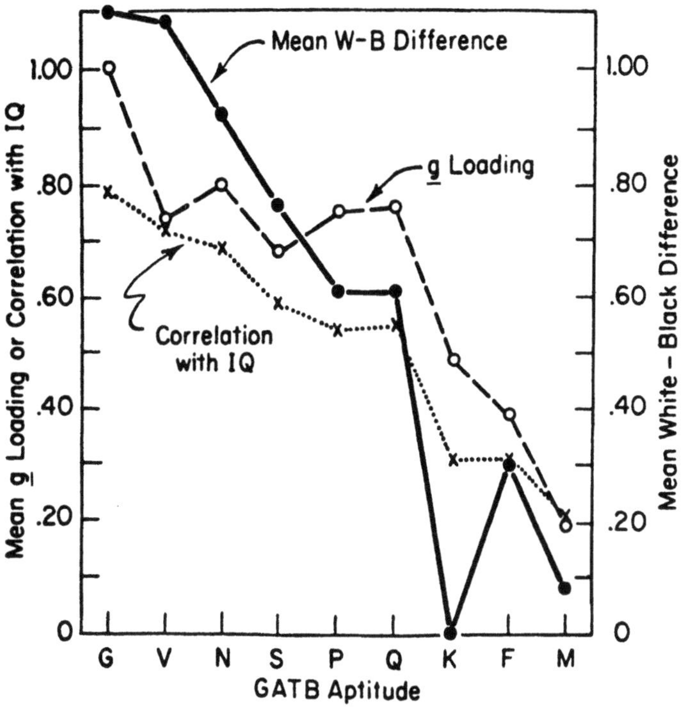

Figure 4. Mean *g* loadings or correlations with IQ and mean white–black differences on subtests of the GATB aptitude test. From Jensen, 1981a, with permission.

are astonishingly high, ranging from .9 to 1.0. Thus complex psychometric theories are capable of resolution by combining the psychometric (factor analytic) approach with direct experimental testing along the lines of criterion analysis.

The implications of the findings on the electrophysiological measurement of IQ go much further than already noted. The contamination of orthodox IQ tests with educational and cultural material has always made difficult the study of questions which are of considerable interest to psychologists, such as the development in young children of intellectual ability, the decline of this ability in old age, and differences in ability between males and females or between different national or ethnic groups. Differences are usually less than the 20% of the variance allocated to environmental factors in the genetic analysis of IQ data, and consequently it can be argued endlessly that they are due entirely to the environmental factors. Having available tests which are independent of environmental factors of an educational or cultural kind would make it possible to resolve these problems in a direct, scientific manner, without having to have recourse to complex statistical arguments which make assumptions which can always be challenged. The only such problem to have been attacked hitherto is that of sex differences (Eysenck, 1983). It was found that among the boys and girls tested there were no mean differences in evoked potential scores, but boys had a larger *variance* than girls, which is in good accord with the literature on orthodox IQ tests. Thus on this important question the evoked potential gives results similar to those previously obtained with IQ tests, but as expected the difference in variance is greater for the EEG measure than for the IQ. Differences in variance on the ordinary IQ would be likely to be lessened by the introduction of cultural and educational factors which would tend to equalize variances; here the argument is exactly the opposite to that presented previously for differences in SES.

Even more important from the theoretical point of view are the repercussions these findings have on our conception of the *nature* of intelligence. It is well known that there is little agreement among psychologists on the definition of intelligence or on its nature; what most definitions appear to agree on, however, is that intelligence is related to problem solving, the induction of relations and correlates, cognitive-type work of one kind or another, the learning of complex and difficult tasks, whether verbal or nonverbal, and so on. All these definitions emphasize culturally determined types of activities which are based in large or small part on education and previous experience of one kind or another. The new theory suggests that the only important variable in intelligence is the individual's innate structure of the CNS enabling this system to propagate messages correctly across axons and synapses. This is an elegant and intuitively appealing notion, but it opens up certain important problems. Thus we would imagine that in solving a problem more would be involved than simply passing along information correctly through the central nervous system; some kind of use would have to be made of this information in order to obtain a correct solution. Yet the observed correlation coefficients leave very little leeway for any additional factors over and above the simple propagation of correct information through the cortex; it would appear that bringing together correct items of information relevant to the problem automatically produces the solution! This is certainly counterintuitive and may deserve some discussion.

Let us note first of all that the information in question is not necessarily only information coming in through the senses and coming from the outside; we would include equally information obtained through memory retrieval, that is, items of information stored in the cells of the CNS and addressed during the information-processing stage. It is equally difficult to think of a solution's being arrived at directly from the free interplay of the bits of information involved as it is to think of a separate problem-solving stage—such a stage easily conjures up the view of a homunculus sitting in the nervous system, obtaining the necessary information and then producing the solution! It has to be said that at the moment the theory has no real contribution to make to this fundamental problem, other than to state that in mathematical terms very little of the variance, if any, is left over for a separate problem-solving stage.

One further theoretical problem requires discussion, namely, that of the relation between the different types of noncognitive tests we have enumerated as being so closely related to cognitive ability. If evoked potentials, reaction time, and inspection time all correlate around .80 with IQ, then clearly these different measures must correlate quite highly together themselves, and the question arises as to which is fundamental. An additional question is whether we can account for the observed findings on reaction and inspection time in terms of what we have found or hypothesized with respect to evoked potentials. The following is thus a brief sketch of a possible solution: If we agree that IQ (in its genetic aspects) is determined completely by the probability in a given CNS that information will be processed more or less accurately, than it would appear to follow that high-IQ individuals should have short reaction and inspection times, because erroneous processing of information would require a large number of redundant messages to be passed before reaction is possible. Similarly, we can account for the fact that variance in reaction times is negatively correlated with IQ in the same terms; when the signal is passed correctly through the cortex, then reaction times will be short; when it is in error, reaction times will be long because further messages will have to be passed before reaction is possible. Thus the greater the error propensity of a given CNS, the more variable will the results be. This explanation also accounts for the fact that even for retardates, the fastest reaction times are just about as fast as the fastest reaction times of highly intelligent subjects. This can be explained in terms of the fact that even in a CNS given to making many errors occasionally these errors will be absent. We can thus see that there is considerable agreement between these different noncognitive tests of cognitive ability, pointing to a common causal theory, namely, that of error-proneness. Obviously all these hypotheses are in an early stage of development and much further work will be required to provide backing and support. However, judging simply on the basis of existing knowledge we now seem to have a unified theory of intelligence based on firm physiological foundations, which enables us to measure intelligence without the usual adulteration provided by socioeconomic, cultural, and educational factors irrelevant to "pure" intelligence (Eysenck, 1984).

This whole chain of reasoning, against its historical background, has been introduced in order to demonstrate most forcibly how the introduction of experimental methods into the correlational field can deliver us from the inherent impossibility of allocating the total variance in a meaningful and convincing manner to different hypothetical causal elements. In order to do this we must formulate hypotheses that can be tested in a more direct manner than is possible through factor analysis itself, that is, along experimental lines. This does not mean, of course, that the elements involved in such hypotheses cannot also be introduced into a factorial design; it should be possible to make precise predictions (e.g., along the lines of criterion analysis) as to the outcome of such studies, and as we have seen the outcome hitherto has been very positive. But the essential point is that such theories and experimental studies have an explanatory value which does not appear in purely statistical "factors" of the kind usually studied by psychometrists. Here, then, we have a demonstration that just as the study of individual differences is indispensable for the experimentalist, so experimental studies are indispensable to the psychometrist. The two disciplines of scientific psychology stand and fall together.

5. Theory or Weltanschauung

Theories in science generally, and in psychology in particular, can of course be of varying types, and it is important to bear in mind the precise degree of generality of a theory. Clearly such theories as those regarding the nature of intelligence or personality, as discussed in previous sections, are very general; they enable a large number of deductions to be made and tested. In experimental psychology theories are usually on a much smaller scale; an example might be the theory of reminiscence (Eysenck & Frith, 1977). But we also have even more wide-ranging and far-reaching types of theory which begin to meld into something more closely akin to a weltanschauung. As an example we might take the doctrine of behaviorism.

Behaviorism, particularly the naive construction of Watson but also the neobehaviorism of later practitioners, clearly had its roots in French empiricism (Condillac, Bonnet, and Descartes properly understood) and French materialism (La Mettrie, Helvétius, and Cabanis). The objections of behaviorists to introspection sound in many ways similar to the French schools' objections to the concept of a soul, that is, a revolt against idealism in any of its forms. This philosophical stance has been the mainspring of behaviorism and is much more important than some of the accidental features which were the product of Watson's and Skinner's particular outlook, such as the denigration of physiology (shared with Condillac) and of genetic factors. Insofar as behaviorism is more than a methodology, it must be regarded as a philosophical doctrine and hence more in the nature of a weltanschauung rather than a scientific theory.

Another type of weltanschauung (or zeitgeist, as Boring prefers to call it) has already been mentioned in connection with work on intelligence; this is the belief in equality, also originating in its modern form in France (Rousseau). This belief has powerfully influenced American psychology in particular and has led to the neglect of genetic studies and the disregard of many empirical findings such as those mentioned in the last section. Egalitarianism of this kind is clearly a philosophical and political belief, not a scientific theory, although it may masquerade as such. The test of a scientific theory, of course, is the possibility of making deductions from it which are capable of being tested. Insofar as egalitarianism denies the importance of genetic factors it might be said to give rise to testable deductions, but clearly the refusal of egalitarians to carry out such tests, or to consider in their writings the results of tests carried out by geneticists, removes their beliefs from disproof. Theirs is not what Lakatos (1968) would call a program of research; their concern is entirely with what Lakatos calls a "protective belt of auxilliary hypotheses" which can be modified when empirical difficulties arise. But for any kind of research program that Lakatos envisages, such a protective belt surrounds a hard core of theoretical postulates and intepretations; it is this hard core that is completely missing in the weltanschauung of the egalitarian. His method is simply to deny the observed facts and to denigrate those whose experiments are responsible for the emergence of these facts.

Consider what Lakatos has to say about his general view:

> One of the crucial features of sophisticated falsificationism is that it replaces the concept of *theory* as the basic concept of the logic of discovery as a concept of *series of theories*. It is the succession of theories and not one given theory which is appraised as scientific or pseudo-scientific. Thus members of such series of theories are usually connected by a remarkable *continuity* which welds them into *research programmes*. This *continuity*— reminiscent of Kuhnian "normal science"—plays a vital role in the history of science; the main problems of the logic of discovery cannot be satisfactorily discussed except in the framework of a methodology of research programmes. (p. 84)

But the weltanschauung of the egalitarian does not give rise to such a series of theories, or such a methodology of research programs. It therefore departs decisively from science altogether, and egalitarianism should therefore not be regarded as a theory in the scientific sense; it is an expression of the zeitgeist, inspired by ideological and political beliefs and hostile to empirical demonstration in scientific discovery.

The degree to which egalitarianism and environmentalism are weltanschauungen is clearly shown in the vagaries of Marxist thought. During Stalin's reign, environmentalism and egalitarianism were fashionable among communists on both sides of the Iron Curtain, and Lysenkov's heresy was widely accepted. Since then there has been a return to a more correct interpretation of the thinking of Marx and Lenin, as shown for instance in the books on intelligence published by Guthke (1978) and Mehlhorn and Mehlhorn (1981) in East Germany, the work of Lipovechaya, Kantonistova, and Chamaganova (1978) on the heritability of intelligence among Russian school children, and many other sources. Mehlhorn and Mehlhorn, to take but one example, warn against the danger of calling scientists reactionary who point to the biological limitations in the development of intellectual functions and call such an attitude "*unmarxistisch*" (p. 7); they quote Marx and Engels' *Die Deutsche Ideologie* to illustrate the importance of genetic factors in Marxist thinking and go on to a quotation from Lenin (1965):

> When one says that experience and reason demonstrate that men are not equal, one understands under equality the equality of intelligence, or the similarity of physical strength and cognitive ability of men. It is of course clear that in this sense men are not equal. No single reasonable man and no single socialist can forget this. *This* equality has *nothing* to do with socialism. (p. 137)

Lenin goes on to say that to extend the notion of equality into this field is an "absurdity," and further that "when socialists speak of equality they always understand *social* equality, i.e., equality of social position but not equality of physical and mental abilities of individual persons" (p. 140). And the Russian psychologist Krutezki (1974, p. 140) concludes after a review of classical Marxist writings: "When it is said, 'from each according to his abilities', then it is clearly indicated that human beings in this respect are not equal." The conclusion Mehlhorn and Mehlhorn draw from their discussion is as follows:

> From the point of view of Marxism it is equally important to uncover the biological foundations and conditions of mental development of human beings as to study the specific influence of environment to the extent that these influence the development of mental capacities. (p. 8)

This view is largely identical with that always presented by western psychologists whose tendency to include genetic as well as environmental influences in their explanation of individual differences in intelligence has often led to their being castigated as anti-Marxian and even fascist. *Sic transit gloria mundi!*

An important consequence of these studies is a revision of the concept of discrimination, as widely used in sociology and politics. It is often suggested that the simple fact that a given group (women, blacks, children of low SES parents) is more or less frequently represented in certain social groups (university students, educationally subnormal classes, etc.) is sufficient evidence for the existence of discrimination of a racial or sexual kind in that field. Boudon (1973) defines inequality of educational opportunity and mechanisms of social immobility in these terms. By simply demonstrating that children of low SES parents are less frequently represented in student populations, he believes that he has demonstrated the existence of discrimination. But of course this would only be true if the children of high and low SES groups had similar abilities which enabled them to profit equally from university education; this is clearly not so. Yet throughout his book Boudon does not even mention the demonstrated existence of such differences, and his argument throughout is based on the unstated hypothesis of complete human equality with respect to intelligence. In a similar way political arguments are often based on the observed differences in proportions of blacks or other racial minorities in various socially defined or educationally relevant groups, but without precise knowledge of the actual IQs of the groups in question no deductions can be made about discrimination.

A recent government-inspired study of school performance of white, black, and Indian children in England demonstrated quite clearly the inferior performance of black West Indian children; the conclusion drawn was that racial discrimination was at the base of this. Yet the Indian children, against whom equal discrimination is shown in England, had school achievement scores actually superior to the white children, if anything! This agrees with the known IQ performance of these three groups of children, but the report made no mention of these IQ differences, nor did it draw attention to the fact that one group of socially disadvantaged and discriminated against children did better than the whites whereas the other did much worse. Thus the definition of discrimination has to be very carefully phrased in terms of known biological differences between groups.

We might say that generalized world concepts like egalitarianism are not even a "degenerating problem shift," to use Lakatos's term, but goes right outside the field of science altogether. What is interesting in much modern debate is this argument between rationality and irrationality, between science and zeitgeist, with the advocates of the latter trying to use the language of science but failing to come up to its spirit.

Such battles are of course not unheard of even in the field of the "hard sciences." Galileo's fight against the inquisition and Darwin's battle against the fundamentalists are still well remembered. A somewhat less well known case is that of atomism, the theory that nature is made up of atoms and their interaction. This theory, originating with Democritus and finally entering modern science through Dalton (Greenaway, 1966), was, as Bernal (1969) points out, not only a scientific but also a philosophical theory introducing materialism and radicalism into science. As he goes on to point out, in the nineteenth century the knowledge of thermodynamics began to reach into chemistry and even into biology, thanks to the work of Le Chatelier and Gibbs, and for a while it seemed that the whole of natural phenomena could be explained in terms of simple observables of mechanical energy and heat. In the hands of philosopher scientists like Mach (Blackmore, 1972), and chemists like Ostwald this produced a new positivism which stated that matter and physical hypotheses such as atoms were no longer necessary and that the whole of science could be deduced directly from elementary observations. Dumas in France and many others followed this lead and denied the existence of atoms until the work of Einstein and Perrin made such recalcitrance untenable. The kinetic theory of heat, evolved by Maxwell in 1866, had indeed implied the existence of atoms,but these were entirely hypothetical, and new evidence was required before these atoms could be accepted as measurable and countable material objects. The whole history of the atomic theory, and its belated acceptance by physicists, is a lesson in the determination of scientific theories by the zeitgeist and the prevailing weltanschauung.

The link between weltanschauung and such an all-embracing theory as behaviorism (if indeed this can be called a theory in the scientific sense) is of course complex and difficult to trace. Behaviorism as a methodology does not carry the implications of behaviorism as doctrine, but it is the latter which has had much the greater impact. After all, behaviorism as methodology simply preaches the virtue of the scientific approach; it adds little which is specific to psychology. Behaviorism as doctrine, however, does make a number of statements which, although not logically implicit in a behaviorist-methodological approach, are in good agreement with the American weltanschauung of the early 1920s. The ability to mold and manipulate people's behavior; the control over responses by controlling stimuli; the specificity of responses, and the absence of organized personality and cognitive structures; the determination of behavior by environment rather than by heredity; finally, the influence of social rather than physiological factors—all these mirror the hopeful and optimistic American outlook characteristic of that period, in which the application of science to psychological and social problems seemed to hold the answer to all the ills of mankind. A similar kind of zeitgeist prevailed in the Soviet Union, where the optimistic hope of being able to create "Soviet man" in the new image favored the conditioning and behavioristic proclivities of Pavlov, who became an honored hero.

Fundamental in this general approach was the notion of *specificity*. This thesis proclaimed essentially that all human behaviors were specific products of learned S–R connections and hence capable of environmental influence and change; this view, of course, goes back to Locke's *tabula rasa* image of the human mind.

Thorndike (1903) already preached this philosophy when he held that:

> There are no broad, general traits of personality, no general and consistent form of conduct which, if they existed, would make for consistency of behaviour and stability of personality, but only independent and specific stimulus–response bonds or habits. (p. 29)

This doctrine of "sarbondism" as McDougall used to call it, with its attending notions of the equipotentiality of the conditioned stimulus, has been decisively disproved in the fields of personality and intelligence (Eysenck, 1970, 1979), but it still persists in the public domain and is supported by ideologically motivated psychologists either unaware of or disregarding the established facts. Court judgments, based on evidence given by professional psychologists in contravention of the established facts, may illustrate the prevalence of this attitude.

First, consider the *Hobson* v. *Hansen* case, which abolished ability grouping in the schools of Washington, D.C., in 1967. Judge J. Skelly Wright stated:

> The skills measured by scholastic aptitudes tests are verbal. More precisely, an aptitude test is essentially a test of the student's command of standard English and grammar. The emphasis or these skills is due to the nature of the academic curriculum, which is highly verbal; without such skills a student cannot be successful. Therefore, by measuring the student's present verbal ability the test makes it possible to estimate the student's likelihood of success in the future. Whether a test is verbal or non-verbal, the skills being measured are not innate or inherited traits. They are learned, acquired through experience. It used to be the prevailing theory that aptitude tests—or "intelligence" tests as they are often called, although the term is obviously misleading—do measure some stable, predetermined intellectual process that can be isolated and called intelligence. To date, modern experts in educational testing in psychology have rejected this concept as "false". Indeed, the best that can be said about intelligence insofar as testing is concerned is that 'it is whatever the test measures.' . . . The IQ tests must be recognized as artificial tools to rank individuals according to certain skills.

In more recent years, in the well-known *Larry P.* v. *Wilson Ryles* case (1979), which resulted in prohibiting the use of standardized intelligence tests for the placement of black and Hispanic pupils in classes for the mentally retarded in California schools, Judge Robert Packham, in his final decision, expressed the specificity doctrine very succinctly: "IQ tests, like other ability tests, essentially measure achievement in the skills covered by the examinations . . . the tests measure the skills tested, and each of the tests subject to this litigation assesses very similar skills." He went on to quote the psychologist Leon Kamin as a witness for the plaintiffs as follows: "IQ tests measure the degree to which a particular individual who takes the test has experience with the particular piece of information, particular bits of knowledge, the particular habits and approaches that are tested in these tests."

The law, of course, as Dickens's Mr. Bumbles observed, is an ass; we do not expect judges to give expert rulings on scientific topics. That psychologists and educationalists should still be found to give evidence along these lines is proof of the overriding strength of the zeitgeist when confronted with indisputable facts; the facts are simply disregarded, as are the theories tested experimentally and giving rise to these facts. If IQ tests measure nothing but specific items of information, how is it that this information correlates so highly with nonverbal and culture-fair tests of the type exemplified by Raven's Matrices? How can items of information, each specifically acquired in isolation, give such high correlations with reaction time, inspection time, or evoked potentials? The specificity theory is not a theory in the scientific sense; it is an expression of a weltanschauung, a child of the zeitgeist, an ideological commitment. As such it is clearly proof against any factual disconfirmation, just as religious beliefs counter Galileo's and Darwin's factual evidence with irrelevant theological arguments. It is important for psychologists to recognize the difference between ideology and theory, the evocation of the zeitgeist and a scientific hypothesis. It is the absence of such recognition which makes the reading of psychological journals such a sad experience.

One of the worst consequences of the invasion of science by weltanschauung has been the interpretation of scientific data along preconceived lines, without even a consideration of alternative hypotheses. The developmental, educational, and clinical literature is full of examples of the *post hoc ergo procter hoc* fallacy, in which theoretically neutral findings are interpreted along the lines of one of several competing theories. Explanations are practically always framed in terms of the specificity-environmentalist hypothesis. Consider a study in which a child shows behavior *Y*, which is correlated with parental behavior *A*. This simple correlation is practically always interpreted causally in terms of the influence of *A* on *Y*; *A* is considered the independent variable, *Y* the dependent variable. However, clearly many other causal chains are conceivable and indeed have strong empirical support.

Suppose that the child's behavior is one of aggression, cruelty, and sadism, and the alleged environmental influence is that the parent cruelly beats the child at various times. Now it is, of course, theoretically conceivable that beatings of this kind may influence the child to behave in an aggressive and cruel manner later; if such a correlation is found, the data certainly do not contradict the hypothesis. But an alternative hypothesis would be that the genes which are partly responsible in causing the parents to beat the child savagely are inherited by the child and produce similarly cruel and sadistic behavior in him. There is much evidence for the genetic determination of aggressive and similar impulses, and the hypothesis certainly has stronger *a priori* support than does the much more widely accepted environmental one.

A third hypothesis is also tenable, of course, namely, that children who behave in an abnormally aggressive and sadistic manner are so difficult to manage that some parents are reduced to beating them savagely in order to establish their authority and preserve some degree of discipline. Other theories are of course also possible, and these three theories in turn are not mutually exclusive—all three causal chains may apply in individual cases, or even in all cases. What is suggested is simply that the existence of a correlation between *A* and *Y* does not *establish* any of these theories as correct; it simply fails to disconfirm any of them but gives no clue as to which (if any) is the correct causal theory—assuming (*pace* Hume) that causal theories are meaningful! In these circumstances, therefore, no causal interpretation can be given, and the fact that such causal interpretations along environmentalist lines are practically always given by the author (and accepted by the referees and the editor, who ought to be acting as guardians of the scientific conscience!) indicates the prevalence of a weltanschauung which renders both authors and judges blind to the realities of the situation. Clearly the theory itself is a scientific one in the sense that it can be disproved; but certainly the influence of the zeitgeist is felt in the *interpretation* of the facts along predetermined lines. Yet such studies are presented as being scientific in the strict sense, that is, an hypothesis is presented, an experiment is conducted to test the hypothesis, and the results are declared to bear out the hypothesis. What is neglected, of course, is an essential element of all scientific theory testing, namely, the consideration of alternative theories and the design of "crucial" experiments to decide between these theories. Scientists nowadays are aware of the difficulty of constructing a truly crucial experiment, but the complete neglect even to consider alternative theories is certainly not in line with the modern philosophy of science.

I have devoted a whole section to this discussion of differences between theories and weltanschauung and the disastrous effect which the latter can have on the testing of the former. Such a consideration is particularly relevant in psychology whereas in the "hard" sciences we no longer suffer from this type of indoctrination, at least to any noticeable degree. In biology the battle between fundamentalism and evolution has again emerged, at least in the United States, but biologists overwhelmingly reject the interjection of theological arguments based on a non-scientific weltanschauung. It is in psychology (and even more so in sociology) that the battle is joined most crucially. It is here that weltanschauung dictates to a large extent the interpretation of results, and even the very kinds of results which are acceptable. Jensen (1981b) gives some horrifying examples of this tendency. This is what he has to say on one case in a very large research project, one of potentially great public importance regarding the educational effects of school busing. He reports that the research was suddenly halted before it was half-completed, and the explanation given to him by a school official was that "The school system is a political unit, not a research institute, and cannot ignore political pressures in the community." Just two weeks before being treated to this shocking announcement, Jensen reports having been in Washington and having been told by a high government official in the White House that he was being overly naive to think that, at that time, he would be allowed to carry out bona fide research on the effects of school busing.

Other points made by Jensen (1981b) concern the funding of research. Referring to some instances, he says:

> They raise the question of the ethics of accepting research funds when there are strings attached as to the possible outcomes of the study, or restrictions on the reporting of results. (p. 15)

In one case, for example, the funding agency said they would consider supporting the research only if they could know beforehand the conclusions that the investigators intended to reach. In another instance, the funding agency would make the grant only if different racial groups than those originally proposed by the researchers were used in the study (whites and Asians had to be used instead of whites and blacks). In yet another case, a granting agency stipulated:

> Although data could be obtained on different racial groups, the researchers could not report group means or standard deviations, or any other statistics that might reveal the direction or magnitude of the group differences in scholastic abilities, but could report only correlations and factor analyses among different test scores. (p. 18)

These and many other examples quoted by Jensen show the grip which the zeitgeist has on research, whether concerning the funding of research, the organization of research, or the interpretation of results. It appears clear that explicit instruction in the differentiation between weltanschauung on the one hand and scientific theory on the other should form a very important part of the training of psychologists and sociologists. Only in this way will the vicious consequences of the disregard of this distinction be avoided.

6. The Biosocial Nature of Man: Theory or Metatheory?

If the very notion of *theory* is somewhat uncertain in the minds of psychologists, that of *metatheory* is probably even less clear. The term has been used by social psychologists who follow Marxian dialectics to denote the hypothesis that scientists are influenced in their work by social and economic factors and by the structure of the society in which they live, but the more critical followers of this view (e.g., Mackenzie, 1981) have acknowledged that while their views can be supported to some extent by arguments drawn fron the consideration of historical developments, theirs is not a scientific theory in the ordinary sense and cannot be falsified in ways that would satisfy Popper. Philosophical notions such as idealism or materialism might be thought of as metatheories, but they are too far removed from the nitty gritty of psychological research to be easily accommodated under that heading; it would be better to retain the title of *philosophical* for views of this kind. The same might perhaps be said about questions such as the body–mind relation; here the connection is closer, but testability is equally remote.

As an example of metatheory in psychology one might perhaps choose the conception of man which informs a psychologist's choice of theories, topic of research, and method of approach. We have already had occasion to refer to various different approaches to the nature of man, such as the respective weight given to genetic or environmental factors, but in an even more general way one might say that psychologists tend to be biophilic or sociophilic—in other words, they tend to prefer concepts of man emphasizing his *biological* nature or his *social* nature. It would here be contended that man is a biosocial animal and that there is no way in which we can understand man or interpret psychological data correctly unless and until we take into account this fundamental conception. This is not properly speaking a theory in the ordinary sense because it would be difficult to prove in a quantitative sense and specific deductions from it are difficult to make because the conception is such a very wide one, embracing the whole of psychology. However, the concept finds uses in directing research along the right channels and may be judged in relation to the extent of its success in doing so. The concept of man as a biosocial animal would certainly put an end to the absurd squabbles between those who embrace a 100% environmentalism and those who favor an equally exclusive biological reductionism.

It might be objected that such extremes are not really to be found in modern psychology, and that to oppose them would be to erect a man of straw. Up to a point, of course, such an objection would be justified; few psychologists are as incautious as Kamin (1974, 1981) who carries environmentalism to extremes by denying that there is any evidence for genetic factors in the causation of individual differences in intelligence. However, as pointed out in the last section, in practice many if not most psychologists certainly act as if they did hold such a view and interpret empirical data which are quite ambiguous in this respect entirely in terms of environmentalist concepts.

Few psychologists would be found at the opposite end of this continuum, but the recent doctrines of sociobiology as advocated by Wilson (1975, 1978), some of the authors in Caplan's book (1978) and many others certainly tend in that direction, although they are certainly less dogmatic and doctrinaire than extreme environmentalists like Kamin (1974). Wilson and his followers have taken up, at a higher level, the arguments of William McDougall originally advanced in his series of debates with Watson. McDougall's theory of instincts, of course, would not now be seriously supported by any biologist in its very primitive form, but in essence he was right and Watson was wrong in the importance attributed to the biological factors in human conduct.

What ended the supremacy of the behaviorists, of course, was the rise of the ethological school in Europe; Tinbergen, Lorenz, and many others demonstrated beyond any doubt the existence, importance, and specificity of mammalian instincts. McDougall had been right, in principle if not in detail, and Watson had been wrong; Watson's success in the argument had been a disaster for psychology, and we shall have to make good the years the locusts ate. With a realization of this sad calamity has come a realization of the importance of individual differences and of biological and genetic factors in psychology. In the treatment of mental disorder, to take but one example, behavior therapy, based on principles of conditioning pioneered by Pavlov, is taking the place of psychoanalysis, demonstrating greatly superior powers of alleviating distress (Eysenck, 1977). Personality theory, relating individual differences to biological factors (limbic system; reticular formation) is again getting into its stride (Eysenck, 1976a). Above all, genetic research into individual differences among human beings is again taking its rightful place, using new and much improved methods that were unheard of even a few years ago (Mather & Jinks, 1971).

These new developments are of crucial importance to any appraisal of sociobiology, although curiously enough this relevance has not hitherto been brought out clearly by Wilson or any of his followers. We may see how this comes about by looking at the three alternative states for our species which Wilson discusses in his foreword to Caplan's (1978) book. Either, he says, natural selection has exhausted the genetic variability underlying social behavior, or else the social genotype is uniform but prescribes a substantial amount of instinct-like behavior; or finally, some variability in human social behavior has a genetic basis and as a consequence at least some behavior is genetically constrained. He concludes:

> The evidence immediately available seems to leave room only for the last conclusion, that human social behaviour is to some extent genetically constrained over the entire species and furthermore subject to genetic variation within the species. (p. 3)

With this conclusion it would be difficult to quarrel (although as Caplan's book of readings shows, many people have managed to do just that!) It rests securely on two legs, one the phylogenetic type of evidence surveyed in Wilson's (1975) book, using evolutionary theory to account for human social behavior, the other the ontogenetic evidence of modern behavioral genetics, using the methods of biometrical genetical analysis to sort out the contributions to phenotypic variance of genetic and environmental factors.

Curiously enough, Wilson relies almost exclusively on the weaker of these two sources and seems to shun the stronger. In his first book he hardly ever mentions biometrical genetics; in his second book hardly more than two pages out of 260 are devoted to a desultory discussion of this evidence, and even this discussion is unsystematic, inaccurate, and not integrated with the remainder of the book. If there is to be a criticism of sociobiology, then I think it must be this failure to see that it stands securely on both feet, rather than totter insecurely around on one foot, with very little help from the other. If Wilson's argument had to rest on one line of evidence alone, then surely he has made the wrong choice; the ontogenetic argument is inherently the stronger because it rests on direct, experimental evidence rather than on brilliant argument from possibly shaky foundations, impossible in the nature of things to prove directly.

The major difference between Wilson's standpoint and mine is brought out very clearly in a sentence in his 1978 book:

> Human social behaviour can be evaluated . . . first by comparison with the behaviour of other species and then, with far greater difficulty and ambiguity, by studies of variation among and within human populations. The picture of genetic determinism emerges most sharply when we compare selected major categories of animals with the human species. (p. 84)

I would suggest that the argument from comparison with other species is beset by far greater difficulty and ambiguity than that from studies of variation among and within human populations; Wilson's own admission that "sociobiological theory can be obeyed by purely cultural behaviour" is ample evidence for this view.

Critics have sometimes suggested, as does Kamin (1974), that biologically oriented researchers favor this view because it supports the status quo, whereas socially oriented researchers favor environmentalism because it permits more freedom for social change. This belief that a person's scientific stance is determined by his political view is not borne out by historical fact. Watson, the archenvironmentalist, was also an archconservative; J. B. S. Haldane, one of the leaders of the genetic-biological camp and a precursor of sociobiology, was one of the leaders of the Communist Party in Great Britain! Noam Chomsky, too, is of the left wing politically, but favors genetic theories. *Argumenta ad hominem* arising from this ancient and often disproved notion should be laid to rest now; even if the correlation were perfect between social views and political affiliation, nevertheless the arguments in favor of either side would still have to be answered—throwing doubts on the scientist's motivation does not disprove his argument.

It is perhaps an ironic comment on the ideological onslaught which the presentation of genetic hypotheses in biology (Wilson, 1975), psychology (Eysenck, 1975a), history (Darlington, 1969), the study of race (Baker, 1974), and other social fields has provoked that ideology itself has been found to have strong genetic roots and to be intimately linked with personality factors genetically determined (Eaves & Eysenck, 1975, 1977; Eysenck & Wilson, 1978). In a large-scale twin study, Eaves and Eysenck (1974) found that radicalism–conservatism had a heritability of 65%; toughmindedness, a factor identifiable with ideological commitment, had a heritability of 54%. The tendency to voice extreme views, irrespective of right- or left-wing bias, had a heritability of 37%. This tendency and toughmindedness were found to be genetically connected with appropriate personality variables. It would thus appear that not only are left-wing ideologues wrong in assuming that scientists hold genetic views because they have been environmentally conditioned to defend the status quo; their own antigenetic views would appear to have a genetic basis! *Difficile est non satiram scribere.*

What is the upshot, substantively, of these considerations as far as the nature of human nature is concerned? In one sense, empirical studies simply support what common sense would unhesitatingly proclaim: man is a biosocial animal, whose aims and motives are shaped in part by his ancestral inheritance, in part by the pressure of the society in which he grows up and has his being. Curiously enough such a generalization would probably be approved by almost all geneticists, psychologists, biologists, sociologists, psychoanalysts, historians, and anthropologists who have given serious consideration to the problem; unfortunately such approval would be little but lip service in the majority. Even so, such lip service is the homage that vice pays to virtue; fundamentally we all know that nature and nurture are but the opposite sides of one and the same coin and that neither could exist without the other. The only real problem is a quantitative one; for particular groups and situations, what is the relative contribution of either? Such quantitative considerations demand a quantitative reply, and at present only the methods of biometrical genetical analysis can give us such an answer—qualified by the smallness of samples, their unrepresentative nature, and the unreliability of our measuring instruments, but nonetheless a first step in the unending quest for more precise information.

The general notion that man is a biosocial animal is of course too general to be of direct use in framing hypotheses. However, it can be rendered more precise by seeking the help of sociobiology and modern physiological research. As an example of how this can be done, consider the work of MacLean (1973) on "a triune concept of the brain and behaviour." As MacLean points out:

> Perhaps the most revealing thing about the study of man's brain is that he has inherited the structure and organisation of three basic types which, for simplification, I refer to as reptilian, old mammalian, and new mammalian. . . . It cannot be over-emphasized that these three basic brains show great differences in structure and chemistry. Yet all three must intermesh and function together as a *triune* brain. The wonder is that nature was able to hook them up and establish any kind of communication among them. (p. 7)

The hierarchy of these three brains is shown in Figure 5. Man's oldest brain is basically reptilian, forming the greater part of the upper brainstem and comprising much of the reticular system, midbrain, and basal ganglia. It is characterized by greatly enlarged basal ganglia which resemble the striatopallidal complex of mammals. The old mammalian brain (paleocortex) is distinctive because of the marked development of a primitive cortex which corresponds to the limbic system. Finally, there appears late in evolution a more complicated type of cortex, neocortex,

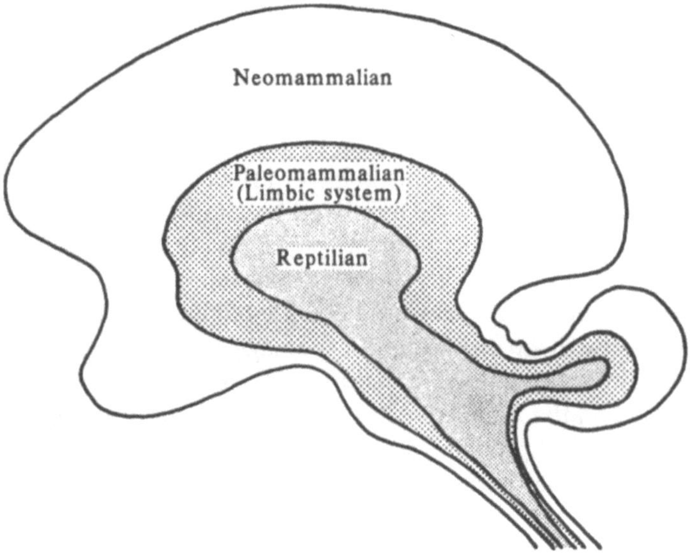

Figure 5. The triune theory of the brain.

which is the cerebral development characteristic of higher mammals and which culminates in man to become the organ mediating reading, writing, and arithmetic.

According to MacLean, the reptilian brain programs stereotype behaviors according to instructions based on ancestral learning and ancestral memories. He suggests that the mammalian counterpart of the reptilian brain is fundamental to such genetically constituted forms of behavior as selecting home sites, establishing territory, engaging in various types of display, hunting, homing, mating, breathing, imprinting, forming social hierarchies, and selecting leaders.

MacLean goes on to point out that the evolutionary development in the lower mammals of a rudimentary cortex appears to represent nature's attempt to provide the reptilian with a better means for adapting to its internal and external environment. In all mammals, most of this primitive cortex is found in a large convolution which Broca called "the great limbic lobe" because it surrounds the brainstem. (*Limbic* means "forming a border around.") This brain, as Papez was the first to show in 1937, plays an important role in elaborating emotional feelings that guide behavior in relation to two basic life principles of self-preservation and the preservation of the species. This limbic system has strong connections with the hypothalamus which plays a basic role in integrating emotional expression. MacLean goes into great detail regarding the structure and function of this system, but it cannot be part of this paper to follow him there. Suffice it to say that his discussion gives ample justification for regarding the limbic system as being concerned with emotion and with conditioning.

The neocortex, of course, is the characteristic mark of man and is concerned with reasoning, language, and number. Because of its great development in man, it has become customary to speak of man as *Homo sapiens*, but of course this is an absurd oversimplification. Man has a triune brain; the three parts are morphologically and functionally differentiated, and communication between them is relatively restricted. The neocortex may tell us how to achieve our wishes and desires, but the source of these is the reptilian brain, and the limbic system. Culture and other social concepts relate almost exclusively to the neocortex, but more primitive forms of behavior, immensely more powerful, relate to the reptilian brain and to the paleocortex. It would be equally wrong to deny the importance of the neocortex as it would be to deny the importance of reptilian brain and the paleocortex; recognizing the importance of all three gives us a picture of man as a biosocial animal, driven by primitive impulses and instincts but also possessing the means of rational adjustment to circumstances, evaluation of rewards and punishments, and powers of rational decision making.

The best example of the relative independence of these different parts of the triune brain is the existence of neurotic disorders. The behavior of neurotic patients is clearly determined by the paleocortex, probably through conditioned emotional reactions of one kind or another, and there is little or no communication with the neocortex. The patient *knows* perfectly well that his phobias, anxieties, and fears are irrational; this knowledge does not help him in the slightest to overcome his difficulties. The whole development of behavior therapy (Kazdin, 1978; Kazdin & Wilson, 1978; Rachman & Wilson, 1981), which has proved immensely more powerful in dealing with neurotic disorders than previous theories and methods, was premised on some such conception of man; insight therapies of the Freudian kind, which have been found seriously wanting, were based essentially on neocortical views of behavior emphasizing the importance of cognitive factors such as insight. Behavior therapy bases itself on laws of conditioning, and it has been shown that concepts such as those of extinction are basic to all forms of therapy insofar as these are successful (Eysenck, 1976a, 1979, 1980). It cannot, of course, be said that these developments prove the correctness of the view of man as a biosocial animal; all that can be claimed is that such a view leads to research which has been found to be eminently useful in this and other fields.

One obvious consequence of regarding man as a biosocial animal, and not looking upon his behavior exclusively from one point of view or the other, is clearly the abandonment of categorical viewpoints and the adoption of a quantitative way of looking at rival claims. We cease to regard behavior as being determined by environment or heredity and adopt rather the viewpoint that clearly all behavior is determined to some extent by both, as are individual differences; the task of science is to assess on a quantitative basis the contribution made by each of these factors. Indeed, analysis must go further and look at different ways of interaction and their contribution to the total variance. Also important is a breakdown of the total genetic variance into additive and nonadditive factors such as dominance, epistasis, and assortative mating. Environmental variance must be broken down into within-family and between-family, and so on. It is interesting to note that such a program of research (which is clearly a "progressive problem shift" in Lakatos's sense, is advocated and implemented by geneticists (Mather & Jinks, 1971). Nothing corresponding to it has been suggested by the environmentalists, who thus demonstrate their one-sidedness, their refusal to look at the total picture, in the most clear-cut fashion of all, that is, by using a methodology clearly inappropriate for the problems it is designed to solve. The conception of man as a biosocial animal, adopted as a guiding principle for research and not as a shibboleth deserving lip-service only, may thus be seen as a vitally important kind of metatheory for psychology (and sociology).

From the practical point of view, too, as has already been mentioned in relation to behavior therapy, such a conception is of extreme importance. Education, industry, government, and all areas of life involving intercourse with people could clearly improve their effectiveness by basing themselves on this principle, conducting research in conformity with it, and acting on the results of such research. To take but one example, consider the work of Hunter and Schmidt (1980), who have looked at methods of estimating the effectiveness of different selection strategies for various jobs. At the moment selection is largely disregarded and judicially frowned upon as leading to discrimination; furthermore, it is clearly a breach of the egalitarian shibboleth. Their analysis is essentially concerned with the probable cost of ignoring the kinds of information that can be provided by psychological tests of intelligence, personality, and so forth for job selection in all spheres of our national economy. They find that employee differences in job proficiency correspond to considerable difference in the actual dollar value of their performance.

Thus in a study of budget analysts, Schmidt and Hunter (1980) estimated that the dollar value productivity of superior performers (top 15%) was 23,000 dollars per year greater than that of low proficiency (bottom 15%). Computer programmers showed a comparable difference. Hunter and Schmidt point out that when these dollar losses are multiplied by the number of employees in an organization and by the number of years they are employed the losses quickly mount into millions of dollars.

In a study of the Philadelphia Police Department with 5,000 employees Hunter (1979) estimated that the abandonment of a general ability test for the selection of police officers would cost a total of $180 million over a 10-year period. The estimated gains in productivity resulting from one year's use of a more valid selection procedure for computer programmers in the federal government range from $5.6 million to $92.2 million for different sets of estimation parameters (Hunter & Schmidt, 1980). For the whole federal government with four million employees Hunter and Schmidt conservatively estimated that optimal selection procedures would save $16 billion per year!

The use of such tests is of course based on the concept of man as a biosocial animal, with genetically determined differences in abilities, personality, and so forth; these differences can be measured with considerable success and predictions made about future behavior on the basis of such tests. Failure to use these tests is based essentially on a picture of man vitally different from that given by modern science, and the resulting loss to society is of a size that should not be tolerated any longer. These figures are a clear indication of the deprivation which society imposes on itself by adopting a radically false picture of man. The concept of man as a biosocial animal may hence be regarded as a psychological metatheory which generates extremely important and testable consequences; it leads us to more specific theories, many of which have been tested and found to be borne out by the facts. Hence the importance of the metatheory and the need for defending it against ideological onslaughts of an unscientific kind.

7. Epilogue

This paper does not fit neatly into one of several categories which might be considered relevant, such as the philosophy of science, the history of science, the sociology of science, or the psychology of science. Having read fairly widely on all these topics, I believe that none of them can properly be studied without reference to the others; the dividing lines are arbitrary, and a proper understanding of science requires all these aspects to be taken into account. As a practicing scientist I have of course been most interested in seeing to what extent the debates, teachings, or pronouncements of these various groups are relevant to my own work and throw some light on the various areas of research in which I have been engaged. This paper is the outcome of such an active interchange between experiences gained in taking part in several scientific "revolutions" (Kuhn) or (hopefully) progressive problem shifts (Lakatos), on the one hand and consideration of the writings of philosophers, sociologists, and historians interested in these various aspects of science on the other.

Inevitably the outcome is somewhat untidy when looked at from the point of view of the professional philosopher, historian, or sociologist. It is my hope, nevertheless,that the paper will add to a better understanding of the place of theory in psychology. Most published works deal with theory in the more advanced sciences, and what is said there is not always relevant to a science which clearly is in a much less advanced stage. It is for this reason that I consider the developmental scheme outlined in Figure 1 to be particularly important and regard the distinction betweek weak and strong theories as potentially useful to research workers in psychology. Many a good theory is cut off in its prime because it does not come up to the standards of strong theories; had it been allowed to develop without having undue demands made upon it, it might have benefitted psychology considerably. Altogether psychology appears to oscillate between two extremes, both of which are equally unacceptable from the scientific point of view. One is the excessive leniency towards theories such as the Freudian; pronouncements along dynamic lines are uncritically accepted, without the usual requirement of any form of acceptable and reproducible proof. This whole area of psychology, which includes projective tests like the Rorschach, interpretations (as of dreams and stories), and therapies (psychotherapy and psychoanalysis), is clearly in a prescientific stage and should not be admitted to the group of scientific disciplines that make up psychology. There are certain minimal requirements which are not fulfilled by these dynamic theories, and until they are it would be wise to exclude theories of this kind from serious consideration.

The other extreme is the narrow exclusiveness often shown by so-called experimental psychologists who believe that strict adherence to relatively arbitrary criteria of methodology is the be-all and end-all of science and suffices to decide the acceptability or otherwise of a given study. Along these lines we have a long series of quite unimportant papers reporting experiments which are technically competent but do not make any valuable addition to psychological theory. Typically what happens is that someone invents a slightly different mode of investigating memory, or motivation, or whatever; this original paper is then followed by large numbers of replications, slight changes, and complex discussions and arguments, without any consideration of whether or not the original paradigm makes any real contribution to psychological theory. If psychology can learn to avoid the excesses of rigorous dogmatism, leading to *rigor mortis*, as well as the delinquencies of overspeculation of a Freudian kind, we may finally succeed in making psychology into a science, rather than, as William James expressed it, the hope of a science.

8. References

Achinstein, P. (1965). The problem of theoretical terms. *American Philosophical Quarterly, 2,* 193–203.
Baker, J. R. (1974). *Race.* London: Oxford University Press.
Bergman, G. (1951). The logic of psychological concepts. *Philosophy of Science, 18,* 93–110.
Bergman, G. (1954). *The metaphysics of logical positivism.* New York: Longman, Green.
Bergman, G. (1957). *Philosophy of science.* Madison, WI: University of Wisconsin Press.
Bergman, G., & Spence, K. (1941). Operationism and theory in psychology. *Psychological Review, 48,* 1–14.
Bernal, J. P. (1969). *Science in history* (4 vols.). London: C. A. Watts & Co.
Blackmore, J. T. (1972). *Ernst Mach.* Berkeley: University of California Press.
Boudon, R. (1973). *Education, opportunity, and social equality.* London: Wiley.
Caplan, A. C. (Ed.). (1978). *The sociobiology debate.* London: Harper & Row.
Carnap, R. (1962). *Logical foundations of probability.* New York: Basic Books.
Carnap, R. (1966). *Philosophical foundations of physics.* New York: Basic Books.
Cohen, M. R., & Nagel, N. (1936). *An introduction to logic and scientific method.* New York: Harcourt, Brace & Co.
Cronbach, L. J. (1957). The two disciplines of scientific psychology. *American Psychologist, 12,* 671–684.
Darlington, C. D. (1969). *The evolution of man and society.* London: Allen & Unwin.
Dashiell, J. F. (1938). Some rapprochements in contemporary psychology. *Psychological Bulletin, 36,* 1–24.
Eaves, L. J., & Eysenck, H. J. (1974). Genetics and the development of social attitudes. *Nature, 249,* 288–289.
Eaves, L. J., & Eysenck, H. J. (1975). The nature of extraversion: A genetical analysis. *Journal of Personality and Social Psychology, 32,* 102–112.
Eaves, L. J., & Eysenck, H. J. (1977). Genotype-environmental model for psychoticism. *Advances in Behaviour Research & Therapy, 1,* 5–26.
Eddington, A. (1920). *Space, time and gravitation.* Cambridge: Cambridge University Press.
Eysenck, H. J. (1939). Primary mental abilities. *British Journal of Educational Psychology, 9,* 270–275.
Eysenck, H. J. (1950). Criterion analysis—An application of the hypothetico-deductive method to factor analysis. *Psychological Review, 57,* 38–53.
Eysenck, H. J. (1957). *Dynamics of anxiety and hysteria.* London: Routledge & Kegan Paul.

Eysenck, H. J. (1959). Serial position effects in nonsense syllable learning as a function of interlist rest pauses. *British Journal of Psychology, 50,* 360–362.
Eysenck, H. J. (1967). *The biological basis of personality.* Springfield: C. C Thomas.
Eysenck, H. J. (1970). *The structure of personality.* London: Methuen.
Eysenck, H. J. (1975a). *The inequality of man.* London: Temple Smith.
Eysenck, H. J. (1975b). The measurement of emotion: Psychological parameters and methods. In L. Levi (Ed.), *Emotions—Their parameters and measurement.* New York: Raven Press.
Eysenck, H. J. (1976a). The learning theory model of neurosis—A new approach. *Behaviour Research and Therapy, 14,* 251–267.
Eysenck, H. J. (1976b). *The measurement of personality.* Lancaster: Medical and Technical Publishers.
Eysenck, H. J. (1977). *You and neurosis.* London: Temple Smith.
Eysenck, H. J. (1979). The conditioning model of neurosis. *Behavioral and Brain Sciences, 2,* 155–199.
Eysenck, H. J. (1980). A unified theory of psychotherapy, behaviour therapy and spontaneous remission. *Zeitschrift für Psychologie, 188,* 43–56.
Eysenck, H. J. (1981). *A model for personality.* New York: Springer.
Eysenck, H. J. (Ed.). (1983). *A model for intelligence.* New York: Springer.
Eysenck, H. J. (1985). The theory of intelligence and the psychophysiology of cognition. In R. J. Sternberg (Ed.), *Advances in the psychology of human intelligence.* New York: Academic Press.
Eysenck, H. J., & Barrett, P. (1985). Psychophysiology and the measurement of intelligence. In C. R. Reynolds & V. Willson (Eds.), *Methodological and statistical advances in the study of individual differences.* New York: Plenum Press.
Eysenck, H. J., & Frith, C. D. (1977). *Reminiscence, motivation and personality.* New York: Plenum Press.
Eysenck, H. J., & Wilson, G. D. (1973). *The experimental study of Freudian theories.* London: Methuen.
Eysenck, H. J., & Wilson, G. D. (1978). *The psychological basis of ideology.* Lancaster: Medical and Technical Publishers. Baltimore: University Park Press.
Feyerabend, P. K. (1975). *Against method.* London: New Left Books.
Greenaway, F. (1966). *John Dalton and the atom.* London: Heinemann.
Grünbaum, A. (1976). Is falsifiability the touchstone of scientific rationality? Karl Popper versus inductivism. In R. S. Cohen, P. K. Feyerabend, and M. W. Wartofsky (Eds.), *Essays in memory of Imre Lakatos.* Dordrecht: R. Reidel.
Grünbaum, A. (1977). How scientific is psychoanalysis? In R. Stern, L. Horowitz, and J. Lynes (Eds.), *Science and psychotherapy.* New York: Haven Press.
Grünbaum, A. (1979). Is Freudian-psychoanalytic theory pseudo-scientific by Karl Popper's criterion of demarcation? *American Philosophical Quarterly, 16,* 131–141.
Grünbaum, A. (1981). Can psychoanalytic theory be cogently tested "on the couch"? In A. Grünbaum & L. Landon (Eds.), *Pittsburgh series in philosophy and history of science.* San Francisco: University of California Press.
Guthke, J. (1978). *Ist Intelligenz Messbar?* Berlin: Deutscher Verlag der Wissenschaften.
Hanson, N. R. (1958). *Patterns of discovery.* Cambridge: Cambridge University Press.
Harré, R. (1975). *Problems of scientific revolutions.* Oxford: Clarendon Press.
Hempel, C. G. (1952). *Fundamentals of concept formation in empirical science.* Chicago: University of Chicago Press.
Hempel, C. G. (1965). Aspects of scientific explanation. In C. G. Hempel (Ed.), *Aspects of scientific explanation and other essays in the philosophy of science.* New York: Free Press.

Hempel, C. G. (1966). *Philosophy of natural science.* Englewood Cliffs, NJ: Prentice-Hall.
Hunter, J. E. (1979). *An analysis of validity, differential validity, test fairness, and utility for the Philadelphia Police Officers Selection Examination prepared by the Educational Testing Service.* Report to the Philadelphia Federal District Court, *Alvaraz* v. *City of Philadelphia.*
Hunter, J. E., & Schmidt, F. L. (1980). Fitting people to jobs: The impact of personnel selection on national productivity. In E. A. Fleishman (Ed.), *Human performance and productivity.* New York: Academic Press.
Jensen, A. R. (1962). Extraversion, neuroticism and social learning. *Acta Psychologica, 20,* 69–77.
Jensen, A. R. (1981a). *Test validity,* g *versus the specificity doctrine.* Invited address, Division 14, presented at the annual convention of the A.P.A., Los Angeles, August 26.
Jensen, A. R. (1981b). Taboo, constraint and responsibility in educational research. *New Horizons,* No. 22, 11–20.
Kamin, L. J. (1974). *The science and politics of I.Q.* London: Wiley.
Kamin, L. J. (1981). In H. J. Eysenck & L. J. Kamin (Eds.), *The intelligence controversy.* New York: Wiley.
Kazdin, A. E. (1978). *History of behavior modification.* Baltimore: University Park Press.
Kazdin, A. E., & Wilson, G. T. (1978). *Evaluation of behavior therapy.* New York: Ballinger.
Krige, J. (1980). *Science, revolution and discontinuity.* Sussex: Harvester Press.
Krutezki, W. A. (1974). Die Entwicklung Leninscher Ideen in der sowjetischen Psychologie der Fähigkeiten. In W. A. Krutezki (Ed.), *Lenins philosophisches Erbe und Ergebnisse der sowjetischen Psychologie.* Berlin: Springer.
Kuhn, T. S. (1962). *The structure of scientific revolutions.* Chicago: University of Chicago Press.
Lakatos, I. (Ed.). (1968). *The problem of inductive logic.* Amsterdam: North-Holland.
Lakatos, I., & Musgrave, A. (Eds.). (1970). *Criticism and the growth of knowledge.* Cambridge: Cambridge University Press.
Lenin, W. I. (1965). Ein liberaler Professor über die Gleichheit. In W. I. Lenin (Ed.), *Werke* (Bd. 20). Berlin: V.E.B.
Lipovechaya, N. G., Kantonistova, N. C., & Chamaganova, T. G. (1978). The role of heredity and environment in the determination of intellectual functions. *Medicinskie Problemy Formirovanija Livcnosti,* 48–59.
Mackenzie, D. A. (1981). *Statistics in Britain: 1865–1930.* Edinburgh: University of Edinburgh Press.
MacLean, P. D. (1973). A triune concept of the brain and behavior: Including psychology of memory and sleep and dreaming. In T. J. Bord & D. Campbell (Eds.), *Brain and behavior.* Toronto: University of Toronto Press.
Marrow, A. (1969). *The practical theorist.* New York: Basic Books.
Mather, K., & Jinks, J. L. (1971). *Biometrical genetics.* London: Chapman & Hall.
Mehlhorn, G., & Mehlhorn, H. (1981). *Intelligenz.* Berlin: Deutscher Verlag der Wissenschaften.
Popper, K. R. (1959). *The logic of scientific discovery.* London: Hutchinson.
Popper, K. R. (1974a). *Conjectures and refutations.* London: Routledge & Kegan Paul.
Popper, K. R. (1974b). Replies to my critics. In P. A. Schilpp (Ed.), *The philosophy of Karl Popper.* La Salle: Open Court Publishing.
Putnam, H. (1962). What theories are not. In E. Nagel, P. Suppes & A. Tarski (Eds.), *Logic, methodology and philosophy of science.* Stanford: Stanford Press.
Quine, W. V. O. (1962). *From a logical point of view.* Cambridge: Harvard University Press.
Rachman, S., & Wilson, T. (1981). *The effects of psychological therapy.* New York & London: Pergamon Press.

Roley, T. B. (1959). An opinion on the construction of behavior therapy. *American Psychologist, 14,* 129–134.
Rosenbaum, G. (1953). Stimulus generalization as a function of level of experimentally induced anxiety. *Journal of Experimental Psychology, 45,* 35–43.
Rosenbaum, G. (1956). Stimulus generalization as a function of clinical anxiety. *Journal of Abnormal and Social Psychology, 53,* 281–285.
Savage, R. D., & Eysenck, H. J. (1964). The definition and measurement of emotionality. In H. J. Eysenck (Ed.), *Experiments in motivation.* London: Pergamon Press.
Schmidt, F. L., & Hunter, J. E. (1980). *New research findings in personnel selection: Myths meet realities in the 1980s.* Public Personnel Administration: Policies and Procedures for Personnel. New York: Prentice-Hall.
Spearman, C. (1927). *The abilities of man.* New York: Macmillan.
Spence, K. W. (1944). The nature of theory construction in contemporary psychology. *Psychological Review, 51,* 47–68.
Suppe, F. (1974). *The structure of scientific theories.* Chicago: University of Illinois Press.
Tarski, A. (1941). *Introduction to logic and to the methodology of deductive sciences.* New York: Oxford University Press.
Tarski, A. (1956). *Logic, semantics, metamathematics.* Oxford: Clarendon Press.
Thorndike, E. L. (1903). *Educational psychology.* New York: Teachers College.
Toulmin, S. (1953). *The philosophy of science.* London: Hutchinson.
Vaucouleurs, G. de (1957). *Discovery of the universe.* London: Faber.
Williams, L. P. (1965). *Michael Faraday.* London: Chapman & Hall.
Wilson, E. O. (1975). *Sociobiology: The new synthesis.* London: Harvard University Press.
Wilson, E. O. (1978). *On human nature.* London: Harvard University Press.

2

The Use of Theory in Psychology

Edward Erwin

Eysenck's nonrealist account of the role of theory in psychology is new and provocative. In what follows, I compare his view to other kinds of nonrealism in psychology and then state a realist's response.

A *realist*, as I will use the term, is someone who holds that scientific theories are literally true or false and should be judged accordingly. (For a somewhat similar definition, see van Fraassen, 1980, p. 8.)

In saying that a theory is *literally* true or false, I mean to exclude the view that scientific theories are nothing more than abbreviations for observation statements. For example, someone who holds that quantum theory is true but that there are no elections—"There are elections" is shorthand for some observation statement—is not a realist in this sense. Popper (1962), then, is a realist, but Eysenck in Chapter 2 is not. Rather, Eysenck is not a realist about "weak" theories, the kind that typically appear in psychology; in his view they should be judged not on the basis of being right or wrong, but on pragmatic criteria.

> The value of a weak theory, to put it briefly, lies in the fact that *it directs attention to those problems which most repay study from a systematic point of view;* in Thomson's words, it defines a policy of action and research. It is by giving rise to worthwhile research, rather than by necessarily being right, that a weak theory makes its greatest contribution to science. (p. 24, Eysenck's italics)

1. Other Nonrealist Accounts

1.1. The "No Theory" View

One kind of nonrealism that is still important in American psychology holds that theories need not be used at all. The view has been associated with B. F. Skinner and has had some impact on those working

Edward Erwin • Department of Philosophy, University of Miami, Coral Gables, Florida, 33124.

in the operant tradition. For example, Biglan and Kass (1977) report that one operant-oriented journal, *The Journal of Applied Behavior Analysis*, will not publish research unless it concerns what is publicly observable (this may not reflect current editorial policy). This stress on the observable is consistent with Skinner's (1950) suggestion that the most rapid progress toward an understanding of learning may be made by research not designed to test theories. His alternative is to collect data showing orderly behavioral changes and to relate the data to manipulable and observable variables, selected for study through a commonsense exploration of the field.

Skinner does not, in fact, reject *all* theories, but merely psychological theories that try to explain an observed fact by appeal to events taking place somewhere else, at some other level of observation, described in different terms, and measured, if at all, in different dimensions. He made clear, however, that all physiological and mentalistic theories of learning are to be excluded; this would rule out several of the theories discussed by Eysenck in the current paper.

It is sometimes objected that Skinner himself is tacitly employing a theory. Why else would he recommend operant research (Skinner, 1950) unless he is conjecturing that learning is caused by such variables as deprivation and changes in reinforcement schedules? Skinner could reply, however, that his theory does not fall within the class of those he finds objectionable; it does not try to explain behavioral changes by appeal to events taking place at some other level of observation. This reply invites another question. Why is the "other level" type of theory not worth pursuing, whereas Skinner's kind of theory is? Stating Skinner's answer is complicated by the fact that he has special objections to mentalistic theories, but he does have at least one criticism of *all* other level theories. That a theory generates research, he points out, does not prove its value unless the research is valuable, but research designed with respect to a theory is likely to be wasteful. Most psychological theories are eventually overthrown, and the greater part of the associated research is then discarded. Skinner is employing here the same criterion as Eysenck in judging theories; their disagreement concerns the usefulness of certain "other level" theories.

The weakness in Skinner's argument, I think, is its reliance on inadequate data. To show that it is likely that no "other level" psychological theory will be valuable, it is not sufficient to look at the poor track record of previously discarded theories. The evidence from past failures may be overridden by the evidence that a current theory is generating fruitful research. Eysenck could reply to Skinner, then, by simply citing the evidence discussed in his paper that certain other level theories have already proved to be of some value in generating worthwhile research.

It may be unfair to critize a position that Skinner elaborated more than 30 years ago, but a more recent paper (Skinner, 1977) suggests that he has not changed his mind about other level theories. He begins by asserting that the variables of which human behavior is a function lie in the environment. If that is true, then his own research program is on the right course, and there is no need of other level theories, whether cognitivist or physiological, to explain human behavior. The issue, then, between Eysenck and Skinner becomes an empirical one concerning the causes of human behavior. On this issue, Eysenck has already replied to Skinner (Eysenck, 1976); the evidence cited in his current paper provides additional, powerful evidence that Skinner is betting on the wrong research program and that his rejection of all other level theories is unjustified.

1.2. Abbreviation Views

A second form of nonrealism that has had some influence in psychology holds that theories may be used but only if they are translatable into observation statements. Sometimes the thesis is extended to all theories and is defended on the grounds that all meaningful theoretical terms must be operationally definable in observation terms. A more restricted version, usually termed *logical behaviorism*, is applied only to cognitivist theories (or any theory that postulates mentalistic causes). The restricted version has been used to question the intrusion of cognitive theories into the field of behavior therapy (see, e.g., Marzillier, 1979; Waters & McCallum, 1973). Eysenck, too, has appealed to this view. In replying to a discussion of mine (Erwin, 1978), he writes:

> Erwin makes much of the fact that behavior therapists talk and write using mentalistic terms; but this is just a shorthand and does not commit them to a mentalistic, cognitive type of theory. To say "Bloody hell!" when you hit your thumb with a hammer does not commit you to a belief in the existence of the nether regions; we have become conditioned to the use of certain verbal formulations, and continue to use them, particularly when the theoretically correct usage would take many more words, and would be intolerably clumsy. (Eysenck, 1979, p. 513)

Eysenck, then, apparently holds two different positions with respect to theories; the positions are logically compatible but reflect very different attitudes. In the view expressed in Eysenck's current paper, psychological theories have a useful role to play: their utility lies in their ability to generate useful research. In the second view, which applies only to cognitivist theories, a theory has no useful role to play except as an aid to concise expression. Cognitivist theories are mere shorthand for statements about observable matters; whatever they say can be said without using them.

The view that cognitivst theories are abbreviations has been criticized often in the philosophical literature, usually under the heading of logical behaviorism; there is no need to discuss the standard criticisms here (see Erwin, 1978, Chapter 2). It might be replied, however, that such criticisms show only that sentences containing mentalistic expressions used in their *ordinary* sense cannot all be translated into a behavioristic language. This leaves open the following question: is the cognitivist language of behavior therapists, which may be used in a technical sense, amenable to such a translation? A proper answer to this latter question depends, I think, on which behavior therapists we are talking about. Some do use sentences containing cognitive terms as a shorthand for observation statements, but some do not. For example, in his discussion of conditioning, Bandura (1974) makes the point that, contrary to mechanistic metaphors, outcomes change behavior in humans through the intervening influence of thought. Whatever other behavior therapists might mean by using the term *thought*, Bandura clearly does not mean *behavior*; he is contrasting his view with that of those behaviorists who contend that the environment affects behavior in an automatic and mechanistic fashion. Also, in explaining the effectiveness of behavior therapy in terms of beliefs about self-efficacy, Bandura (1977) is not using an abbreviation for a statement about behavior; whether his theory is useful or not, it does postulate a certain kind of cognitive cause. Again, in research reports on systematic desensitization (Paul, 1966) and aversive imagery (Lichstein & Lipshitz, 1982), behavior therapists often refer to mental images in explaining how these therapies work. They are not merely talking about verbal reports of mental images but are apparently referring to mental images in the same sense that certain experimentalists do (e.g., Kosslyn, 1980). There is no reason to think that all such statements about mental imagery are mere shorthand for statements about behavior.

1.3. Other Nonrealist Views

There are other types of nonrealism discussed in the philosophy of science literature. For example, on one account, theories can be eliminated by appeal to Craig's theorem. Another view, called *constructive empiricism*, is, I believe, consistent with Eysenck's view of weak theories, but it is supported by quite different arguments; it also applies to all theories, whereas Eysenck restricts his account to weak theories. Neither of these kinds of nonrealism, as far as I can tell, has had any impact in psychology so far, so I will not discuss them here. (For a useful discussion of Craig's theorem, see Nagel, 1961; constructive empiricism is defended by Van Fraassen, 1980). I turn now to Eysenck's view.

2. Eysenck's View

One of Eysenck's most useful points is that we must distinguish between different kinds of theories. Philosophers of science tend to take their main examples from physics,but such theories may differ significantly from those typically employed by psychologists. It is possible, of course, to retreat to a very high level of abstraction where the differences disappear, and for some purposes this is desirable; but to a psychologist the differences may be important. Eysenck proposes that one important difference is that psychological theories tend to be weak in contrast to strong theories that sometimes appear in physics. A strong theory, such as Newton's theory of gravitation, tends to have the following features: (1) It is based on a very large number of accurate observations, made over many years by large groups of people. (2) It brings together a number of subfields in which quantitative laws have already been discovered and verified. (3) The phenomena covered by the theory are relatively clear-cut and unambiguous; in particular, they are not embedded in or entangled with groups of other phenomena. (4) The mathematical relations tend to be not of a very complex order. (5) Largely as a result of having the preceding features, the theory's predictions are straightforward and precise; verification and confirmation of deductions do not give rise to special problems. (6) The postulates are interdependent; it is not possible to change one without changing the rest.

Weak theories tend to have the opposite characteristics, even if the differences are often a matter of degree. They tend to be supported by relatively few observations; there are few, if any, established quantitative laws in subfields; the nature of the phenomena is not clear-cut; changes in one part of the theory can be made without changing the rest, and so forth.

The value of a weak theory, as noted earlier, is its ability to direct attention to those problems which most repay study from a systematic point of view. Its main purpose is not to generate predictions that would confirm or disconfirm it, but to generate research concerning certain problems which, but for the theory in question, would not have arisen in that particular form.

It seems to me that most of what Eysenck says about theories in psychology is true. A realist, then, should be prepared to make certain concessions, some of which are required by Eysenck's arguments and some of which should be made for other reasons. First, if we want to say something relatively concrete about theories in psychology, then we need to investigate theories actually employed by psychologists. Second, one virtue of a psychological theory is its ability to generate useful research, and it may possess that virtue even if it is false. Third, to focus primarily on determining the truth or falsity of a theory, especially in the early stages of research, may result in premature rejection of a potentially fruitful theory. Fourth, although Eysenck does not stress this point, a realist should agree, as most do, that even when truth is important, we are primarily interested in whether or not a theory is *approximately* true. Fifth, in certain cases, a term in a theory may denote an ideal entity or process and not something that actually exists. (For example, as psychologists become interested in economic behavior, they, like their economist counterparts, may talk about behavior under conditions of perfect competition, even though such conditions rarely if ever exist.) Having made these concessions, a realist might then make the following points, some of which may reflect only a difference in emphasis from Eysenck's account.

1. Eysenck is concerned about premature rejection of potentially fruitful theories. An example he gives is that of his own theory, developed in his *Dynamics of Anxiety and Hysteria,* which predicted that extraverted people would have greater reminiscence effects after massed practice than would intraverted people. All of the investigators who tested the prediction using a practice period of ninety seconds failed to confirm it, but all who used a practice period of five minutes verified it. At the time the studies were done, knowledge of the subfield of reminiscence was too restricted to specify a required practice period. Had investigators used only a ninety-second practice period, disconfirmations would have been the rule and Eysenck's theory might been rejected. Eysenck sees this problem of premature rejection as arising from an inappropriate concern with truth and falsity, but will not the problem also arise if we use a pragmatic criterion? Whether or not a theory is judged useful for research purposes depends on several factors, one of which is the availability of information concerning subfields. A theory judged not very useful at one time may be seen, with justification, to be extremely useful after new knowledge is acquired. Conditioning theories of treatment for neuroses were available as early as the 1920s but, to judge by the paucity of subsequent research, were not thought by many to be useful during

the 1930s and 1940s. In the 1950s, as new evidence was obtained concerning the absence of symptom substitution, the weak data base for psychodynamic therapies, the apparent usefulness of systematic desensitization, and so forth, conditioning theories of treatment were judged to be useful for research purposes. Another example is the use of theories of mental imagery. During the 1930s and 1940s, research on mental imagery flagged for several reasons; one of these was probably lack of knowledge about how to test relevant theories without relying entirely on introspective reports. As more sophisticated experimental techniques became available through the work of several investigators (e.g., Kosslyn, 1980; Paivio, 1971) research on theories of mental imagery began to flourish. I am not saying that exactly the same theories about mental imagery that were proposed fifty years ago are beng employed today, but rather that the rejection of earlier theories, whether or not they were potentially useful, was due partly to a lack of information about how to investigate them. Other examples are theories concerning genetic contributions to schizophrenia and intelligence. Some psychologists dissatisfied with traditional studies of twins reared apart have been skeptical about the usefulness of such theories (see, for example, Ullmann & Krasner, 1969); they have been dubious about the prospects of tearing apart genetic and environmental causative factors. Such skepticism has become less warranted as new research techniques have become available, including the adoptee method used by Kety, Rosenthal, and others in studying schizophrenia (Rosenthal, 1971) and some of the methods cited by Eysenck for studying intelligence.

Some of the above examples require more discussion, but what I hope they illustrate is that the adoption of a pragmatic criterion is no guarantee that useful theories will not be prematurely rejected.

2. If good psychological theories too often die young, bad theories often persist in the literature far too long. One example that Eysenck mentions is psychoanalysis. Some philosophers (e.g., Karl Popper) would claim that it is untestable and not a scientific theory, but Grünbaum (1979) and others, I think, have refuted this claim. It is difficult but not impossible to test large portions of Freud's theory, and Freud and some of his followers made some efforts to do so. One problem has been the overreliance on clinical evidence, which has proved inadequate at least for confirmation (Grünbaum, 1984). The interpretation of the experimental evidence is still controversial (see Fisher & Greenberg, 1977), but I agree with Eysenck and Wilson (1973) that much of it is of poor quality. My point, then, is that psychoanalytic theory has been (and in some quarters still is) accepted—with profound effects on the development of

clinical psychology—because not enough was done to determine the truth or falsity of Freud's hypotheses. To take one important example, had the symptom substitution hypothesis been tested and rejected earlier, the development and acceptance of behavior therapy techniques might have been much more rapid. This is only one example, but the realist has a more general concern: To put aside efforts to determine the truth or falsity of weak theories is likely to result in the persistence of error, with all its harmful consequences. I may be unduly extending Eysenck's position; he does not say that we should never ask whether a weak theory is true. My point is relevant, however, to anyone who recommends using only a pragmatic criterion in judging theories (e.g., Yates, 1970, p. 396).

3. Farrell (1981) has recently argued that we still need Freudian theory because of its heuristic value; he means its ability to stimulate research. In support of his claim, consider a couple of examples. Blum (1954) makes clear that his research on perceptual defense was inspired by Freudian theory. As a second example, consider the Sarnoff and Corwin (1959) study of the relationship between castration anxiety and the fear of death, a topic also suggested, as the authors make clear, by Freudian theory. In each case, and in many others, research was done on certain problems which, except for the use of Freudian theory, would not have been done in that particular form. It looks, then, as if Freudian theory is not so bad if we use Eysenck's criterion for judging a theory. It might be argued, however, that an additional condition must be met; the research generated by the theory must, as Eysenck says, be "worthwhile." But how are we to judge what is worthwhile research? There is more than one way to judge a study, but one relevant question to ask is this: are the assumptions that prompted the study true, or at least plausible? In the Blum (1954) study it was assumed that repression is an important causal factor in perception. If that assumption is false, or without warrant, there is little point to using the particular experimental paradigm that Blum used, although there would still be reason to investigate perceptual defense effects in some other fashion. The Sarnoff and Corwin (1959) investigation assumed the existence of castration anxiety; it also presupposed a certain theory about the Blacky Test, that the test does measure castration anxiety. If either the theory about the Blacky Test or the Freudian assumption that castration anxiety occurs is false, the point of doing the study is lost. I am not saying that there was no reason to do either of the above two studies—that depends partly on the evidence available to the investigators at the time of the study—but rather, looking back, we can judge whether or not the research proved to be worthwhile, and one way to do this is to ask about the truth of

certain weak theories that were being tested or which were presupposed by the researchers. If this is right, then questioning the truth or falsity of weak theories will at least sometimes be important in judging the research value of a theory. To forgo asking questions about the truth of weak theories is to abandon an important tool for judging their utility.

4. Many nonrealists tend to stress the importance of prediction and control and to downplay the role of explanation in science. Eysenck, however, does not do this, and I think he would agree that explanation is important; but then he has a reason to be concerned about the truth or falsity of certain weak theories. Consider a theory which he discusses here, that the only important variable in intelligence is the individual's innate structure of the CNS enabling this system to propagate messages correctly across axons and synapses. If we want to explain what makes an individual intelligent, then we should want to know if this theory is correct, or approximately correct. My point is that someone who agrees with the realist in valuing illuminating scientific explanations has a reason to be interested in the truth or falsity of at least some weak theories. There are also practical reasons to be concerned about truth or falsity. One already mentioned is that determination of correctness is an aid to evaluating research. Another is the need to predict the usefulness of a theory in guiding research in new areas. For example, operant conditioning theory proved to be useful in generating research on token economy programs, which themselves have proved useful (Paul & Lentz, 1977). Some operant theorists, however, have believed that operant theory can also explain verbal behavior (Skinner, 1957), schizophrenic behavior (Ullmann & Krasner, 1969),and indeed almost all voluntary behavior. There is a good practical reason to examine the evidence for such beliefs: If operant theory when extended to the aforementioned areas can be shown to be implausible, then the wasting of very many research hours can be avoided. I think these points become obvious once one grants that even weak theories are true or false and that explanation is important, but I think the realist should highlight what is sometimes overlooked by nonrealists. Eysenck, however, might agree to these points but argue that in the case of weak theories there are overriding reasons to place more emphasis on utility than on correctness. I turn, finally, to these reasons.

5. I think Eysenck sees two main problems in being concerned about correctness: (a) It is very difficult to disprove a weak theory, and (b) many good (i.e., useful) theories are rejected prematurely because they do not come up to the standards of strong theories. A realist must explain how these problems can be avoided without abandoning realism.

One might begin by contrasting a simplistic view of falsification with the picture developed by Eysenck. In the simplistic view, widely rejected now by philosophers of science, a theory has empirical content, consisting of the totality of observation statements it implies. Falsification consists of deducing one of these observation statements and then running an experimental test to determine its truth or falsity. If the observation statement, that is, the prediction, is falsified, then the theory is falsified and should be abandoned. One problem with this picture is that the determination of the falsity of a prediction is sometimes not a simple matter; it may involve the use of other theoretical assumptions that have not been firmly established. The problem is compounded when the theory being tested already explains a good deal of data and no other plausible explanation is available; the evidence in favor of the theory may then outweigh the evidence that the prediction is false. Another problem that Eysenck mentions is that theories generally do not entail observation statements without the aid of auxiliary assumptions. Consequently, even when we know that the prediction is false, we can infer only that either the theory or one of the auxiliary assumptions is false. The trouble may lie with the latter rather than with the theory. Finally, theories are often conjunctions of many hypotheses; the falsification of the entire conjunction does not imply that each of its components is false. When these points about falsification are combined with Eysenck's points about weak theories, it appears that falsification of weak theories is very difficult. In particular, the auxiliary assumptions used in testing weak theories, at least in psychology, are often not well established. Also, the parts of weak theories are often not interdependent; consequently, even if the theory as a whole conflicts with established facts, one part of it may be deleted in favor of some new assumption. The entire theory need not be abandoned. Consider the following example. Brewer (1974) has produced what appears to be a very powerful indictment of operant theory; his arguments seem not only to undermine the supporting evidence but also to falsify the theory. Dulaney (1974), however, has shown how an operant theorist might reply by adopting certain auxiliary assumptions that nullify Brewer's criticisms. The Brewer-Dulaney exchange nicely illustrates, I think, how difficult it is to refute a theory like operant theory. The difficulty increases when methodological and philosophical assumptions are also at issue (see, for example, MacCorquodale's (1970) reply to Chomsky's review of *Verbal Behavior*).

If falsification of weak theories is difficult or, as Eysenck claims, impossible, that does not mean that all attempts at assessing correctness should be abandoned. We may not be able to establish, without fear of revision, that a theory is false, but we may be able to do something weaker: to render it implausible given all current evidence. In the case of operant theory, we can assess the plausibility of the auxiliary assumption invoked by Brewer or the epistemological theses appealed to by MacCorquodale; if the protective assumptions can be shown to be if not conclusively false at least implausible, then, the negative empirical evidence will tip the balance, at least for the present, against the theory (for an extended discussion of this point, see Erwin, 1978, Chapter 3).

Consider one more example. Eysenck claims that Freudian theory has been falsified by the successful "symptomatic" treatment of neurotic disorders without symptom substitution or relapse On his own account of falsification, the issue should not be so clear cut; it is not. A crucial assumption in deriving the prediction that symptom substitution or relapse must follow purely symptomatic cures is this: Neurotic symptoms persist because of the existence of repressed wishes. Some Freudians, however, have denied this assumption. Weitzman (1967, p. 307) even quotes a passage in which Freud questions whether the repressed wish must persist in the unconscious when neurotic symptoms remain. Again, we have what appears to be a clean refutation, but the objection is blocked by the invocation of a "new" assumption (even if Freud did accept this assumption, it is new to most Freudians). This example illustrates Eysenck's point that when the assumptions of a theory are relatively independent one can be replaced without abandoning the entire theory. However, this is not the end of the matter. The cost to the Freudian of adopting the new assumption is great: the new version of Freudian theory not only fails to predict that successful symptomatic treatment will inevitably be superficial, but it also no longer explains the persistence of neurotic symptoms or provides a rationale for using psychoanalytic therapy. The attempt to refute Freudian theory, then, may not have been decisive, but it is still very important. (Strictly speaking, one version of Freudian theory has been refuted.)

The realist, then, can agree with Eysenck that failed predictions of a theory (plus auxiliary assumptions) do not necessarily constitute a falsification and that falsification of weak theories is difficult and perhaps impossible. It can still be argued that something weaker can be done and is worth doing: a weak theory can be shown to be more or less plausible than its rivals.

What of Eysenck's other point, that a concern with correctness, using standards appropriate for strong theories, will result in the premature rejection of useful weak theories? The way to avoid this difficulty, it seems to me, is for more psychologists to take seriously what Eysenck says about the complexities of falsification. A failed prediction may be a sign that something is wrong, but given that it will rarely, if ever, constitute a decisive refutation of a weak theory, it provides insufficient warrant for abandoning altogether what may be a promising theory.

3. Conclusion

It might be useful to sum up some of the points of agreement and disagreement between Eysenck and the realist. Both agree, in contrast to Skinner and the abbreviationist, that theory has an important role to play in psychology. They can also agree that an important criterion for judging a theory is its fruitfulness in generating worthwhile research and that emphasizing attempts at falsification, especially in the early stages of research, may kill a potentially useful theory. Even when Eysenck and a realist disagree, there may be mainly a difference of emphasis. Eysenck does say that for a theory to be taken seriously it must show some evidence that it can mediate experimental predictions. Once that is accomplished, the early failures of *some* predictions should be regarded as a challenge to discover the causes of the failure, rather than as necessarily disproving the theory. With this, too, the realist should agree, for the failed predictions may constitute only *apparent* falsifications. The realist will add, however, that despite the difficulties involved in confirmation and disconfirmation, there is still reason to try to answer the following question of any interesting theory, be it weak or strong: Is it true? I think Eysenck might agree even with this last comment, provided that his warnings about bogus falsifications are taken seriously and that, consequently, concern with correctness does not interfere with the development of fruitful theories.

4. References

Bandura, A. (1974). Behavior theory and the models of man. *American Psychologist, 28,* 849–869.

Bandura, A. (1977). Self-efficacy: Toward a unifying theory of behavioral change. *Psychological Review, 84,* 191–215.

Biglan, A., & Kass, D. (1977). The empirical nature of behavior therapies. *Behaviorism, 5,* 1–15.

Blum, G. S. (1954). An experimental reunion of psychoanalytic theory with perceptual vigilance and defense. *Journal of Abnormal and Social Psychology, 49,* 94–98.

Brewer, W. (1974). There is no convincing evidence for operant or classical conditioning in adult humans. In W. Weimer & D. Palermo (Eds.), *Cognition and the symbolic processes.* Hillsdale, NJ: Lawrence Erlbaum.

Dulaney, D. (1974). On the support of cognitive theory in opposition to behavior theory: A methodological problem. In W. Weimer & D. Palermo (Eds.), *Cognition and the symbolic processes.* Hillsdale, NJ: Lawrence Erlbaum.

Erwin, E. (1978). *Behavior therapy: Scientific, philosophical, and moral foundations.* New York: Cambridge University Press.
Eysenck, H. J. (1976). The learning theory model of neurosis—A new approach. *Behavior Research and Therapy, 14,* 251–267.
Eysenck, H. J. (1979). Behavior therapy and the philosophers. *Behavior Research and Therapy, 17,* 511–514.
Eysenck, H. J., & Wilson, G. D. (1973). *The experimental study of Freudian theories.* London: Methuen.
Farrell, B. A. (1981). *The standing of psychoanalysis.* Oxford: Oxford University Press.
Fisher, S., & Greenberg, R. (1977). *The scientific credibility of Freud's theories and therapy.* New York: Basic Books.
Grünbaum, A. (1979). Is Freudian psychoanalytic theory pseudo-scientific by Karl Popper's criterion of demarcation? *American Philosophical Quarterly, 16,* 131–141.
Grünbaum, A. (1984). *The foundations of psychoanalysis: A philosophical critique.* Berkeley: University of California Press.
Kosslyn, S. (1980). *Image and mind.* Cambridge, MA: Harvard University Press.
Lichstein, K., & Lipshitz, E. (1982). Psychophysiological effects of noxious imagery: Prevalence and prediction. *Behavior Research and Therapy, 20,* 339–345.
MacCorquodale, K. (1970). Chomsky's review of verbal behavior. *Journal of the Experimental Analysis of Behavior, 8,* 109–125.
Marzillier, J. (1979). Cognitive therapy and behavioural practice. *Behaviour Research and Therapy, 18,* 249–258.
Nagel, E. (1961). *The structure of science.* New York: Harcourt, Brace & World.
Paivio, A. (1971). *Imagery and verbal processes.* New York: Holt, Rinehart & Winston.
Paul, G. L. (1966). *Insight vs. desensitization in psychotherapy.* Stanford, CA: Stanford University Press.
Paul, G. L., & Lentz, R. J. (1977). *Psychosocial treatment of chronic mental patients.* Cambridge, MA: Harvard University Press.
Popper, K. (1962). *Conjectures and refutations.* New York: Basic Books.
Rosenthal, D. (1971). *Genetics of psychopathology.* New York: McGraw-Hill.
Sarnoff, I., & Corwin, S. (1959). Castration anxiety and the fear of death. *Journal of Personality, 27,* 374–385.
Skinner, B. F. (1950). Are theories of learning necessary? *Psychological Review, 57,* 193–216.
Skinner, B. F. (1957). *Verbal behavior.* New York: Appleton-Century-Crofts.
Skinner, B. F. (1977). Why I am not a cognitive psychologist. *Behaviorism, 5,* 1–10.
Ullmann, L., & Krasner, L. (1969). *A psychological approach to abnormal behavior.* Englewood Cliffs, NJ: Prentice-Hall.
Van Fraassen, B. (1980). *The scientific image.* Oxford: Clarendon House.
Waters, W. F., & McCallum, R. N. (1973). The basis of behavior therapy, mentalistic or behavioristic? A reply to E. A. Locke. *Behavior Research and Therapy, 11,* 157–163.
Weitzman, B. (1967). Behavior therapy and psychotherapy. *Psychological Review, 74,* 300–317.
Yates, A. J. (1970). *Behavior therapy.* New York: Wiley.

2

Theory, Metatheory, and Weltanschauung

Stephen P. Stich

"The Place of Theory in a World of Facts" is vintage Eysenck: erudite, acute, wide-ranging, and provocative. There is much in it that I endorse with enthusiasm. The distinction between strong and weak theories appears to me to be a valuable one, and Eysenck's emphasis on the utility and importance of weak theories is timely and right on target. During the last decade the Popperian picture of science has become widely known and quite influential in scientific circles. Unfortunately, the popularized version is a relatively crude Popperianism which wields the criterion of falsifiability like a cudgel not a scalpel. Eysenck rightly notes that the premature application of Popperian strictures to weak theories may choke off innovative theories before they have had the chance to demonstrate their value. Also to be applauded is Eysenck's brief for the integration of experimental and correlational metatheories. He has mounted a compelling case for the claim that theorists in either metatheoretical tradition neglect theories in the other only at the risk of missing important generalizations and trivializing their own findings.

But enough. Agreements and endorsements make for dull reading, and there is much in Eysenck's paper which I think is quite wrong. Since space is limited, I will focus my critique on two themes: his account of the metatheories of contemporary psychology and his discussion of the egalitarian weltanschauung.

Stephen P. Stich • Committee on the History and Philosophy of Science, University of Maryland, College Park, Maryland 20742.

1. The Metatheories of Contemporary Psychology

As Eysenck uses the term, a *metatheory* is somewhere along the continuum between a theory and a weltanschauung. It is a general perspective of a domain which may "inform a psychologist's choice of theories, topic of research, and method of approach" (p. 59). If I understand him properly, the notion of a metatheory is a close relative of Kuhn's notion of a paradigm. Eysenck mentions a total of five metatheories which play a role in modern psychology: the experimental and the correlational, the biological view of man, the social view of man and the biosocial view of man. Perhaps some of these are best viewed as versions or variants of others, but my concern here is not to shrink Eysenck's catalogue of metatheories. Rather, I wish to expand it. What is striking in his discussion is the total absence of any mention of what has clearly become the ascendant paradigm or metatheory in psychology, *viz.*, the *computational view of the mind.*

The computational view differs in at least two major ways from the experimental and correlational views that dominated the scene two decades ago.[1] First, it postulates complex, interacting systems of mental mechanisms which underlie observed behavior. These typically include a language-processing system (which is itself broken up into interacting subsystems to handle phonological analysis, parsing, semantic analysis, etc.), various perceptual systems, a memory system, an inference system, and more. The liberal postulation of complex perceptual and cognitive mechanisms sets the computational metatheory in sharp contrast to the experimental metatheory which, as Eysenck sketches it, is concerned to find relatively simple functional relationships between dependent and independent variables, with the variables themselves generally being some overt, observable feature of the organism's environment, history or behavior.

Eysenck rightly notes that the experimental metatheory is motivated, in part, by an analogy with elementary physics, and a bad analogy at that. For physics often takes account of such factors as the modulus of elasticity. This is an "individual difference" between samples of different composition and corresponds to the sort of individual differences that correlational methods in psychology reveal. But even when the importance of individual differences is duly noted, the analogy between

[1]To avoid a proliferation of terminology, I will continue to use the term *experimental* for the metatheory that Eysenck characterizes under that heading in the third section of his paper. However, there is a certain inappropriateness to the term since the computational metatheory is also committed to testing theories by appropriate experiments.

physics and the resulting hybrid experimental-correlational metatheory is still a bad one. It is only at its simplest and most primitive that physics is concerned with functional laws and constants reflecting "individual differences." When physics gets sophisticated it begins to postulate unobservable entities and systems—clouds of molecules, atoms, subatomic particles, and the like. It is in terms of the laws governing these postulated systems of entities that the behavior of the observable, macroscopic physical world is to be explained. The computational metatheory postulates unobservable mental mechanisms in much the same spirit. It marks a clean break with the discredited spirit of verificationism which lingers on in both the experimental and the correlational metatheories.

The second distinguishing feature of the computational metatheory is that the mechanisms or subsystems it postulates embody computational processes. It is no easy matter to say exactly what a computational process is.[2] To a first approximation, the idea is to treat the inputs and outputs to a system as *representations*—quasi-linguistic entities with a specifiable formal or syntactic structure. The way a system processes an input is a function of the formal properties of that input. The guiding metaphor here is the digital computer. And, indeed, many of the most impressive computational theories are written as computer programs.[3]

It is tempting to view the advent of the computational metatheory as a Kuhnian revolution. The new paradigm has plainly won over a substantial fraction of the best young minds now working in psychology. Whether the computational revolution will bear fruit commensurate with its early promise remains to be seen. But it is clear that part of the excitement the new paradigm has generated can be traced to the new questions it encourages us to ask. Consider, for example, the study of intelligence which is the recurrent illustration in Eysenck's paper. After sketching the Hendricksons' work and the "error-proneness" theory of intelligence, Eysenck writes, with evident enthusiasm, that "we now seem to have a unified theory of intelligence" (p. 49). But from the computational perspective, this claim would appear to be almost absurdly exaggerated. The skills or abilities that differentiate intelligent people from dull ones—reasoning, problem solving, verbal abilities, spatial abilities, etc.—are at the very core of our notion of intelligence. To have a "theory of intelligence" seriously worthy of the name we should have

[2]For a pair of recent attempts, cf. Fodor (1981) and Haugeland (1981).

[3]See, for example, Anderson (1976), Schank and Abelson (1977), and Newell and Simon (1972).

to have an account of the mechanisms underlying these abilities. How is it possible to do sums quickly, to note the ambiguity of a sentence, to recover relevant information from memory, to solve a chess problem or a spatial transformation puzzle? The answer, on almost all fronts, is that we have only the dimmest idea. But without some reasonable theory about how these things can be done it is something of a joke to suggest that we have "a unified theory of intelligence." I think it is closer to the mark to say that we have almost no theory of intelligence at all. What we do have, perhaps, is a theory about the *variance* in IQ scores in terms of the error proneness of the CNS.

To see just how far this is from a theory of intelligence, an analogy may be useful. The Apple computer on which I am writing this paper can, when properly programmed, do some things that, prior to the invention of digital computers, could only be done by people with considerable intelligence. It can play a remarkably good game of chess, for example, or solve the "Towers of Hanoi" problem in a twinkle. My neighbor across the road has the same model Apple, but his is rather the worse for wear. The pins in its cables and memory boards have been inserted and removed so many times that they are no longer reliable. Thus my neighbor's computer is more prone to error than mine. If we set mine to play chess with his, using the same program in each, mine will generally win. It is "smarter" because it is less error-prone. I suppose that with some effort we could even produce a systematic theory explaining the variance in "Apple intelligence" as a function of the number of times various components had been snapped in and out. But note that this would not, in any serious sense, be a "unified theory of Apple intelligence." It would not tell us the first thing about how the Apple succeeds in doing such clever things as playing first-rate chess. To explain *that* we would need an account of the program the Apple is running and an explanation of how it succeeds in implementing the program. It is just this sort of account that computational theories of the mind seek to provide. Note, finally, that when contrasted with an account of the Apple's program and architecture, the error-proneness theory of Apple intelligence variance appears utterly uninteresting, although it might be thought to have some practical application. I am inclined to think that much the same is true for studies of IQ variance, including those that flow from the hybrid experimental-correlational metatheory that Eysenck endorses. But if this is right it leaves us with a puzzle. Why have theories of so little scientific significance attracted so much interest? Before I attempt an answer, I will have to back up a bit and get a running start by looking at Eysenck's remarks on the egalitarian weltanschauung.

2. Egalitarian Weltanschauung

A weltanschauung, for Eysenck, is a very general way of looking at the world. The egalitarian weltanschauung "should not be regarded as a theory in the scientific sense; it is an expression of the zeitgeist, inspired by ideological and political beliefs and hostile to empirical demonstration" (pp. 51–52). I think Eysenck is quite right here. The egalitarian weltanschauung is a fundamentally moral or political conviction with roots reaching back to Rousseau and earlier. At its core is a view about how wealth, power, and privilege ought and ought not to be distributed in society. Like many fundamental moral perceptions, egalitarianism is difficult or impossible to capture as an explicit principle or a set of principles. Rather, I think, it is best viewed as a collection of *moral prototypes*. Certain fairly specific situations are viewed as morally desirable, others as morally repugnant. New cases are located in moral space by their similarity or dissimilarity to the moral prototypes. Of course, judgments of similarity are multidimensional, vague, and subject to much dispute. Yet in certain core cases egalitarians will be in agreement: Enslaving people or denying them political rights on the basis of their race is morally repugnant; a hereditary aristocracy whose members are accorded political power and wealth because of their parentage is morally unacceptable; a political system in which the weak of mind or body do not have the material means to lead a decent and fulfilling life is not a morally acceptable political system.

It cannot be denied that certain writers have overgeneralized the core egalitarian prototypes, sometimes in quite silly ways. To buttress their egalitarianism, some have gone so far as to deny that there *are* inherent differences among people. But as Eysenck's discussion of egalitarianism in Marxism makes clear, egalitarians need make no such claim. The core idea of the egalitarian is that people ought to be accorded equal political and economic rights *even if they differ* in strength or intelligence or race or parentage. I join with Eysenck in lamenting the often unconscious generalization of egalitarianism into the domain of empirical science. Where I part company with him, however, is in the fact that I see much more pernicious consequences flowing from the unconscious effects of a rival weltanschauung. And I detect ample signs of this unconscious seepage in Eysenck's paper.

For want of a better term, I will call the rival weltanschauung *antiegalitarianism*. In its oldest versions, antiegalitarianism is the view that different people have different natural places in society, places determined by their birth or parentage and ultimately, perhaps, sanctioned by God. A nobleman is born a nobleman; he has noble blood; he rules by the grace of God. The good society, for the antiegalitarian, is the society in which each person accepts his natural station. Antiegalitarianism is a far more ancient view than its rival, with roots extending back before the beginning of recorded history. But modern versions have had to make some changes to cohere with the modern zeitgeist. Appeals to God are out of favor, as are appeals to noble ancestors. They have been replaced by a more scientific sounding story—an appeal to better genes. We noted above that egalitarianism is basically a moral and political conviction and that, despite the unfortunate excesses of some of its advocates, it is quite compatible with the existence of substantial inherent differences among people. Antiegalitarianism is also a moral and political conviction. But, unlike egalitarianism, modern antiegalitarianism *requires* that there be substantial differences in those abilities which qualify people for a leading role in society. Moreover, for antiegalitarianism to make sense, these differences must not be rooted in environmental differences; they must be innate, biological differences which assign a person to his natural place in society quite independently of the quality of the environment in which he was reared. With God's grace and noble parentage ruled out as justification of special privilege, the antiegalitarian is in trouble unless he can point to some biological differences which assign people to their natural place.

All this should make us suspicious. If a deeply held political conviction *requires* that the facts turn out in a certain way, there will be strong subconscious pressures to stack the deck and help the facts out a bit. These suspicions have been borne out in the tragic history of Sir Cyril Burt, who, it now appears, invented studies and even research assistants to buttress his antiegalitarian views. Nor is this an isolated example. In a recent book Stephen Gould (1981) documents case after case of the unconscious influence of antiegalitarianism on science. Gould's history is a profoundly disquieting one, since the men whose scientific judgment was jaundiced by their antiegalitarian convictions included some of the finest scientists of their time.

I have claimed that I see the symptoms of an antiegalitarianism in Eysenck's essay as well. Let me conclude by making good on this charge. The principal symptom I would note is Eysenck's pervasive use of the language of biological determinism. He talks of the "genetic determination" of aggressive impulses (p. 57); he suggests that the "innate structure of the CNS" is "the only important variable in intelligence" (p. 48); he claims that evoked potential is "an almost pure measure of innate ability" (p. 44). This is combined with a negative attitude toward efforts, like Head Start, which attempt to change a person's IQ by modifications of the environment. The impression Eysenck conveys is that such efforts are folly since innate structures are the only important variable in intelligence. But this talk of innate abilities and genetic determination is quite literally nonsense, and in more careful moments it is clear that Eysenck knows it. What we inherit from our parents are *genes*, period. These genes interact with the prenatal and postnatal environment to produce our phenotypic characteristics. It makes no sense to ask whether a phenotypic characteristic is determined by genes or environment, since the answer is always both. Nor does it make any sense to ask for a quantitative assessment of the contributions of genes and environment in producing a phenotypic characteristic in a given individual. There is simply nothing to measure. Thus it does not make sense to say that a phenotypic feature in a given individual is genetically determined, or even that it is mostly genetically determined. There is no phenotypic feature that *must* be displayed by the adult resulting from a given fertilized egg, regardless of the environment in which the egg comes to maturity. Change the environment radically enough and you can *always* change the phenotype. Indeed, it is generally quite easy to accomplish this feat (think of thalidomide); what is hard is producing a desirable phenotype.

None of this should be taken to suggest that we cannot measure the *heritability* of a characteristic. But, as Eysenck clearly states, "heritability estimates are always population estimates" (p. 40). The very same trait may have a high heritability in one population and a low heritability in another. And from the fact that a trait is highly heritable in a population it does not even begin to follow that it cannot be radically altered by environmental manipulation. Yet in the same sentence in which Eysenck notes that heritability estimates are always population estimates he goes on to say, "hence tests of this type are inherently incapable of telling us very much about the specific genetic endowment of a given person." The suggestion is that we need better tests of the specific genetic endowment. But that is nonsense. Unless we are literally talking about the DNA which a person receives from his parents (and clearly Eysenck is not), the notion of the "specific genetic endowment of a given person" makes no sense. How could a scientist as acute as Eysenck be lead to such silliness? The culprit, I suggest is the antiegalitarian weltanschauung which requires that there be biologically determined differences in ability among individuals.

Let me close by linking this section with the previous one. At the end of the previous section we had begun to wonder why there was so much interest in the intrinsically uninteresting question of how variance in intelligence is to be explained. Once again I am inclined to see the influence of the antiegalitarian weltanschauung. Unless we think that an explanation of IQ variance will buttress (or undermine) the antiegalitarian weltanschauung, it is hard to see why the question should be of any interest to anyone.

3. References

Anderson, J. R. (1976). *Language, memory and thought*. Hillsdale, NJ: Lawrence Erlbaum.

Fodor, J. A. (1981). "Something on the state of the art." In J. A. Fodor, *Representations*. Cambridge, MA: MIT Press.

Gould, S. J. (1981). *The mismeasure of man*. New York: W. W. Norton.

Haugeland, J. (1981). "Semantic engines: An introduction to mind design." In J. Haugeland, (Ed.), *Mind design*. Cambridge, MA: MIT Press.

Newell, A., & Simon, H. A. (1972). *Human problem solving*. Englewood Cliffs, NJ: Prentice-Hall.

Schank, R., & Abelson, R. (1977). *Scripts, plans, goals and understanding*. Hillsdale, NJ: Lawrence Erlbaum.

2

The Place of Psychology in a Vacuum of Theories

Alexander Rosenberg

Eysenck's paper perceptively directs us to several of the most crucial issues in the assessment and improvement of contemporary psychological theory. In this comment I employ his lucid exposition of these issues to explain why a *very old* science, like psychology, has not become a more advanced one, like physics. Science, Eysenck writes,

> gradually . . . pulls itself up by its bootstraps, a process which may not be pretty but which has received rather less attention from philosophers of science than have the later, more coordinated stages. Nevertheless, from the point of view of a very young science (like psychology) it is precisely the earlier stages that are of much interest, and it is doubtful whether the psychologist can receive much help from the philosophical conceptualizations of the more advanced sciences. (p. 21)

But the fact is that the very useful "help" for psychology that Eysenck's very paper provides constitutes reflections on and inferences from philosophical conceptions of the more advanced sciences and in particular from mechanics; furthermore, the insights Eysenck has gleaned and deployed from mechanics are the very ones reflected in the earlier, less coordinated stage of physics, to which philosophers have devoted considerable attention; finally, the suggestion that psychology is any younger a science than physics is only a disguised way of saying that it is far less successful than physics, for as human endeavors both go back at least to the Greeks. The difference between physics and psychology is that the latter has continued to tug at its own bootstraps without appreciable movement whereas for the last 400 years physics has been making giant strides. This is a fact which requires explanation and on whose

Alexander Rosenberg • Departments of Philosophy and Social Science, Syracuse University, Syracuse, New York 13210.

explanation a really successful psychology might be forthcoming. Eysenck's paper provides the tools with which to put together this explanation.

The tools for an explanation of why psychology is a "young" science are to be found in Eysenck's distinction between "strong" and "weak" theories in science. Such theories seem to be distinguished by Eysenck in two different ways, between which the relationships are not clear. On the one hand the distinction seems to be drawn by appeal to the content and internal structure of theories of the two types; on the other hand, the context within which they are employed seems to fix the distinction. Thus, Eysenck writes:

> In strong theories the different postulates are *interdependent*; it is not possible to change one without changing the rest, indeed without throwing overboard the whole theory. In weak theories, such interdependence is much less marked, and changes in one part of the theory are quite permissible without the necessity of altering other parts as well. (p. 28)

This is a distinction that admits of degree and reflects the claim that "there is a continuum ranging all the way from the *weak* to *strong* theories" (p. 23). But the other, really crucial distinction Eysenck makes between strong and weak theories is not one of degree but one of kind and is orthogonal to the issue of intratheoretical interdependence. It is the presence or absence of auxiliary assumptions: "Strong theories are elaborated on the basis of a large, well-founded, and experimentally based set of assumptions, . . . so that the results of new experiments are interpreted almost exclusively in terms of the light they throw on" (p. 27) the strong theory itself, and not the "experimentally based" set of assumptions. The assumptions in question, it is crucial to see, are not those of the theory under examination but are components and consequences of other theories, statements about conditions, measuring instruments, and the absence of interfering forces, that are guaranteed by other theories, independent of the strong theory in question. These assumptions are known in philosophy of science (and outside it) as auxiliary assumptions. It is the availability of such auxiliary assumptions that distinguishes strong from weak theories and also explains differences in criteria on which we assess them. The existence of auxiliary assumptions for the test of a given theory is not a matter of degree; either there are such resources or there are not. And such resources may be available for a theory quite independently of whether its postulates are interdependent or not. Accordingly, this way of drawing the strong-weak distinction among theories will produce different divisions from the other ways in which Eysenck fixes the distinction. And this way of doing so, in terms of the presence or absence of auxiliary assumptions, is the crucial one for explaining what the matter is with psychology.

For, as Eysenck writes, "in psychology *K* [the set of auxiliary hypothesis] is infinitely less strong than it is in physics and consequently theories in psychology lie toward the weaker pole" (p. 27).

It is the presence or absence of such auxiliary assumptions that explains why a weak theory is prized for its few positive predictions, in spite of its vast burden of falsified ones, whereas a strong theory is rejected on the strength of a few crucial falsifications, no matter what its positive track record (provided, of course, that there is another theory in the wings which can do as well or better). The auxiliary hypotheses' role is to determine whether the conditions under which a theory is tested are the ones to which the tested theory is in fact relevant. It enables us to establish that the antecedent conditions specified in the theory under test obtain, so that the presence or absence of the consequent conditions provides a real test of the theory. Thus, in physics, testing the gas law $PV=rT$ requires that we acquire an accurate measurement of the values of the variables, pressure, temperature, and volume, and a correct determination of the parameter r (Ryberg's constant). These values we can determine only by employing apparatus, thermometers, manometer, meter sticks and so forth, the behavior of which is *independently* warranted. That is, other theories are required to construct, calibrate, and correct the instruments we employ to test the gas law. If we employ an alcohol thermometer and the temperature of the gas exceeds the boiling point of alcohol, then our test of the gas laws will produce a falsification which redounds to the discredit of our theory of thermometers and not to the law under test itself. If we cannot measure the pressure of the gas at all, we cannot test the theory; and if we cannot measure the pressure of the gas except by employing the equation $PV=rT$ then our test will be a sham at worst and only a very weak confirmation at best. Of course, most scientific theories begin with such weak tests, because they are elaborated in the absence or ignorance of the auxiliary hypotheses needed to provide revealing tests of their claims. This is the respect in which, as Eysenck notes, science "gradually . . . pulls itself up by its bootstraps" (p. 21). Weak theories, that is, theories propounded in the absence of auxiliary assumptions, or of very well grounded and precise ones, are insulated from rejection through falsification, just because of the lack of confidence surrounding the auxiliary assumptions implicitly employed or explicitly recognized in their test. The defect in any prediction is as likely due to these auxiliaries as to the theory under test. By contrast, when the auxiliaries are established, a disconfirmation threatens only the strong theory under test.

Eysenck writes that weak theories "usually imply the absence of precise, trustworthy data, and it is interesting that in the history of science most strong theories started out, in fact, as weak theories; their very existence stimulated the accumulation of precise data which later transformed these theories into strong ones" (p. 29). Of course, it is not the theories which imply the absence of trustworthy data; this want is due to the lack of auxiliary hypotheses assuring us the data we collect is relevant to the theory we hope to test and improve. Moreover, the historical claim may well be false; by and large, in natural science the relevant auxiliary hypotheses antedate the theories the testing and improvement of which requires them. Thus, the rudimentary theory of thermometers and manometers antedates the gas laws. Indeed, because auxiliary assumptions must sustain greater confidence than the theories with which they are associated, they perforce will be older and more well established than the theories on which they bestow "strength." What the history of science shows is that as a theory is developed it comes increasingly to serve as the basis for new auxiliary hypotheses that can be employed in the elaboration of still further theories. So, for instance, the gas laws enabled scientists to construct, calibrate, and correct new temperature measuring devices which could be used to measure temperature below and above the freezing and boiling points of the thermometers employed to test the gas laws and so enabled us to extend and strengthen thermodynamics throughout the course of the nineteenth and twentieth centuries. One generation's strong theory is another generation's auxiliary hypothesis. And occasionally, in periods of revolutionary science, a previous generation's auxiliary hypothesis becomes a current generation's falsified strong theory. Such was the fate of Euclidean geometry; treated as a description of physical space, it was the auxiliary hypothesis of mechanics from Newton's day to Einstein's. But after the advent of general relativity it came to be falsified in tests the auxiliary hypothesis of which included relativity.

The upshot for the science of human behavior should be obvious. There is no theory, either in psychology or outside it well enough established to serve as the auxiliary hypothesis in the elaboration of any other theory. *A fortiori*, there are no strong theories in psychology. Psychology is not a *new* science but an *old* one, insofar as we have been searching for and propounding explanations of human behavior farther back than recorded history. But in all that time there has been no real improvement in our abilities to explain and predict behavior: Plato was no worse at explaining the actions of his fellow Athenian in the Agora than is today's social psychologist. And the reason is that we have never been able to acquire the sort of auxiliary assumptions required to turn the weak theory embedded in common sense and implicitly embraced by psychology into a strong one. There is a systematic reason for this failure to propound such a theory. Recall that the function of an auxiliary hypothesis is to measure the initial and consequent conditions under which a strong theory is tested and to do so independently of the theory's own claims; equally, improvements in the predictive power of a theory are attained by improving the independent measures of the causal conditions and the effects predicted by the theory. But it is just these sorts of independent measures of the determinants and effects systematized in psychological theories that have never been available and which the conceptual presuppositions of psychological theory hitherto precludes.

Psychology, both experimental and correlational, has by and large been a resolutely *intentional* science. *Intentional* is employed here in the sense first propounded by Brentano: Psychology has been an intentional science because it has identified the mental states, dispositions, and events which it treats as the effects and causes of stimuli and behavior in terms of their contents, propositions believed and ends desired. Because we have no direct access to the propositions a subject believes or the ends he desires, we cannot identify his mental states directly but must appeal to indirect measurement of his psychological states. The only two sources for such indirect measurement are behavior and neurology. Behaviorism is of course the doctrine that observing behavior provides more than just an indirect measure of psychological states but enables us to identify them as unambiguously as any observable property of a system under study. Unfortunately, behaviorism is false. We cannot unambiguously identify intentional states by ennumerating their non-intentional causes and/or effects. For any intentional state is identifiable only against the background of other intentional states, and this is a circle out of which we cannot break: My verbal behavior does not reveal my beliefs unless you can assume that my desires are sincerely to express these beliefs; my overt goal-seeking behavior will not identify my desires unless you know what my beliefs are about my environment. All our intentional terms are interconnected in a circle into which behavior alone cannot break. But this means that there is no identification of the intentional variables of a psychological theory independent of that theory. That is, no auxiliary hypotheses that could convert a weak intentional theory into a strong one are possible. That is, none are possible unless we can ground such variables in auxiliary hypotheses drawn from the neurosciences. If there is any avenue along which we can provide the auxiliary hypotheses that will convert weak theories to strong ones in psychology, it is to be found in research on the central nervous system. But the prospects are dim. It is not that the prospects for neuroscience are dim, not at all; what is impossible is that the expectation be fulfilled that such work will provide the correlates for the sorts of intentional states, events, and dispositions that figure in intentional psychology. A variety of considerations make improbable any reduction of the intentional to the neurological. For the empirical psychologist the most salient of these is the fact that intentionally characterized mental states are functional ones and that the physical realizations of any functional state are almost always extremely heterogeneous. For example, the Mendelian gene is identifiable only with a vast and unmanageable disjunction of molecular mechanisms. A psychological state cannot be expected to be correlated with any less complex disjunction of neurological states. The upshot, of course, is that neuroscience cannot provide the auxiliary hypotheses of psychology because it does not provide a means of improving our measurements of the theoretical variables of psychology. This conclusion is illustrated in Eysenck's lucid account of the research of Hendrickson and Hendrickson on the relation between evoked potential in the central nervous system and information. The correlation that Eysenck reports between the complexity of a curve measuring evoked potential and IQ is striking and extremely important. Its subsequent elaboration may tell us much about behavior, differences in it, and their neurophysiological determinants, but it cannot tell us anything about "information processing" in the sense in which that term functions in an intentional theory of cognitive processes. For to describe a system as registering information is to make a claim about the suitability of an entire afferent-efferent network to its immediate and mediate spatiotemporal environment (see Dennett, 1966, Chapter 3), and an EEG is obviously limited to the recording of ordinal quantitative differences on a two-dimensional scale from a small number of loci indirectly connected to that afferent-efferent channel. The point is not that neuroscientific methods are immaterial to psychology, but that they cannot be expected to improve the conventional theories of contemporary psychology because they cannot be employed to measure the variables of those theories to improving degrees; indeed, they cannot discriminate between states which conventional theory identifies by their intentional content.

It must be emphasized that the point is not that behavioral approaches or neurological ones cannot succeed in advancing our understanding of human behavior. Quite the contrary, only these approaches will work, but they cannot be expected to do so by providing bridge-principles that connect conventional psychological theory to the other theories required for any improvement in the explanatory and predictive powers. For the conceptual apparatus of conventional theory cannot be manageably linked up with these more productive theories that might be thought to provide their auxiliary hypotheses. Of course, any particular mental state, event, or disposition is the instantiation or realization of a particular brain state and eventuates from and to stimuli and responses. But the intentional types of states, events, and dispositions into which particular instances are classified cannot be expected to link up with a manageably small number of types of neural realizations, stimuli, and/or responses identified in a behavioral or neuroscientific theory.

Eysenck's appeal that experimental and correlational approaches to behavior must be integrated is crucial. Without such an integration, no progress can be expected on either front. In fact, not only is the very separation of research strategies into a functional and a constitutional approach foreign to most nature sciences, but where it has succeeded, the separation has been relatively short-lived, as in twentieth-century genetics, and the success has been due to our ability to identify and hold constant functional or constitutional variables while studying the ones allowed to vary. The successful pursuit of separate strategies presupposes a much better understanding of underlying mechanisms and functional determinants than exists in contemporary psychology. The cause of the failure hitherto to link experimental and correlational psychology, and therefore of the explanatory and predictive weaknesses of both, has been that both are isolated from auxiliary hypotheses that can strenth them and knit them together in a fruitful way. It is not enough simply to deplore the persistent split in psychology between experimental and correlational approaches, or to recommend that the variables with which they deal be integrated. This integration must itself be the consequence of scientific successes and not of an agreement to lump heterogeneous categories together. Of course, Eysenck cannot be accused of making such a superficial suggestion. His concern for integration goes far deeper than any half-baked attempt at papering over important differences between theories and research strategies. But his recognition of the importance, and of the *absence* of, auxiliary hypotheses in either area of psychology must be brought to bear on a diagnosis of the difficulties and requisites of the theoretical integration he urges.

The obstacle to the integration necessary for progress is in fact a weltanschauung, but one even more thoroughgoing than the ideological dogmatism that Eysenck instances and attacks. Eysenck writes:

> It is important for psychologists to recognize the difference between ideology and theory. . . . One of the worse consequences of the invasion of science by weltanschauung has been the interpretation of scientific data along preconcieved lines, without even a consideration of alternative hypotheses. (p. 56)

He goes on to suggest that "in the 'hard' sciences we no longer suffer from this type of indoctrination, at least to any noticeable degree" (p. 58).

Some philsophers will of course reject these claims, arguing that the distinction between theory and weltanschauung, or paradigm, ideology, or conceptual scheme, is impossible to draw and that every theory rests in an extrascientific context which cannot itself be shaken by experimental or even theoretical considerations. These views appear to me to be far too extreme, mainly because they cannot account for the facts: the occurrence of rational scientific change or technological and predictive improvements that are the effects and the causes of substantial change in weltanschauung. The weltanschauung does, as Eysenck says, preselect the nature and interpretation of scientific data, but this influence is unavoidable and not always nefarious. What is crucial is to recognize when the failures of a research program must be attributed to the weltanschauung and not, for example to the research methods. In the case of psychology the fundamental ideological commitment, one shared by almost all parties to the lively debates in contemporary psychology, is to the *intentional stance,* to the treatment of the subjects of psychology as *agents* with *beliefs* and *desires* who undertake *actions*. This stance lies behind even those theories which self-consciously substitute neologisms for these terms, to the extent that the states they recognize are identified by their propositional or teleological contents. The intentional weltanschauung stands in the way of psychological theory even more forcefully than the psychologically irrelevant egalitarianism of many contemporary psychologists, for it blocks the provision of the auxiliary hypotheses that could convert conventional psychology from a body of weak and weakly confirmed theories into a collection of strong and decisively falsified theories, the falsification of which could pave the way for the neuroscientific approach in which our understanding of human behavior would really be improved.

1. References

Dennett, D. (1966). *Content and consciousness.* London: Routledge & Kegan Paul.

2

The Place of Theory in a World of Facts

Reply to Commentators

H. J. Eysenck

Entering the lion's den with trepidation, this particular Daniel appears to have escaped the wrath of philosophers with fewer wounds than might have been anticipated. I have never shared the belief, endemic among many scientists, that philosophy has nothing to teach us; the evils wrought by a wrong philosophy of science are too obvious for one to make such lighthearted comments, which indicate little but that the writer is self-taught and a good example of the horrors of unskilled labour. Nevertheless, my contribution was not intended to be the kind of composition that a professional philosopher might have put forward; I tried as best I could to join together philosophy of science, history of science, sociology of science, and psychology of science and to illustrate my conclusions by reference to some of my own studies. This was perhaps too ambitious an undertaking for a chapter and in due course may lead to the preparation of a book; however, the kindly welcome my efforts have received from three professional philosophers encourages me to believe that some, at least, of the major points I was trying to make may have been in the right direction.

Let me begin with Erwin's points regarding the contrast between nonrealist and realist accounts of theory. Erwin classes my account as nonrealist, as contrasted with his own realist view. I would hesitate to adopt a simple qualitative distinction, feeling as I do that my own position would not really fit into either of these two views. I would rather think of a continuum, with weak theories being largely but not entirely

H. J. Eysenck • Institute of Psychiatry, University of London, De Crespigny Park, London SE5 8AF, England.

nonrealist and strong theories being much closer to being realist. This continuum would therefore be collinear with the continuum shown in Figure 1 (p. 20). I cannot believe that many of the theoretical concepts shoring up psychologists' weak theories have any pretention to being "realist," but I would be quite prepared to believe that many of the concepts appearing in the strong theories of physics and astronomy have such pretentions. As Hegel used to believe in thesis and antithesis, giving rise to synthesis, so I tend to believe that in most situations qualitatively distinct and categorically opposed positions are often not so representative of reality as would be a continuum leading from one to the other. Hence I would not like to regard myself as either a nonrealist or a realist, but rather as a partial realist, depending on the particular concepts and theories I am writing about. I can see that such an attitude would be rather painful to philosophers trained to draw a distinction which would be as clear, definite, and consistent as possible, but in this case I do not believe that the distinction is as clear-cut as is often made out, and I think that a continuum represents the facts of scientific life rather more clearly.

Erwin makes the point that I see the problem of premature rejection as arising from an inappropriate concern with truth and falsity, and he goes on to say that the problem would also arise if we used a pragmatic criterion. However, I do not think that this is as likely to happen, and the examples he gives are usually from a different direction, namely, the invocation of a different weltanschauung. Theories concerning genetic contributions to schizophrenia and intelligence, which he cites, probably come into that category; so do conditioning theories of treatment for neurosis. Neither could have been rejected using a pragmatic criterion. The truth would appear to be that the too early application of a realist criterion to theories, resulting in their rejection, is only one of several causes of inappropriate rejection; opposition by a dominant weltanschauung, as I pointed out, is another one. So, incidentally, is the curious faddism that runs through much of American psychology, that is, the tendency to pick up a theory, develop it for a few years with great enthusiasm, and then suddenly drop it, regardless of its promise. This happened, for instance, to Lewin's *level of aspiration* work; it happened to Helson's *adaptation level*; it happened to Festinger's *dissonance theory*; and it seems to be happening to Atkinson's *achievement motivation*. These were all excellent progressive research programs, and none of the reasons hitherto mentioned caused them to be abandoned. Faddism is the only explanation, but faddism itself probably needs an explanation in psychological terms.

As regards Freud's hypothesis, I agree with Erwin that "not enough was done to determine the truth or falsity of Freud's hypotheses." This is exactly the point I was making in my paper; at an early stage of development positive research results are required to make the theory reasonable and sufficiently acceptable to go on working with it. Only once a theory has established a reasonable batting average should it be taken seriously and other scientists be asked to continue working on it. Freud's theory, I think, was accepted because it agreed with the generally *geisteswissenschaftliche* approach to psychology, that is, a literary and subjective approach, against which behaviorism and scientific psychology generally had to fight hard in order to establish themselves. On my criterion Freud's is a weak theory which might have been forgiven a fair number of experimental failures but which should never have been forgiven a complete lack of positive results! This lack in my view is so fatal that even the fact that the Freudian theory has inspired some research should not be used to reprieve it—particularly as the research so inspired has usually been of rather poor quality and of little general importance.

I think on the whole that Erwin and I do not really disagree on most of these points to any great extent; I would recognize that my theory of weak and strong theories is itself a weak theory and hence will inevitably show inconsistencies and a failure to fit reality in all places. I think most of the difficulties can be overcome by taking seriously my suggestion of a continuum between realism and its opposite; philosophers seldom use the concept of a continuum, but there are greys as well as blacks and whites in life, and indeed neither a perfect black nor a perfect white is really achievable.

I do of course agree with Erwin that although a weak theory cannot be shown to be false it can be shown to be more or less plausible than its rivals. I think this is what we in fact try to do all the time, but of course neither such attempts, nor even the actual attempts at falsification of a strong theory, can guarantee the outcome, that is, the abandonment of "wrong" theories, and the continuation of work on "correct" theories. Ultimately there can be no rules about this, and the quality of an experimentalist is ultimately judged on the basis of his ability to judge the right point when contrary evidence is sufficient to abandon a wrong theory or when it is insufficient to discontinue work with a correct theory. My main point is simply that applying rules that on the whole are reasonable in connection with strong theories may not be reasonable with weak theories and that psychologists would do well to bear these points in mind in judging theories in their fields.

Stich points out:

> It is only at its simplest and most primitive that physics is concerned with functional relations and constants reflecting "individual differences." When physics gets sophisticated it begins to postulate unobservable entities and systems. . . . It is in terms of the laws governing these postulated systems of entities that the behavior of the observable, macroscopic physical world is to be explained. (p. 89)

And he advocates the use of what he calls the *computational view* of the mind, which

> typically includes a language-processing system (which is itself broken up into interacting subsystems to handle phonological analysis, parsing, semantic analysis, etc.), various perceptual systems, a memory system, an inference system, and more. The liberal postulation of complex perceptual and cognitive mechanisms sets the computational metatheory in sharp contrast to the experimental metatheory which, as Eysenck sketches it, is concerned to find relatively simple functional relationships between dependent and independent variables, with the variables themselves being some overt, observable feature of the organism's environment, history or behavior. (p. 88)

My point, of course, was precisely that the stage of development of psychology is that of "physics at its simplest and most primitive"; it is here that we find a reasonable analogy, not by going to physics at its strongest and most highly developed! The computational view of the mind may sound good and much more embracing, but the question arises of whether it can make any more accurate predictions than can the simpler type of experimental metatheory with which I was concerned. I doubt it, and I am not aware of any experimental evidence that this is so. As Stich said, "Whether the computation revolution will bear fruit commensurate with its early promise remains to be seen" (p. 89). Exactly. I remain to be convinced.

Stich goes on to say that from the computational perspective my claim that work on reaction time, evoked potential, and intelligence seems to have given us a unified theory of intelligence "would appear to be almost absurdly exaggerated" (p. 89). In one sense he is of course right; causal explanations in the sense criticized by David Hume the theory does not give us. But, as I pointed out in my *Model for Intelligence*, the theory does tie together in a unique fashion the huge body of empirical evidence which previously seemed to have no connecting links within it at all. In that sense it certainly is a *unified* theory, and I think one would be justified in asking how the computational metatheory would explain the hundreds of facts described in my book? How would such a theory predict and/or explain the fact that the highest correlation with intelligence of all the reaction time scores is obtained when we use intraindividual variances? Our theory does both predict and explain it;

until and unless computational metatheory can do the same, I beg to suggest that from the scientific point of view ours is distinctly the better theory. I have only given one example; many others could be given. To my mind computational metatheory, like much of what is called cognitive psychology, is just a promissory note, drawn on a bank with no reserves, and on an account which is, if not in the red, at least insufficient to cover the costs.

What Stich has to say about the weltanschauung underlying my own work is, I think, based on a misunderstanding. If I have a negative attitude toward efforts like Head Start "which attempt to change a person's IQ by modifications of the environment," this is not because I disapprove of the aim but because knowledge of past psychological investigations, and behavior genetic ones also, convinces me that the money will be largely wasted and that the outcome will be, as it has indeed been, largely negative. I would have preferred more promising lines of research, such as efforts to use physiological, biochemical, and general biological methods to influence the individual's CNS, such as the use of glutamic acid (Eysenck, 1973). There is some interesting evidence to show that this can increase the "intelligence" of both low IQ children and dull rats, while leaving the IQ of average or above-average children (or rats) unaffected. I feel strongly that if even a small proportion of the money wasted on Head Start had been used to investigate glutamic acid or other pharmaca which have been linked experimentally with improvements in mental performance, we might have got a long way further in our attempts to reduce the biological equality which makes it so difficult to achieve social equality.

I have never said, as Stich seems to imagine, that intelligence cannot be radically altered by environmental manipulation; the possibility always exists. But the realistic point is that at the moment we simply lack any knowledge of how to achieve it. There are, of course, many people who pretend to know, but clearly, as Firkowska and her colleagues (Firkowska-Mankiewicz & Czarkowski, 1982) have shown, even the most egalitarian regime using the most drastic methods of enforcing environmental equality does not seem able to affect biological inequality in the slightest! Of course we should be on the watch for any method which might enable us to gain control over intellectual performance; all I am concerned to say at the moment is that we have hitherto failed conspicuously to do anything of the kind. This is not an antiegalitarian sentiment; it is simply a statement of fact.

An example may make clear just precisely what is intended. The shape, size, and consistency of the female bosom is determined almost entirely by genetic factors (except when famine and other disasters interfere with ordinary living conditions). Yet recent advances in plastic surgery, silicone injections, and hormonal treatment bid fair to make environmental manipulation much more important than genetic factors, and it is conceivable that in 50 years' time a study in California might find an almost zero heritability for size, shape, and consistency of the female bosom. It is impossible to deny that similar advances might be made in intelligence, but the crucial point is that they have not yet been made and that we can talk only about conditions as they are. The available evidence suggests that most efforts to improve human intelligence have been along the wrong lines, paying little heed to what is known about the biological determination of cognitive abilities and preferring rather to go along educational lines which can indeed make up for educational deficits and deprivations but do not seem to be able to increase IQ beyond this very limited stage.

I do not think Rosenberg can really be serious when he says that psychology is as old as physics, and that "the difference between physics and psychology is that the latter has continued to tug at its own boot straps without appreciable movement, whereas for the last 400 years physics has been making giant strides" (p. 95).

I think the answer lies in the old observation: "Psychology has a long past but a short history." In other words, psychological *problems* have always agitated the minds of poets, dramatists, philosophers, military men, and indeed the man in the street as well; in this sense psychology has a long past and is indeed as old as physics. But it is only recently that psychology has become a *science*, or that psychologists as a group have begun to exist. Archimedes was already a recognizable experimental physicist; Plato and Aristotle were not psychologists in any commensurate sense.

Rosenberg goes on to make a rather startling claim, starting with the repetition of the one just discussed:

> Psychology is not a *new* science but an *old* one, insofar as we have been searching for and propounding explanations of human behavior farther back than recorded history. But in all that time there has been no real improvement in our abilities to explain and predict behavior. (p. 98)

As a statement, this is delightfully definite and clearly false. Giving youngsters of five years old an IQ test enables us to predict with very great accuracy how well they will do in school in their final examination at the age of 16½ (Yule, Gold & Busch, 1982); this is a considerable achievement which certainly would not have been possible even 100 years ago! Similarly, during World war II the usual methods of selecting officers for the British army had completely broken down around 1940; the use of psychological selection methods produced a tremendous improvement in the accuracy of the selection and avoided a potentially very dangerous situation (Eysenck, 1953). No such methods were available 100 years ago.

My theory of personality has made it possible to predict with considerable accuracy the ways in which introverted and extraverted children respectively react to new methods of teaching, like automated teaching or the use of the discovery method, and the way in which they react respectively to success and failure, praise and blame (McCord & Wakefield, 1981). This knowledge was not available 100 years ago. Psychology has indeed acquired a great deal of factual knowledge, much of which can be used in a practical manner already and is being so used. This should not make us complacent, of course, but any current textbook of psychology would show that to say that "there has been no real improvement in our abilities to explain and predict behavior" is simply untrue.

Even a few years ago, obsessive-compulsive neurosis was almost impossible to cure. Rachman, using the theoretical explanation developed by Eysenck and Rachman (1965), adapted the model of flooding with response prevention from earlier animal work to the treatment of his patients and showed that they could be cured in a relatively short period of time in something like 90% of all cases (Rachman & Hodgson, 1981). To say that this is "no real improvement in our abilities to explain and predict behavior" is patently wrong; we can not only explain and predict behavior but also control and change it.

Rosenberg goes on to exaggerate the lack of auxiliary hypotheses in psychology. I have pointed out in my chapter that these auxiliary hypotheses are not nearly so strong as they would be in physics; I never said that they did not exist at all. Weak theories are surrounded by weak auxiliary hypotheses; gradually in interaction they succeed in strengthening each other, just as they have done in physics over the past 2,000 years. Some auxiliary hypotheses, indeed, are quite strong now; recall the use of Hick's law by Jensen in his work on a relationship between reaction time and intelligence (Jensen, 1982).

Rosenberg's critique of Brentano's view of psychology as an *intentional* science is well taken and indeed was of course the target of behaviorism's early onslaught. Whether, as he says, "behaviorism is false" is another matter. Behaviorism by now has become a weasel word which means one thing to one person, another to another; it is almost impossible to discuss it without lengthy explanation of what we mean by the term. It is no doubt true that "we cannot unambiguously identify intentional states by enumerating their nonintentional causes and/or effects" (p. 99). We are certainly much closer than we used to be to establishing reasonably firm relations between intentional states and nonintentional causes and/or effects, whether behavioral or neurological. Pavlovian conditioning theory, as applied to neurotic disorders, certainly helps us to understand the origin of "intentional" mental content as well as the observed phobic behavior, and we can manipulate both through Pavlovian extinction methods. Obviously there still is a long way to go, but as my example of obsessive-compulsive neurosis and its treatment by Rachman shows, we are getting a grip on the relations involved. Rosenberg seems to feel that something more drastic is needed; unfortunately he does not tell us what exactly such a change would imply.

An example may make clear what I have in mind. Expectancy is one of the many mental contents which psychology is expected to study. Event-related brain potentials such as the CNV (contingent negative variation of the EEG) can be seen as a close psychophysiological analogue of introspectively reported expectancy. The CNV is one of the family of slow cerebral event-related potentials, the term referring to the slow surface-negative shift in voltage found during the warning period in a signal reaction-time task. The shift in negative potential starts to build up shortly following presentation of a conditional or warning stimulus (S_1) and reaches maximum amplitude just prior to presentation of the unconditional or imperative stimulus (S_2) which typically cause some response, either overt or covert, from the subject. Upon delivery of S_2 and performance of the response, which is usually a button-press movement, the slow negative shift normally resolves to the baseline level. Walter, Cooper, Aldridge, McCallum, and Winder (1964), who first described the phenomenon, invoked expectancy as the hypothetical process underlying CNV development, since the latter was said to reflect subjective probability or relative certainty that S_2 will follow S_1. Other hypotheses have also been put forward, relating the phenomenon to motivation, conation, motor readiness, and arousal/attention, but as Howard, Fenton, and Fenerick (1982) show, there is much support for the original conception of the CNV as a measure of expectancy. Clearly this research, like much else in psychology, is in its early stages, but the possibility certainly appears to exist that by suitable manipulations of the experimental situation we should be able to use the CNV as a close parallel to the mental content of expectation.

Another way of studying expectancy has been by looking at the preparatory interval (PI) in reaction time experiments, that is, the interval between the warning or preparatory signal (*PS*) and the *RS*. RT is an increasing function of the preparatory interval and this fact can be thought of in terms of the *PI* contributing directly to the uncertainty involved in having more than one *RS*. (It is of course well known that increasing the number of *RS*s increases reaction time according to Hick's law, that is, *RT* is a linear function of the bits of information presented by the *RS*s). Schafer and Marcus (1973) succeeded in demonstrating the neurophysiological counterpart to expectancy, which they controlled by having subjects administer a stimulus, as contrasted to automatic presentation at random intervals, while the subject's average evoked potential (*AEP*) for the stimulus was recorded. Self-stimulation, implying foreknowledge of the exact moment of arrival of the stimulus and hence a reduction in uncertainty, resulted in shorter latency and smaller amplitude of the *AEP* to both visual and auditory stimuli. The percentage reduction in amplitude under the self-stimulation condition as compared with the condition in which the subject has no control over the timing of the stimuli was termed the "self-stimulation effect." This expectancy index was found to be related to intelligence, with people who gave larger than average evoked potentials to unexpected stimuli and smaller than average *EP*s to stimuli whose timing they knew as a result of self-stimulation tending to have higher IQs (Schafer, 1979; Jensen, Schafer, & Crinella 1981). Here, then, we have an alternative approach to the neuropsychological measurement of "psychic content" which gives a sound biological foundation to a mental event.

These examples illustrate one reason why psychology must be regarded as a *new* science, even though speculation has been rife regarding psychological phenomena for thousands of years. If, as I firmly believe, psychological processes are intimately bound up with psychophysiological events, then clearly it has only become possible recently, through technological developments of the EEG and other psychophysiological measures, to investigate directly this underlying set of events.

In very much the same way astronomy was almost completely restricted to elementary planetary events and had nothing to say about the stars, other than providing a description of their actual positions. Stellar astronomy as a science took off only with the invention of the telescope, and in the same way a truly scientific and biologically oriented psychology could not have been developed until quite recently when new technological developments made the direct study of neural events possible. Up to date we have had to make do with what amounts essentially to an intriguing preview of the main feature in this field, but continuing technological and methodological improvements make it likely that we will not have to wait very long for the main feature to begin.

Rosenberg may be right in saying:

> The intentional weltanschauung stands in the way of psychological theory even more forcefully than the psychologically irrelevant egalitarianism of many contemporary psychologists, for it blocks the provision of the auxiliary hypotheses that could convert conventional psychology from a body of weak and weakly confirmed theories into a collection of strong and decisively falsified theories, the falsification of which can pave the way for the neuroscientific approach in which our understanding of human behavior would really be improved. (p. 102)

I have no doubt that reductionism will ultimately lead to unification of physiology and psychology, the borders between which have always seemed to me rather insecure and artificial; whether such a reductionism will be quite as drastic as forecast by Rosenberg I cannot tell. In any case, such a discussion would take us a little beyond the limits of my paper.

What can we conclude? I think psychology is beset by a number of dangers and enemies which prevent it from developing into the kind of science many of us would like to see, one which would be able to look physics and astronomy in the eye. One major reason is the literary, humanistic, and generally idiographic predilection of many psychologists who do not want the subject to become truly scientific in the usual sense of that term, who do not like quantification, and who shy away from testable theories and verifiable (or falsifiable) predictions. As long as most clinical psychologists hold these views, and as long as they constitute the majority of members of such bodies as the APA or the BPS, psychology will not develop into a proper science.

A second burden we have to bear is the fact that many of our findings are relevant to weltanschauungen with which they may be incompatible. When this is so, many people reject the facts and the theories based upon them and prefer to retain the weltanschauung. As we used to say during the war: "Don't confuse me with facts, my mind is made up." This unfortunate predilection augurs ill for any development of a scientific social psychology.

A third difficulty, and one to which I devoted a good deal of space in my chapter, is concerned with the fact that many scientifically minded psychologists fail to realize the difference between weak and strong theories and use methods of refutation appropriate to strong theories in relation to weak theories presented by psychologists, thus making it impossible for these to develop from weak into strong theories. I am convinced that the evolutionary approach to scientific theories is of fundamental importance in an understanding of the methods optimally suited to each stage of the development of a science, and I do not believe that by being hypercritical we will advance any faster than we will by being completely uncritical (which is the natural stance of many psychoanalysts vis à vis their own discipline.) The rigor of scientific research and theorizing should match the development of the science; otherwise, it is only too likely to turn into rigor mortis! The best, as always, is the enemy of the good; if we aim for perfection and will not rest content with anything less, we will certainly never reach perfection. This is not a plea for substandard research; it is simply a plea for a recognition of the state of development of psychology at the present time.

I cannot say that the comments of the three philosophers who were kind enough to write rejoinders to my paper have made me change my mind, but they have certainly brought up points which made me think very hard about problems of consistency in my own metatheorizing, the nature of psychological theories, and the content of such theories. That, of course, is the aim of theoretical papers; they are not likely to give scientific proof of the correctness of the views expressed by their authors, but one hopes they will make readers more conscious of the complexities of the task they as scientific psychologists have undertaken and will make them ponder ways and means of improving the present rather sad state of psychology. The views expressed are not likely to be true or false, but helpful or irrelevant. It is my hope that they are the former, and I am glad to say that the three commentators do not seem to disagree on this point.

1. References

Eysenck, H. J. (1953). *Uses and abuses of psychology*. Harmondsworth: Pelican.

Eysenck, H. J. (1973). *The measurement of intelligence*. Lancaster: Medical and Technical.

Eysenck, H. J., & Rachman, S. (1965). *Causes and cures of neurosis*. London: Routledge & Kegan Paul.

Firkowska-Mankiewicz, A., & Czarkowski, M. P. (1982). Social status and mental test performance in Warsaw children. *Personality and Individual Differences, 3*, 237–247.

Howard, R. C., Fenton, G. W., & Fenerick, P. B. C. (1982). *Event-related brain potentials in personality and psychopathology: A Pavlovian approach*. New York: Research Studies Press.

Jensen, A. R. (1982). Reaction time and psychometric *g*. In H. J. Eysenck (Ed.), *A model for intelligence*. New York: Springer.

Jensen, A. R., Schafer, E. W. P., & Crinella, F. M. (1981). Reaction time, evoked brain potentials and psychometric *g* in the severely retarded. *Intelligence, 5*, 179–197.

McCord, R., & Wakefield, J. (1981). Arithmetic achievement as a function of introversion–extraversion and teacher-presented reward and punishment. *Personality and Individual Differences, 2*, 145–152.

Rachman, S., & Hodgson, R. (1981). *Obsessions and compulsions*. New York: McGraw-Hill.

Schafer, E. W. P. (1979). Cognitive neural adaptability: A biological basis for individual differences in intelligence. *Psychophysiology, 16*, 199.

Schafer, E. W. P., & Marcus, M. M. (1973). Self-stimulation alters human sensory brain responses. *Science, 181*, 175–177.

Schafer, E. W. P., & Marcus, M. M. (1973). Self-stimulation alters human sensory brain responses. *Science, 181*, 175–177.
Walter, G., Cooper, R., Aldridge, V.J., McCallum, W. C., & Winder, A. L. (1964). Contingent negative variation: An electric sign of sensorimotor association and expectancy in man. *Nature, 203*, 380–384.
Yule, W., Gold, R. D., & Busch, C. (1982). Long-term predictive validity of the WPPSI: An 11-year follow-up study. *Personality and Individual Differences, 3*, 65–72.

3

From Mindless Neuroscience and Brainless Psychology to Neuropsychology

Mario Bunge

Abstract. Three main strategies for the study of behavior and mentation are examined: behaviorism, mentalism, and psychobiology. Behaviorism is found wanting for eschewing most of the problems that traditional psychology posed but left unsolved. Two kinds of mentalism are distinguished: traditional and cognitivist (or information-theoretic). Both are found wanting for ignoring the nervous system and begging the question, since they postulate the mind instead of explaining it. Only the psychobiological (or neuropsychological) approach, which regards the mind as a collection of brain functions, is found promising for studying that which guides behavior and does the mentation, namely, the brain. It is also shown to have the advantage of promoting the union of psychology with biology and of bridging psychiatry to neurology, neurophysiology, and neurochemistry. It is argued that this approach is the only fully scientific one of the three approaches discussed in the paper.

Psychology is, of course, the scientific study of behavior and mentation. It studies bodily movement, sensation, perception, and imagination; emotion, motivation, and attention; learning, memory, and forgetting; intuition, reasoning, and intellectual creativity—and a host of other mental states and processes.

There are several approaches to the study of the problems of behavior and mentation. They can be grouped into three main strategies: *behaviorism*, *mentalism*, and *psychobiology*. Behaviorism is brainless and mindless, in that it ignores the nervous system (except as a stimulus

Whiting lecture, Winter Conference on Brain Research, Keystone, Colorado, 24 to 31 January 1981.

Mario Bunge • Foundations and Philosophy of Science Unit, McGill University, Montreal, Quebec H3A 1W7, Canada.

transducer) and is not interested in mental events: it restricts its attention to overt or observable behavior. On the other hand, mentalism is mindful but brainless: it is concerned with the perennial problems of psychology but has no use for neuroscience because it can explain everything very easily in terms of the immaterial (and possibly also immortal) mind or soul. Finally neuropsychology (or biopsychology) is both brainy and mindful, for it endeavors to identify mental events with brain events, and it studies the development of the psyche as one aspect of the development of the nervous system and the evolution of the nervous system and its abilities as one aspect of biological (and social) evolution (see Figure 1).

We shall examine the three approaches, but first we shall clarify the rather hazy notion of approach or strategy and the more special concept of scientific approach. Then we shall settle our accounts with behaviorism and mentalism. Next we shall sketch the guiding ideas of the psychobiological approach. And finally we shall discuss the problem of whether psychobiology effects the reduction of psychology to neurophysiology, or rather a merger of biology and psychology.

1. Approach

An *approach* to problems of some kind is of course a way of conceiving and treating them: it is a problem-solving strategy. One and the same problem can often be approached in different ways, although not

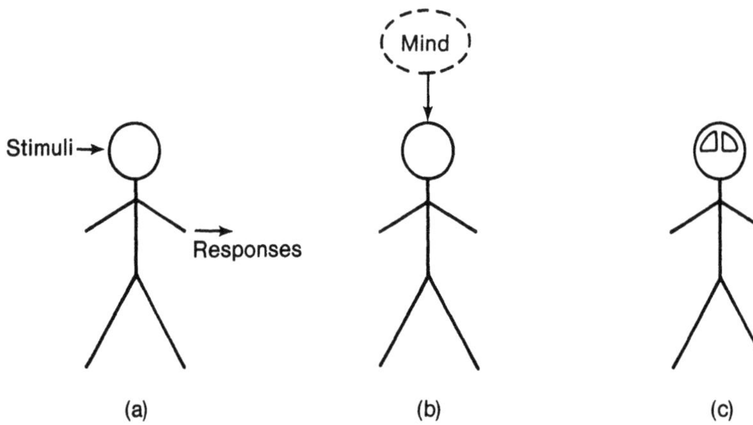

Figure 1. Three models of the organism. (a) Behaviorism: responses are effects of stimuli alone, and there is no subjective experience. (b) Animism: the immaterial mind controls the body. (c) Psychobiology: the mind is a collection of brain processes, and behavior is controlled by both the central nervous system and the environment.

necessarily with the same success; and one and the same approach can be applied to several problems. In particular, the problem of the nature of mind can be approached theologically, philosophically, in the manner of traditional (mentalistic) psychology, or else neurobiologically. And the neurobiological approach, like the other strategies, can be applied to a variety of questions regarding mind, from the origin of mentality to the effects of mental events on other bodily processes. An approach suggests hypotheses and theories without sponsoring any in particular. On the other hand, every approach writes off whole bunches of theories. For example, the theological approach to mind is incompatible with the science of mind, for it endorses an unscientific and ready-made view including the thesis that the mind of a person is an immaterial and immortal substance detachable from the brain and inaccessible to experiment.

The concept of an approach can be analyzed as a *general outlook,* or conceptual framework, together with a *problematics* (problem system), a set of *aims* or goals, and a *methodics* or set of methods (not to be confused with methodology, or the study of methods). The scientific approach is a very particular kind of approach or strategy. It is the one characterized by the following quadruple:

- *General outlook:* (*a*) a naturalistic ontology, according to which the world is composed of concrete changing things (nothing ghostly); (*b*) a realistic epistemology, on which we can frame fairly true representations of things with the help of experience and reason (no supernatural or paranormal cognitive abilities); and (*c*) the ethos of the free search for truth (no reliance on authority, no cheating, and no withholding information)
- *Problematics:* all the cognitive problems that can be posed against the above background (nothing is irrelevant)
- *Aims:* the description, explanation, and prediction of facts with the help of laws and data (not just mere description and not just any old explanation)
- *Methodics:* the scientific method and all the scrutable, checkable, and justifiable techniques or tactics (no crystal ball and no inkblot tests)

The specialist may balk at my assertion that the scientific approach has a bulky philosophical component made of a world view, a theory of knowledge, and a code of conduct. Yet that general outlook does guide the choice of problems, the goals of research into them, and the means or methods. We may not realize this because we take that general outlook for granted: to absorb it is part of our scientific apprenticeship.

To show that this is indeed the case, imagine cases in which the general outlook is not adopted.

Suppose someone held a supernaturalistic world view. In this case he would count on supernatural agencies instead of restricting his theorizing and experimenting to certifiably or putatively real entities. Again, suppose our imaginary scientist were a subjectivist or a conventionalist: in either case he would hardly care for experimental tests. And if he believed in paranormal modes of cognition—such as revelation, intuition, telepathy, precognition, or what have you—he would rely on them instead of checking his hunches. Finally, suppose someone did not adopt the code of scientific ethics. In this case he might feel free to bypass problems that could embarrass the powers that be, he might feel inclined to make up his data, withhold information, or cheat in theorem proving; he might plagiarize, and he would hardly be interested in trying to prove himself wrong. So, the general or philosophical background is an indispensable component of the scientific approach. This is particularly evident in the case of psychology, so much of which is nonscientific, or protoscientific, precisely because it does not partake of the general outlook of the older and harder sciences.

Presumably the other three components of the concept of scientific approach—namely, the problematics, aims, and methodics—will raise no eyebrows, although admittedly every one of them can be the subject of lengthy methodological disquisitions (see, e.g., Bunge, 1983). Indeed we all appear to agree that scientific research is characterized by a peculiar set of problems, aims, and methods. On the other hand, we often forget that an exclusive training in methods, to solve well-defined problems with narrowly circumscribed goals, makes technicians, not necessarily scientists. Hence our emphasis on the first component of the outlook–problematics–aims–methodics quadruple.

Let us now catch a glimpse of the approaches characterizing behaviorism, mentalism, and psychobiology, in order to assess their respective scientific merits.

2. Behaviorism

To begin with, let me caution against the mistaken yet not uncommon confusion between behaviorism and the study of behavior. Behavior can be approached in a number of ways, among them in a behaviorist, a mentalist, or a biological fashion. From Watson (1925) to Skinner (1938) and their followers, behaviorism is not just the study of behavior. It is the approach consisting in leaving out everything else, in particular the so-called neural substrate or correlate of behavior as well as the whole of subjective experience. To put it positively, behaviorism studies the responses of organisms to various stimuli configurations, as if animals were black boxes.

More precisely, the behaviorist approach can be characterized as follows in terms of our characterization of the concept of scientific approach in section 1. First, its general outlook includes a limited naturalistic ontology. It is naturalistic because it discards disembodied entities, and it is limited because it discounts nonbehavioral phenomena such as emotion, imagination, and ideation. Second, behaviorism adopts an immature realistic epistemology. It is realistic because it endeavors to account for an aspect of reality (namely its skin), the existence of which it recognizes the moment it demands that research be objective. However, this epistemology is primitive because it shuns hypothetical constructs such as desire and reasoning and thus does not allow one to pose the problem of whether such constructs represent objective properties, states, or processes of the organism. Behaviorism can make do with a primitive epistemology because it avoids deep (nonphenomenological) hypotheses and theories, which are the ones that pose the toughest problems in the theory of knowledge. (Typically, a behaviorist learning model revolves around the concept of the probability of an organism's making a given response at the *n*th presentation of a stimulus of a certain kind. It deals with whole events and it makes no reference to mental states (see Figure 2); it pretends that the latter do not matter, at least to science.) Third, behaviorism adopts a strict code of scientific conduct. Indeed, we should be grateful to behaviorists for having introduced this code into the field of psychology, where illusion and deception (unwitting or deliberate) are not uncommon. So much for the general outlook of behaviorism, which turns out to be scientific but narrow.

As for the problematics of behaviorism, it must be conceded that it is extremely narrow and that it eliminates the most interesting problems of psychology, namely, all those about mental states and processes as well as their so-called neural basis or correlates. Such elimination is unsatisfactory. We all want to know not only how a woman in pain, love, or deep thought moves about. We also wish to know what pain, love, and thought are, that is, what are the neural processes we call *pain, love,* and *thought.* When watching two boxers exchanging blows or two psychologists exchanging views, we are not satisfied with being told that each is responding to the blows of his adversary or to the verbal behavior of his interlocutor. We want to know what motivates them to start the fight or the argument and what makes them go on or stop. When watching a neuroscientist operating on a rat, inserting electrodes

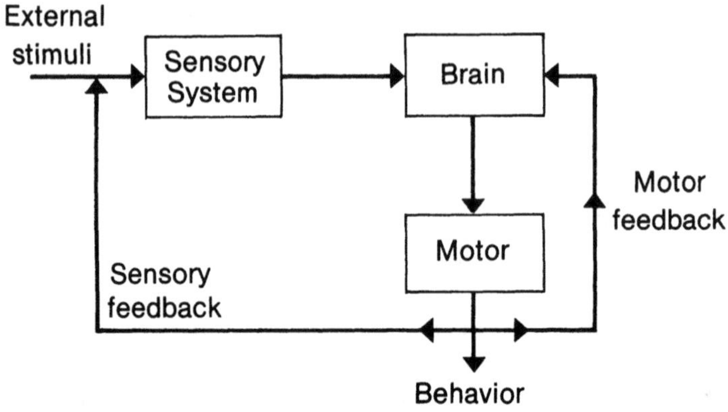

Figure 2. The motor output of an animal is controlled by the brain, the activity of which is modulated (but not fully caused) by external stimuli as well as by the outcome of behavior. In higher vertebrates brain activity is partly prewired ("motor tapes" and other "programs") and partly creative, but in either case is largely autonomous (inner-determined or self-started).

in its brain, or taking instrument readings, we wish to know what makes him tick: what are his problems, hypotheses, and goals, not to speak of his doubts and hopes. By discarding motivation, affect, and ideation, the behaviorist gives a superficial and therefore unilluminating account of behavior; it is like a silent movie without titles. In his extreme ontological, epistemological, and methodological asceticism, the behaviorist denies that such problems are accessible to the scientific approach and thus leaves us in the lurch. In this manner he hands the field over to the mentalist and his wild speculations. The self-denial of the scientist becomes the self-indulgence of the pseudoscientist.

The aim of behaviorism is scientific but, again, narrow. It is scientific because it proposes to describe and predict behavior. (Actually it describes it superficially because it ignores the internal states of the animal, and therefore it cannot predict behavior accurately.) And it is narrow because it adopts the positivistic proscription against tackling unobservable facts and trying to explain. These limitations are crippling, for we cannot attain a satisfactory description of behavior unless we frame hypotheses about the underlying neural processes. After all, it is the brain that makes the stimulus–response–reinforcement process occur (Pribram, 1971). Trying to understand behavior on the sole strength of the observation of behavior is like trying to understand motion without masses, forces, and strains, or radio without electrons and electromagnetic waves, or evolution without natural selection, or history without social forces.

Finally, the methodics of behaviorism too is scientific but narrow. It is scientific because it employs observation, measurement, and controlled experiment. But it is narrow because it makes no full use of its empirical findings; indeed, it minimizes and sometimes even denies the role of theories and in particular that of mathematical models and consequently has no occasion to face the problems of testing them experimentally. And because behaviorism ignores neuroscience it is spared the intriguing problem of checking its S–R hypotheses with neuroscientific data and hypotheses. As for behavior modelling, to be sure there are a few behaviorist models, most of them in the area of learning (cf. Luce, Bush, & Galanter, 1963–65). However, all these models are phenomenological; they deal with appearances or externalities and are therefore shallow (for one thing, they do not bear on neural processes; for another, they ignore cognition and motivation). Worse, they are basically mistaken in assuming the Aristotelian conception of change, according to which the cause (input) alone determines the effect (response) regardless of the inner organization and state of the system. Needless to say, this view is at variance with modern physics, chemistry, and biology, all of which study not only external circumstances but also internal structures and processes. Whether the object of study is an atom or a person, the effect of a stimulus (be it a photon or a word) depends not only upon the kind and strength of the stimulus but also on the internal state of the object—a state that must be conjectured since it cannot be observed directly. Moreover, what happens inside the system—for example, the process that led to the internal state at the time the stimulus impinged—is just as important and interesting as the commerce of the system with its environment. For one thing, every concrete system, from atom to DNA molecule to cell to ecosystem, is in flux; yet according to behaviorism such internal flux does not matter. For another, every concrete system has some spontaneous activity: it may initiate changes (e.g., spontaneous radioactive decay, spontaneous self-assembly, spontaneous formation of a melody) without environmental prodding: not every output is a response to some input. Not so according to behaviorism: all we ever do is respond to stimuli. In short, the methodological strictures of behaviorism, modern though they may appear to be, bind behaviorism to obsolete science—to say nothing of the boredom induced by observing externalities without having a clue as to their mechanism (for further criticisms see Bandura, 1974).

The upshot of our examination of the behavioristic approach is that behaviorism adopts a narrow scientific approach: it is nearer to protoscience than to full-fledged or mature science. Because of its narrowness behaviorism has remained stagnant since the mid 1950s. For the same reason it has been incapable of stemming the mentalist tide—nay, it has in part provoked it. In historical perspective behaviorism may be regarded as the mother of scientific psychology. But it is an unwed mother that has refused to marry the father, namely neurobiology. Like every other mother, behaviorism deserves our love; like every mother, it must be prevented from hindering the advancement of its progeny. When criticizing the limitations of behaviorism, therefore, let us not forget that it was the beginning of a science; let our criticisms not be an excuse for reviving mentalism, which is definitely prescientific, not protoscientific.

3. Mentalism

Mentalism is the approach that focuses on mental events, endeavors to explain them, as well as behavior, by further mental events, and relies mainly on introspection, that is, ordinary intuition. Thus the mentalist holds that he feels, perceives, thinks, and wills with his mind, not with his brain. He emphasizes that mind is immaterial and autonomous with respect to matter. And, of course, he regards his own view as conclusive refutation of naturalism or materialism.

Two main variants of mentalism are popular today. One is the old vulgar idea that mind is a special immaterial (and possibly immortal) substance. As such, mind is inaccessible to the scientific approach even though it may interact in mysterious ways with the brain (cf. Popper & Eccles, 1977; Eccles, 1980). How such interactions between an entity and a nonentity are to be conceived is not explained. The second variant of mentalism, popular among contemporary workers in cognitive science (in particular artificial intelligence practitioners), as well as among philosophers, is that the mind is a set of programs: that it is software, structure, organization, or information, not a matter of hardware or stuff (cf. Fodor, 1975, 1981; MacKay, 1978; Putnam, 1975; Pylyshyn, 1978). We shall call these varieties of mentalism *substantialist* and *functionalist* respectively.

Substantialist mentalism matches with ordinary (i.e., fossil) knowledge and has the blessing of theology. It is a view rather than a theory and it contains no technical concepts, let alone mathematical ones, so anyone can understand it. In fact, substantialist mentalism has still to be formulated in precise terms, and it is doubtful whether it can ever become a testable theory. Consider, for example, the hypothesis formulated by Saint Thomas Aquinas and adopted by Eccles (1980, p. 240) that the mind or soul of an individual is "infused" into it by God at some time between conception and birth, so that every fruitful marriage is actually a *ménage à trois*; or Eccles's claim (1980, pp. 44–45) that the self-conscious immaterial mind scans and reads out the activity of the cortical modules or columns; or his postulate (1980, p. 232) of "the existence of some conscious experiences *prior* to the appearance of the counterparts in the specific modular patterning in the neocortex." Try and formulate these wild speculations in exact terms; try and design experiments to check them; and try to render them compatible with neurophysiology, developmental psychology, or evolutionary biology. If you fail in at least one of these attempts, confess that substantialist mentalism is anything but scientific.

The functionalist (or structuralist or information-theoretic or computational) variety of mentalism is slightly more sophisticated than substantialist mentalism. It is advertised as being neutral between spiritualism and materialism, but in fact it is good old mentalism in new garb, for it holds that form or organization is everything whereas matter or stuff is at most the passive support of form—oh, shades of Plato! To the functionalist mentalist almost anything, from computers to persons to disembodied spirits, can have or acquire a mind: "We could be made of Swiss cheese and it wouldn't matter" (Putnam, 1975, p. 291). According to this view a psychological theory is nothing but "a program for a Turing machine" (Fodor, 1981, p. 120). So why bother studying the brain? And why bother studying the peculiarities and interrelationships of perception, motivation, and cognition? An all-encompassing and stuff-free theory is already in hand: it is the theory of automata. Psychology can learn nothing from neuroscience, and it can expect no theoretical breakthrough.

Although mentalists of the functionalist or computational variety are very critical of behaviorists, their approaches are similar insofar as both are externalists and ignore the nervous system. In fact,functionalist mentalism can be regarded as the complement rather than the opposite of behaviorism. Take, for instance, Turing's criterion for telling—or rather not telling—a human from a computer, namely, not to open them up but to record and analyze the net responses of the two regardless of the way they process the incoming information, that is, irrespective of the stuff they are made of (Turing, 1950). This criterion is behaviorist as well as functionalist. And it will not do, because every theory of machines, in particular Turing's own theory, contains a theorem to the effect that, whereas behavior can be inferred from structure, the converse is false. (Similarity of internal structure implies similarity of behavior but not the other way round.) This is obvious to any psychologist or ethologist. Thus the foraging bee,the migrating swallow, and the human navigator are good at orienteering, yet each "computes" the desired path in its own peculiar fashion.

To be sure the search for similarities, and the accompanying construction of metaphors, is useful—but it cannot replace the investigation of specifics. Trivially any two things are similar in some respects and dissimilar in others. The question is to ascertain whether the similarities weigh more than the differences, so that both things can be grouped into the same species. Functionalist mentalists hold that this is indeed the case with regard to persons, computers, and disembodied spirits. This claim is not only offensive to parents; it is also outrageously false and misleading.

To begin with, the theory of Turing machines is far too poor to account for any real system, if only because Turing machines have a denumerable set of states, whereas the states of any real system form a nondenumerable set. Not even the neutrino and the electron, possibly the humblest things in the universe, are describable as Turing machines: they are far more complicated than that. (A Turing machine is describable by a table exhibiting the properties of its next state function. Neutrinos and electrons are describable by complicated systems of partial differential equations and other complex formulas.)

Second, the human nervous system is far more complicated than a computer, if only because it is composed of variable components capable of some degree of spontaneous activity and creativity—the last thing we want in a computer. Third, computers are artifacts, not organisms with a long evolutionary history. Fourth, computers are designed, built, and programmed to solve problems, not to find them; to process ideas, not to originate them; to supplement the brain, not to replace it; to obey, not to command. It follows that computer science can advance provided it learns from neuroscience, whereas neuroscience will stagnate if it becomes the caboose of computer science. Computers imitate brains, not the other way round.

Another feature that mentalism, whether functionalist or substantialist, shares with behaviorism is that all three regard neuroscience as being irrelevant to psychology, in the case of substantialist mentalism because psychological problems are allegedly solved by resorting to the old philosophical-theological dogma known as mind–body (or psychoneural) dualism and in the case of functionalist mentalism because those problems are said to be solved by decreeing (not proving) that we are Turing machines or at any rate information processors and nothing else. In both cases the solutions are proposed *a priori* and are not checked experimentally. In neither case is the brain necessary, except perhaps to keep neurscientists busy. And in both cases telekinesis, telepathy, reincarnation, and resurrection are possible—nay, they are sometimes the whole point of the exercise.

The scientific status of mentalism can be appreciated by checking the way it matches or mismatches the scientific approach (section 1). To begin with, the general outlook or philosophical background of mentalism involves an ontology countenancing disembodied minds (or souls, egos, superegos, etc.) or stuff-free programs, or energy-independent information—and sometimes also supernatural beings. Accordingly, adopting this ontology turns psychology into the anomalous research field: the only one in which states are not states of concrete things and events are not changes in the state of concrete things. The accompanying epistemology is uncritical, for it relies on intuition and metaphor, which to the scientists are just props. And the ethics of mentalism is dubious because it resorts to authority or chooses to ignore the evidence for the biological view of mind accumulated by physiological, developmental, and evolutionary psychology.

On the other hand, the problematics of mentalism is its *forte* and a reason for its popularity. Indeed, instead of writing off most of the classical problems of psychology, mentalism makes a point of showing that it can handle them and in this it satisfies—alas, ephemerally—our yearning to understand our subjective experience or mental life. This, then, is the one and only merit of mentalism: namely, that it acknowledges the full problematics of psychology. A pity, that it does not approach it scientifically.

As for the aims of mentalism, they are mixed. On the one hand, it endeavors to understand the mind, but on the other, it refuses to do so with the help of laws, or at any rate in terms of laws linking variables some of which are accessible to objective (not introspective) observation or measurement. In fact, the vagaries of the putative immaterial mind are untraceable with the help of scientific instruments, we can speculate about them but cannot check experimentally such speculations as long as we are restricted to introspection. As for the assertion that every psyche, whether fathered by ourselves or by IBM, is a Turing machine or some more complicated computer, it is a dogma, not a hypothesis, and it is one involving no commitment to any scientific laws. After all, every scientific law is stuff-dependent, for laws are nothing but the invariant patterns of being and becoming of concrete things, whereas functionalism claims that mind is stuff-free. In sum, mentalism does not endeavor to account for mind with the help of any laws of matter, and on this count its goals are not scientific.

Finally, the methodics of mentalism is clearly unscientific. In fact, mentalism is typically speculative, metaphorical, dogmatic, and nonexperimental. There is nothing wrong with speculation as long as it is fertile and testable in principle—or at least entertaining. But the mentalistic speculations are untestable because they involve disembodied entities, that is, nonentities. As for the metaphors of mentalism—"The soul is like the pilot of a boat" and "The mind is like a computer program"—they are not intended to be put to the test, for they are not scientific hypotheses. To be sure these superficial analogies do hold in some respects, but so what? Analogies can be heuristically fertile but in this case they are barren and misleading, the pilot metaphor because it ushers people into the theologian's cell and the computer metaphor because it advises them to study machines instead of brains. (Of course one may decide to study only that which all information systems have in common. However, this arbitrary decision does not prove that nervous systems are nothing but information processors and therefore accountable solely in terms of computer science. One may focus one's attention on information rather than, say, the complex neurobiological processes whereby information is transmitted—and generated and destroyed. But this does not prove that information can be transferred without energy; every single signal is carried by some physical process. To be sure, information theory is interested only in the form of signals and thus ignores matter and its properties, among them energy. But this only shows that it is in an extremely general theory—so general, in fact, that it can explain no particular fact.) We conclude that the methodics of mentalism is nonscientific.

The upshot of our examination of mentalism is clear. Of the four components of the mentalistic approach, only one is acceptable, namely, its problematics—and this on the charitable assumption that mentalists pose their problems in a way that is susceptible to scientific treatment, which is not always the case. The other three components of mentalism are not congenial with science. The verdict is that mentalism is nonscientific: it is just the old philosophical psychology, even if it sometimes uses fashionable terms such as *software, program,* and *information* (for further criticisms, see Bindra, 1984).

4. Neuropsychology

The central thesis of neuropsychology (or biopsychology, or psychobiology) is that a mind is a collection of special brain functions (Bindra, 1976; Bunge, 1980; Hebb, 1949). On this thesis the mind is not a separate substance (substantialist mentalism), nor is it a stuff-free program (functionalist mentalism). Instead,the mind is a collection of peculiar activities or functions of systems composed of numerous neurons (or their homologues in minding beings on other planets).

This neurobiological view of mind is not a stray philosophical opinion or a dogma entrenched in theology but is part and parcel of every naturalistic (materialist) world view. As such it is abhorred by all those who seek to explain the world in terms of the unworldly. The neurobiological view of mind transforms the mystery of mind into a problem, or rather a problem system (problematics), that can be approached scientifically. The thesis denies the autonomy of mental life as well as the independence of psychology: it renders our inner life a part of our ordinary life and turns psychology into a branch of biology and, more particularly, of neuroscience. In other words, in the neurobiological perspective psychology is the branch of neuroscience that investigates the specific functions or activities of neuron assemblies that, in ordinary parlance, we call *perceiving, feeling, learning, imagining, willing, evaluating, reasoning,* and so on.

Neuropsychology is revolutionary but it does not ignore the contributions of alternative psychological currents. For one thing, it adopts the methodological rigor of behaviorism; for another, it accepts the genuine problems posed by mentalism. Neuropsychology thus retains whatever is valuable in those two approaches while going far beyond them. In fact, it adopts a very different approach. Let us take a look at it.

To begin with, the general outlook or philosophical background of neuropsychology is a naturalistic (materialist) ontology free from nonscientific (in particular theological) strictures. The ontology underlying neuropsychology is naturalistic because it is concerned with organisms and it does not hypothesize disembodied minds or stuff and energy-free information flows. The epistemology of neuropsychology is realistic and mature. It is realistic because it endeavors to account for mental and behavioral reality, not just for introspective or behavioral appearance. And it is mature because, by not setting limits upon theoretical constructs, it cannot evade the problem of finding out how such theories represent reality. Finally, the ethics of neuropsychological research is supposed to be that of science in general: no argument from authority, no holding back information, no recoiling before problems, methods, or hypotheses that may shock some ideology, and so on.

The problematics of neuropsychology is the entire realm of behavior and mentation: it does not reject any problem that can in principle be investigated scientifically. In particular, it includes the problem of

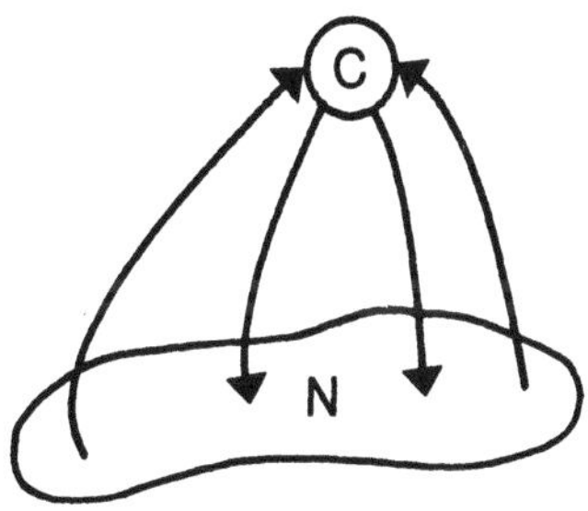

Figure 3. Consciousness: hypothetical neuron system C monitors the activity of neural system *N*. Subject is aware of activity in *N*, or in muscles innervated by *N*, just in case *N* stimulates C or C controls *N*.

accounting for consciousness, perhaps in terms of the self-monitoring of the brain or, better, of the monitoring of the activity of one part of the brain by another part of it (see Figure 3). Thus the problematics of neuropsychology has a considerable overlap with that of mentalism. However, it drops some of the problems of mentalism, reformulates others, and adds some that mentalism cannot pose. For example, neuropsychology rejects as nonscientific the problem of finding out where the mind goes in deep sleep or coma, or at death. It reformulates the problem of finding out at what point in its development a human embryo gets its soul "infused," asking instead whether its brain can be in mental states at all and, if so, from what stage on. And it adds the entire problematics of biological evolution: at what evolutionary level did mind start, what may have been the mental abilities of the hominids, what is the origin and function of lateralization, when may language have started, and so on. In short, the problematics of neuropsychology is far richer than that of behaviorism and far more precise and challenging than that of mentalism.

The aims of neuropsychology are fully scientific: not just the description of overt behavior, nor merely the accumulation of introspective reports—let alone of old husbands' tales about the ghostly—but the explanation and if possible also the prediction of behavior and mentation in terms of the laws of neuroscience. The production of (relevant) data is of course indispensable to this end: no factual science without observation, measurement, and experiment. Still, data are bound to be superficial or even irrelevant in the absence of theories. Besides, the ultimate goal of scientific research is not the accumulation of facts but their understanding. And scientific explanation is achieved only with the help of theories and models—not just descriptive theories summarizing and generalizing the data, but theories explaining them in terms of mechanisms, such as neural processes.

Finally, the basic method of neuropsychology is the scientific method, even though, of course, neuropsychologists add to it their own special techniques, from voltage clamping to surgical ablation. Unlike behaviorism, which shuns theory, and mentalism, which shuns experiment, neuropsychology makes full use of the scientific method: problem, hypothesis (or better, model), logical processing, empirical operation, inference, evaluation of hypothesis, new problem, and so on. In particular, neuropsychology can control and occasionally measure mental variables because it identifies them with certain properties of the brain.

In sum, the neuropsychological approach to the mental is scientific and moreover the only one fully so. The birth of neuropsychology in recent years constitutes a scientific revolution, for it adopts a new approach to an old problematics; it elicits an explosive expansion of this problem system; it is bound to succeed where the alternative approaches failed; and it promotes the merger of previously separate disciplines, from neurophysiology, neuroendocrinology, and neurology to psychology, ethology, and psychiatry. Compare this revolution with the exhaustion of behaviorism and with the mentalist counterrevolution.

Note the following peculiarities of the research strategy in neuropsychology. First, in studying behavior it proceeds centrifugally, from the nervous system to overt behavior. For example, it attempts to explain voluntary movement in terms of specific activities of certain neuron modules presumably located in the frontal lobe and connected with the motor cortex. To be sure, some investigators still try to "read off" brain processes from behavior (in particular, speech) or from global electrophysiological data (in particular, EEG tracings). This centripetal strategy is unlikely to bear fruit because one and the same behavior pattern can be produced by many alternative neural mechanisms. The only strategy that can succeed is to try and explain the global data in terms of neurophysiological hypotheses consistent with our current knowledge of the brain.

Second, rather than regarding the nervous system as a mere information processor restricted to transducing (encoding) external stimuli, neuropsychology is learning that the central nervous system of the higher vertebrate is constantly active and that this activity is largely spontaneous. In other words, the nervous system activity is modulated by environmental stimuli instead of being uniquely determined by them. Hence we must speak of information generation and destruction in addition to information transmission.

Third, neuropsychology does not restrict itself to talk of information processing, for this notion is far too generic; it attempts to understand the peculiar properties of the generation, destruction, and transmission of neural information. Nobody begrudges the information theorist the privilege of talking about information in general, but he cannot explain the functioning of the brain in information-theoretic terms anymore than he can explain the generation, propagation, and detection of electromagnetic waves. Similarly, the general systems theorist is entitled to theorize about systems in general, regardless of the stuff they are made of and consequently regardless of the special laws they satisfy. Not so the specialist in the nervous system of the higher vertebrate: he deals with a unique system possessing properties that no other system in the world has, such as lateral inhibition, synaptic plasticity, spontaneous activity, and the possibility of knowing itself.

Fourth, neuropsychology possesses an all-encompassing theory of behavior and mentation serving as a foil for the construction of partial theories (models) each devoted to a particular kind of behavior or mentation. The general framework is used as a basis and guide for the special theories, much in the same way as general mechanics is the basis for the special theories of the spring, planetary motion, and so on. And, whether extremely general, half-way general, or extremely specific, such theories tend more and more to be couched in mathematical terms, for only mathematics gives precision, deductive power, and conceptual unity.

Fifth, in building neuropsychological theories we can choose among three styles: *holistic* (whole brain), *top-down* (reduction to cellular components) or *bottom-up* (synthesis from components). We need theories at all levels and theories relating the various levels, for the nervous system happens to be a multilevel system, composed of a number of subsystems—the more numerous, the lower the level concerned. Holism is defective because, although it insists that the whole possesses emergent properties absent from its parts, it resists the attempt to explain emergence in terms of composition and structure. Microreduction correctly emphasizes the importance of the composition of a system but is blind to emergent systemic properties, such as the ability to form an image or a concept. Consequently this purely analytic approach is bound to fail in the attempt to explain the mental, which is presumably a collective or mass activity resulting from the activity and the interactions of myriads of neurons. We are left with the synthetic or bottom-up method, which endeavors to reconstruct a system from its components and the interactions among them as well as with the environment. This third strategy has the virtues of the alternative methods: it is concerned with emergent wholes and their composition. And it lacks the defects of its rivals: it does not reject analysis and it does not ignore the multilevel reality of the nervous system. For these reasons it is the most ambitious, the hardest, and the most promising of the three. This is also why it is seldom tried, so that so far only a few precise and reasonably realistic bottom-up models of the activity of neural systems are available (see e.g., Bindra, 1976; Cooper, 1973; Cowan & Ermentrout, 1979; Malsburg, 1973; Pellionisz & Llinás, 1979; Pérez, Glass, & Shlaer, 1975; Wilson, 1975).

The bottom-up modeling strategy is systemic and integrative. Therefore it is the one capable of accounting for the systemic and integrative properties of the various neural systems. By the same token it is the strategy capable of bringing together all the studies in neuroscience and psychology. It has been well tried in physics and chemistry. Thus the solid-state physicist builds a mathematical model of his crystal structure in order to explain the molar behavior of a lump of conducting material.

And the quantum chemist tries to understand the properties of compounds and reactions with the help of hypotheses and data concerning atomic composition (in both cases a knowledge of some global properties is needed and in each case a set of new hypotheses must be added to atomic physics). Similarly, the neuroscientist studying a particular neural system, such as the brainstem, is expected to make a contribution to our understanding of the specific emergent functions of that system in terms of its peculiar composition, structure, and environment—such as its being in charge of awareness and wakefulness.

So much for the salient features of neuropsychology and its scientific worth by comparison with those of behaviorism and mentalism. Let us now place neuropsychology in the scheme of science.

5. Reduction or Integration?

Of the three approaches we have examined, the third is the most promising for being the most scientific. Its peculiar trait is that it reformulates most psychological problems in neurobiological terms (the exception is formed by the problems in social psychology, on which more below). In this way it succeeds in going farther and deeper than its rivals. For example, instead of restricting himself to describing the behavior of an animal engaged in solving a problem, the neuropsychologist will endeavor to identify the neural processes consisting in problem solving and will attempt to trace the evolutionary ancestry as well as the ontogeny of such behavior pattern. In short, the neuropsychologist identifies the mental with certain neural functions, and in this regard he performs a reduction. However, such reduction of the mental to the neurophysiological does not transform neuropsychology into a chapter of physiology. There are several reasons for this.

First, neuropsychology (or psychobiology, or biopsychology) includes not only the neurophysiology of behavior and mentation but also the study of the development and evolution of neural systems and their functions, and therefore the investigation of the genetic and environmental determinants of behavior and mentation.

Second, neuropsychology and even neurophysiology, when studying the mechanisms of behavior and mentation, are guided by findings in traditional psychology. Thus the study of a perceptual system involves knowing what it does (what its specific functions are), how such activity is related to the motor centers, and in what ways it serves or disserves the whole organism.

Third, unlike neurophysiology, which studies individual animals detached from their social environment, neuropsychology cannot ignore social behavior. That is, it must mesh in with sociology to constitute social psychobiology.

In other words, neuropsychology is not a mere logical consequence of neurophysiology. But it does result from enriching neurophysiology with new hypotheses concerning behavior and mentation conceived in a biological fashion. The reduction is then *partial* or *weak* rather than total or strong, to employ terms elucidated elsewhere (Bunge, 1977).

In summary, neuropsychology identifies mental events with neural ones, but it is not a branch of neurophysiology. Instead, it is a merger of neuroscience and psychology with a pinch of sociology. This synthesis has, like every synthesis, properties of its own not found in its components taken in isolation. The main feature of the new synthesis is that it is capable of tackling the most formidable of all scientific-philosophical problems, namely the age-old mind–body problem, instead of either ignoring it or leaving it in the hands of theology or traditional philosophy. And the main advantage of the new synthesis is that it overcomes the fragmentation of the disciplines dealing with behavior and mentation—a lack of cohesion that has often been lamented and sometimes seemed unavoidable (Koch, 1978). The frontiers between the various sciences of behavior and mentation are artificial and bound to be erased.

6. References

Bandura, A. (1974). Behavior theory and the models of man. *American Psychologist, 29,* 859–869.

Bindra, D. (1976). *A theory of intelligent behavior.* New York: Wiley Interscience.

Bindra, D. (1984). Cognitivism: Its origin and future in psychology. In J. R. Royce & L. P. Mos (Eds.), *Annals of theoretical psychology* (vol. 1, pp. 1–29). New York: Plenum Press.

Bunge, M. (1977). Levels and reduction. *American Journal of Physiology,* 233(3), R75–R82.

Bunge, M. (1980). *The mind–body problem.* Oxford: Pergamon Press.

Bunge, M. (1983). *Epistemology and methodology,* vols. 5 and 6 of the *Treatise on basic philosophy.* Dordrecht and Boston: Reidel.

Cooper, L. (1973). A possible organization of animal memory and learning. In B. Lundqvist & S. Lundqvist (Eds.), *Collective properties of physical systems* (pp. 252–264). New York: Academic Press.

Cowan, J. D., & Ermentrout, G. B. (1979). A mathematical theory of visual hallucination patterns. *Biological Cybernetics, 34,* 137–150.

Eccles, J. C. (1980). *The human psyche.* New York: Springer International.

Fodor, J. A. (1975). *The language of thought.* New York: Crowell.

Fodor, J. A. (1981). The mind–body problem. *Scientific American, 244*(1), 114–123.

Hebb, D. O. (1949). *The organization of behavior.* New York: Wiley.

Koch, S. (1978). Psychology and the future. *American Psychologist, 33,* 631–647.

Luce, R. D., Bush, R. R., & Galanter, E. (Eds.). (1963–1965). *Handbook of mathematical psychology* (3 vols.). New York: Wiley.
MacKay, D. M. (1978). Selves and brains. *Neuroscience, 3*, 599–606.
Malsburg, C. von der (1973). Self-organization of orientation-sensitive cells in the striate cortex. *Kybernetik, 14*, 85–100.
Pellionisz, A., & Llinás, R. (1979). Brain modeling by tensor network theory and computer simulation. *Neuroscience, 4*, 323–348.
Pérez, R., Glass, L., & Shlaer, R. (1975). Development of specificity in the cat visual cortex. *Journal of Mathematical Biology, 1*, 275–288.
Popper, K. R., & Eccles, J. C. (1977). *The self and its brain*. New York: Springer International.
Pribram, K. (1971). *Languages of the brain*. Englewood Cliffs,NJ: Prentice-Hall.
Putnam, H. (1975) *Mind, language and reality*. Cambridge: Cambridge University Press.
Pylyshyn, Z. W. (1978). Computational models and empirical constraints. *Behavioral and Brain Sciences, 1*, 93–99.
Skinner, B. F. (1938). *The behavior of organisms: An experimental analysis*. New York: Appleton-Century-Crofts.
Turing, A. A. (1950). Can a machine think? *Mind NS, 59*, 433–460.
Watson, J. B. (1925). *Behaviorism*. New York: The People's Institute.
Wilson, H. R. (1975). A synaptic model for spatial frequency adaptation. *Journal of Theoretical Biology, 50*, 327–352.

3

On Being Brainy

M. C. Corballis

Although I am generally sympathetic to Bunge's theme, the first thing that strikes me about his trinity of psychological strategies is that probably the majority of academic psychologists would not identify with any of them. Mentalism, at least in the sense that Bunge calls *substantialist* mentalism, has long since been relegated to the fringes of psychology, although it may well have a greater following among philosophers. Whether what Bunge calls *functionalist* mentalism is truly mentalism is a matter I shall take up below. I think that behaviorism, at least in the strong sense, has also very largely disappeared from the academic scene, although behavioristic *methods* of course remain useful in both basic and applied research.

The approach that supplanted behaviorism as the mainstream of academic psychology is what has come to be termed cognitive psychology. That term was the title of Neisser's influential book, published in 1967, which crystallized the cognitive approach for North American psychologists. Cognitive psychology owed something to the revolution in linguistics, and perhaps especially to Chomsky's (1959) famous review of Skinner's (1957) book *Verbal Behavior*, which was the most ambitious of behaviorist enterprises. It also owed something to the development of artificial intelligence (AI). Its main roots in psychology probably came from England, where behaviorism had a lesser influence than in North America, and particularly from Cambridge through a lineage that included Bartlett (1932), Craik (1943), and Broadbent. Indeed, it was Broadbent's *Perception and Communication* (1958) that established the "information-processing" approach that has dominated much of the theorizing in cognitive psychology, and in many respects this was the prototype for subsequent works in cognitive psychology.

M. C. Corballis • Department of Psychology, University of Auckland, Auckland, New Zealand.

Cognitive psychology is not behaviorist and indeed derived much of its impetus from a reaction against behaviorism, but it retains the behaviorist emphasis on objective, observable, and indeed very largely behavioral evidence. It is not mentalist. It uses mental terms such as attention, imagery, mental rotation, even consciousness, but it refutes any nonmaterial bases for mental concepts. It is not psychobiological. Neisser's 1967 book contained remarkably little reference to neurophysiological or neurological knowledge, and the same is very largely true of more recent cognitive theory, although one may discern a growing convergence of cognitive psychology and neuropsychology. It might also be noted that Neisser (1976) has recently rejected his own pioneering approach in favor of one that is more "ecological" in tone. Although one might see this in the same light as Bunge's plea for a more psychobiological orientation, Neisser in fact continues to make very little appeal to biological or neurophysiological ideas.

Cognitive psychology does owe a good deal to the computer. The information-processing models that were especially popular in the late 1960s or early 1970s (e.g., Shiffrin & Atkinson, 1969) were based on the sorts of flow diagrams used in computer programming, although in truth one does not have to know much about computers in order to understand them. There have, of course, also been much more explicit attempts to model aspects of human cognition on the computer; these include Newell and Simon's (1963) General Problem Solver, Winograd's (1973) SHRDLU, a simulation of the understanding of language, Newell's (1973) production systems, and many others. The field of AI was at first somewhat removed from that of cognitive psychology, but the two have recently merged in the new discipline known as cognitive science.

1. Strong and Weak AI

Searle (1980) has distinguished "strong" from "weak" AI. Weak AI refers to the use of the computer as a tool for exploring the nature of mental processes, a strategy that neither Searle nor Bunge takes exception to. It should perhaps be stressed, however, that the computer is not merely a tool, but also a metaphor—the mind is likened to a computer or robot, and the language of computers provides a natural medium for articulating cognitive theories. Thus cognitive psychologists speak of such processes as scanning, filtering, coding, storing, retrieving, transforming. Psychology has always relied heavily on mechanical metaphors, such as those based on hydraulic systems, telephone switchboards, balances, or holograms, and the computer offers a degree of complexity and appropriateness so far unrivalled. Metaphor is necessary to psychology because there is still a large gap between the processes and units of mental functioning and those of neurophysiology. If the majority of psychologists still eschew "physiologizing," it is not entirely because they believe that the brain is irrelevant to an understanding of mental processes; rather, it is because they believe that we do not yet have any real idea of the neurophysiological bases of most psychological phenomena. Hebb (1949) made a brave and imaginative attempt to develop a neurophysiologically based theory of mental processes, but I think that his book was more important as a morale booster than as a serious, verifiable psychological theory. It made a strong case for supposing that mental processes might *in principle* be accounted for in neurophysiological terms, but it went only a very small distance toward actually providing such an account.

Modern theories of memory illustrate the power of the computer metaphor over neurophysiologically based theorizing. I do not believe that any neuroscientist has any real idea as to how memory actually works at the neurophysiological or neurochemical level—that is, as to how events are physically recorded in the brain. Yet psychological theories of memory have grown enormously in sophistication over the past quarter of a century, largely through the influence of computer-based cognitive models (see, e.g., Baddeley, 1976). We know quite a lot about the practical side of memory, including the use of mnemonic devices, the effects of different coding strategies, and so forth, and we can frame theories about memory with sufficient precision to test them empirically and modify them if necessary. This is not to say that a purely psychological theory of memory is all that we need ever want. I have little doubt that if the memory code can be "cracked" in the way that the genetic code was, then the implications for psychology will be huge; in the meantime, psychological theories are doing quite well even in the absence of neurophysiological or neurochemical input, just as genetics was a useful and viable science before the genetic code was understood at a biochemical level.

According to strong AI, the computer is more than just a tool or a metaphor for the study of mental processes; rather, a computer programmed to simulate cognitive states can actually be said to *possess* those states. That is, computers can in principle be made to understand, feel, know, perceive, and so on, in the same way that human beings can. To most cognitive psychologists, this assertion is probably unexceptionable; even Neisser (1967), by no means a hard-core cognitivist, argued that cognitive psychology was concerned with the software rather than with the hardware of mental processes. Hence it does not really matter whether a cognitive "program" is realized in a person, or a digital computer, or a sequence of water pipes, or a set of wind machines, or even, as cunningly proposed by Weizenbaum (1976), a device constructed of a roll of toilet paper and a pile of small stones (Searle, 1980).

It is this strong version of AI that both Searle and Bunge take exception to. It seems to me ironic, however, that both authors identify strong AI with mentalism. In Bunge's logic, strong AI implies that anything, even toilet paper and stones, can have a mind, and so mind is disembodied. Searle (1980) writes similarly:

> Strong AI only makes sense given the dualistic assumption that, where the mind is concerned, the brain doesn't matter. In strong AI (and in functionalism, as well) what matters are programs, and programs are independent of their realization in machines. . . . The single most surprising discovery I have made in discussing these issues is that many AI workers are quite shocked by my idea that actual human mental phenomena might be dependent on actual physical-chemical properties of actual human brains. But if you think about it a minute you can see that I should not have been surprised; for unless you accept some form of dualism, the strong AI project hasn't got a chance. (p. 423)

The irony is that, to its advocates, strong AI appeals precisely because it is materialistic rather than mentalistic or dualistic. For instance, Pylyshyn (1980) writes as follows:

> What I wish to focus on here is what I take to be the most fundamental reason why cognition ought to be viewed as computation. That reason rests on the fact that computation is the only worked-out view of *process* that is both compatible with a *materialist* view of how a process is realized and that attributes the behavior of the process to the operation of rules upon representations. In other words, what makes it possible to view computation and cognition as processes of fundamentally the same type is the fact that both are *physically* realized and both are governed by rules and representations. (p. 113, italics added)

It is indeed hard to accept that strong AI is mentalistic or dualistic. I suspect that the mistake that Bunge and Searle make is to treat mind as a thing rather than as a process. I do not think that anyone would seriously claim that a computer program has a mind, but a computer operating that program might possibly be credited with understanding, thinking, recognizing patterns, or whatever. Thus Bridgeman (1980) replies to Searle as follows:

> Searle's accusation of dualism in AI falls wide of the mark because the mechanist does not insist on a particular mechanism in the organism, but only that "mental" processes be represented in a physical system when the system is functioning. A program lying on a tape spool is no more conscious than a brain preserved in a glass jar. (p. 427)

2. Limitations of Neurophysiology

However, I think that the charge of mentalism is probably just a diversionary tactic and that Bunge's real objections are elsewhere. One argument is that the brain is too complicated to be modeled by a computer (or a Turing machine). But there is a touch of irony here, too, because those who have resorted to computers have done so precisely because of the complexity of mental processes. In the attempt to explain the human ability to understand stories, for instance, Schank (1980) writes, "The range of the phenomena to be explained is too broad and detailed to be covered by a theory written in English. We can only know if our theories of understanding are plausible if they can work by being tested on a machine" (p. 446). In any event, it is surely extremely unlikely that human understanding or human language or human imagination can be elucidated by neurophysiology, at least at the present time. Geschwind (1980) points out, for instance, that "there is not a single neuron in the central nervous system of any *mammal* whose connections are fully specified" (p. 188, his italics). Computer science is indeed still a primitive medium for the elucidation of human cognition, but it is not nearly so primitive as neurophysiology.

For all that, there must surely be important insights to be gained from neurophysiology. Geschwind also recalls that von Neumann, in one of his last lectures, spoke of invertebrates whose nervous systems contained fewer units than were possessed by the calculating machines of the period, and yet the invertebrates were the more complex in their functions. Presumably there are principles of design and programming even in simple organisms that are not yet understood and that will be discovered only by neurophysiological investigation. One must agree with Bunge, and indeed with Geschwind, that computer science must learn from neuroscience if it is to progress.

It is in the domain of perception, and most particularly visual perception, that neurophysiological evidence has proved most valuable. The work of Hartline and others on lateral inhibition, Hubel and Wiesel on the functional properties of single cells in the visual cortex, the more recent discoveries about spatial frequency analysis in the cortex and the distinction between sustained and transient cells—these and other basic findings, although derived from work with species other than humans, have provided unexpected insights into the nature of human perception. I do not think that these insights would have emerged from attempts to simulate perception on a computer, or even from purely psychological research, although psychologists have been quick to see their relevance and to adapt their techniques and theories accordingly. Even Hebb's

(1949) unusually insightful speculations about the neurophysiology of visual perception failed even remotely to predict the actual course of events.

Yet we should not be lulled into a false sense of achievement. The insights gleaned from single-cell recordings in the visual systems of crabs, frogs, cats, and monkeys come nowhere near to providing a theory of visual perception in these species, let alone in human beings. It is surely folly to extrapolate the neurophysiological evidence to the point of identifying, at least notionally, the single cell responsive to purple Volkswagens, as some earlier enthusiasts were apt to do. Even the lowly pigeon appears to be able to tell whether or not a photographic slide it has never seen before depicts some natural concept, such as water, or trees, or even a particular human being, but we have simply no idea how it does it (Herrnstein, Loveland, & Cable, 1976). I think that there has been more progress in the field of perception than in most other areas of psychology and much of this progress is related to discoveries in neurophysiology, but we cannot even begin to construct an adequate theory in terms of neural networks.

In the development of neuropsychology over the past 20 years, however, we can identify a more global sense in which mental processes can be linked to the study of the brain. Here the concern is not with individual neurons or specified neural networks, but rather with particular areas of the brain or with relatively global indices of neural activity, such as those provided by the electroencephalogram, or regional cerebral blood flow, or more recently by positron emission tomography. In at least some respects, human neuropsychology has provided the neurobiological basis for cognitive science that Bunge calls for; for instance, there is increasing convergence between neuropsychological and cognitive theories of memory (e.g., Cermak, 1981), and even though little is known of the physical basis of memory there is growing understanding of the brain areas involved. Yet a good deal of the effort in neuropsychology has been toward the crude localization of function, which does relatively little to advance cognitive *theory*, although there are signs of a more sophisticated functional approach (e.g., Kelso & Tuller, 1981; Warrington & Weiskrantz, 1982).

3. Conclusion

If we exorcise the (holy?) ghost of mentalism, Bunge's trinity boils down to behaviorism, AI, and psychobiology. These define the vertices of a triangle, and most psychologists occupy the area in the middle.

Behaviorism will continue to provide the acid test for cognitive theory (as in the Turing test) and will no doubt continue to serve as the empirical basis for psychological research, especially with animals. It will also continue to serve as a counter to what appears to be a natural human tendency to resort to mentalistic interpretations. AI is also here to stay; it is difficult to see how one can approach certain problems in cognitive psychology, such as the way people play chess or understand stories, without the help of the computer or some equivalently complex analogy. Yet there are indeed ways in which computers do not resemble brains, as Bunge notes. I suspect that all of them can *in principle* be overcome; for instance, one *can* inject spontaneous activity into a computer, or (paradoxically) program it to command rather than obey, but the result might be a very roundabout simulation of what brains do naturally. There are theorems that seem to say that *anything* can be simulated, but a simulation of a biological process might be very misleading indeed.

I agree with Bunge, however, that psychologists have been overly reluctant to face the biological side of their discipline. My favorite example has to do with cerebral lateralization, where the research emphasis has been primarily on the different modes of consciousness, or cognitive styles, associated with the left and right cerebral hemispheres. Given that lateralization is fundamentally a biological phenomenon, a biological approach has been curiously lacking, except perhaps over the last few years (e.g., Corballis, 1980). Even in the post-Darwinian era, we seem to be reluctant to admit to the biological aspect of ourselves.

4. References

Baddeley, A. D. (1976). *The psychology of memory*. New York: Basic Books.

Bartlett, F. C. (1932). *Remembering*. Cambridge: Cambridge University Press.

Bridgeman, B. (1980). Brains + programs = minds. *Behavioral and Brain Sciences, 3,* 427–428.

Broadbent, D. E. (1958). *Perception and communication*. New York: Pergamon.

Cermak, L. S. (Ed.). (1981). *Human memory and amnesia*. Hillsdale, NJ: Erlbaum.

Chomsky, N. (1959). Review of B. F. Skinner's *Verbal behavior*. *Language, 35,* 26–58.

Corballis, M. C. (1980). Laterality and myth. *American Psychologist, 35,* 284–295.

Craik, K. J. W. (1943). *The nature of explanation*. Cambridge: Cambridge University Press.

Geschwind, N. (1980). Neurological knowledge and complex behaviors. *Cognitive Science, 4,* 185–193.

Hebb, D. O. (1949). *The organization of behavior*. New York: Wiley.

Herrnstein, R. L., Loveland, D. H., & Cable, C. (1976). Natural concepts in the pigeon. *Journal of Experimental Psychology: Animal Behavior Processes, 2,* 285–311.

Kelso, J. A. S., & Tuller, B. (1981). Toward a theory of apractic syndromes. *Brain and Language, 12,* 224–245.

Neisser, U. (1967). *Cognitive psychology*. New York: Appleton-Century-Crofts.

Neisser, U. (1976). *Cognition and reality*. San Francisco: Freeman.
Newell, A. (1973). Production systems: Models of control structure. In W. G. Chase (Ed.), *Visual information processing* (pp. 463–526). New York: Academic Press.
Newell, A., & Simon, H. A. (1963). GPS, a program that simulates human thought. In A. Feigenbaum & V. Feldman (Eds.), *Computers and thought* (pp. 279–298). New York: McGraw-Hill.
Pylyshyn, Z. (1980). Computation and cognition: Issues in the foundation of cognitive science. *Behavioral and Brain Sciences, 3,* 111–132.
Schank, R. C. (1980). Understanding Searle. *Behavioral and Brain Sciences, 3,* 446–447.
Searle, J. R. (1980). Minds, brains, and programs. *Behavioral and Brain Sciences, 3,* 417–424.
Shiffrin, R. M., & Atkinson, R. C. (1969). Storage and retrieval processes in long-term memory. *Psychological Review, 76,* 179–193.
Skinner, B. F. (1957). *Verbal behavior*. New York: Appleton-Century-Crofts.
Warrington, E. K., & Weiskrantz, L. (1982). Amnesia: A disconnection syndrome? *Neuropsychologia, 20,* 233–248.
Weizenbaum, J. (1976). *Computer power and human reason*. San Francisco: Freeman.
Winograd, T. (1973). A procedural model of language understanding. In R. Schank & K. Colby (Eds.), *Computer models of thought and language* (pp. 152–186). San Francisco: Freeman.

3

Is Neuropsychology Something New?

P. C. Dodwell

Although I am in broad general agreement with Bunge's aims and commentary on the state of psychology, I find much to criticize in detail in his paper. In general his description of the nature and practice of science today is too clean and idealized to be plausible; in particular his distinction between the way neuropsychology is conducted as a scientific discipline, compared to other branches of contemporary psychology in this regard, will not bear detailed scrutiny.

The concept of a scientific approach and its analysis into the four components of *general outlook, problematics, aims,* and *methodics* is very useful and undoubtedly can help us to understand the nature of modern scientific psychology. In applying his analysis to behaviorism and mentalism, however, Bunge has perhaps overstepped the bounds of cautious and careful reasoning which must form the basis of any sound appraisal of scientific work.

1. Behaviorism

Consider first his treatment of behaviorism; much of what he has to say is true of the historical movement which had its heyday more than 50 years ago. In discussing the strict code of scientific conduct of the behaviorists he states, "We should be grateful to behaviorists for

P. C. Dodwell • Department of Psychology, Queen's University, Kingston, Ontario K7L 3N6, Canada.

having introduced this code into the field of psychology, where illusions and deception (unwitting or deliberate) are not uncommon" (p. 119). It is well recognized that behaviorism introduced a stricter set of standards of scientific reasoning into psychology which was of lasting benefit; however, we do need to bear in mind that this happened well over half a century ago. To imply that the rest of psychology was (or is) subject to illusions and deceptions rather suggests that the lessons of behaviorism have not been learnt. For most of the current fields of psychological research and theory I think this simply is not true.

Although we can agree that the problematics of behaviorism is extremely narrow, this does not in itself mean that it is invalid or currently worthless. Indeed much of what we now know and value in the study of the behavior of animals, especially their abilities to learn, stem from behaviorist concepts married to the behavioral and physiological knowledge gained from Pavlov's reflexology. Similarly, there are contemporary situations in which the methodics of behaviorism is entirely appropriate; the question we have to ask is whether these situations yield useful scientific knowledge. Anyone familiar with the application of behavioral concepts in modern forms of control and treatment for maladaptive behavior could scarcely deny that they have such value, at least in an applied sense. Bunge is quite right to criticize behaviorism's claim to provide comprehensive explanations for every form of behavior and mentation which might be of interest to the psychologist, but that grandiose claim was made by Watson, Hull, and their contemporaries and is seldom if ever heard these days. The exception is Skinner, an influential but controversial latter-day behaviorist. He is not, however, a typical modern psychological theoretician.

One might almost say that Bunge's characterization of behaviorism is something of a caricature, or at least an anachronism. Is there, according to his criteria, any "pure" behaviorist left? If not, one should cease to flog a dead horse. Bunge throws a neat compliment behaviorism's way, in view of its influence on the launching of scientific psychology. Let us now add to that compliment the indubitable fact that some of behaviorism's strengths still sustain major parts of our discipline.

The behaviorists made a valuable conceptual clarification between learning and performance, for example. This has carried over into a general principle which is usefully applied to the understanding of development, language, memory, and other areas of cognition. Thus it is not true to say that behaviorism has made no lasting contributions to the broad theoretical structure of modern scientific psychology.

2. Mentalism

In castigating mentalism Bunge makes more sweeping generalizations. In terms of his characterization, one must ask again: are there pure mentalists left in scientific psychology? Surely it is scientific psychology that he means to deal with. To strike a comparison: no serious critic of cosmology and astronomy would venture to ostracize them as sciences on the ground that astrology bears some superficial resemblance to them, in its field of study. Similarly, although it is true that some wild mentalistic speculations have and are being made, they are not part of what most of us would regard as current scientific psychology. However, Bunge seems to equate much of modern cognitive psychology with that old mentalist tradition.

Two questions demand consideration: First, is there a sense in which mentalism can be reconciled with modern scientific psychology? And second, to what extent are modern cognitive psychologists "mentalists"? Bunge assumes that the label fits perfectly, but that is an oversimplification.

Some very respectable psychologists would answer yes to the first question, given a reasonable definition of mentalism. For example J. J. Gibson (1966) whose views have so heavily influenced the development of perceptual theory, and stemming from that certain other branches of cognitive psychology (e.g., Shaw & Turvey, 1982), is a psychologist who believed profoundly that to understand perception it was not necessary to study the brain in detail. To that extent, he certainly was a mentalist. The new Gibsonians have gone further than he did in propounding a notion of psychological explanation that is not contingent on understanding how the brain works. Their reasons for holding these views are, of course, not merely perverse, but follow from their understanding of what psychological explanation should be about. That understanding differs sharply from the behaviorists' model of explanation, as it does from that of the neuropsychologist. Is it not possible that such other attempts at explanation and understanding can be examined and validated in their own right? In my view it is not good enough merely to dismiss them as mentalistic, without further analysis. I must add that some serious attempts to analyze Gibson's theories, such as that of Fodor and Pylyshyn (1981), have discovered other grounds for thinking that those theories are, if held in their strict form, unsound. But those grounds have nothing in particular to do with the lack of neuropsychological hypotheses in Gibson's work. I myself believe that a lack of willingness to concede the importance of physiological coding mechanisms in a visual theory is a serious weakness, but it does not have to be—as Bunge so strongly implies—in itself a fatal weakness.

Thus there are still some mentalists in scientific psychology, but I think it is quite wrong to characterize most cognitive psychologists in this way. Most of them are intensely interested in learning to understand the mechanisms of the mind, which obviously at one level involves knowledge of its neuropsychological properties. What Bunge interprets as a lack of interest in the neuropsychological substrate is, I believe, something rather different. In order to understand the mind—or indeed any other object worthy of scientific investigation—it is necessary at a fairly early stage to ask the question: What is it for? or What does it do? Without such an analysis of its functioning and purpose, there surely would be little point in trying to understand how it is put together, develops, or "emerges." This is a point to which I shall return shortly. In concentrating on functional analyses of cognitive behavior cognitive psychologist are not so much denying the importance of the "wetware" of the brain as they are treating it like a black box. Surely the history of science tells us that a certain methodics may be perfectly valid within a particular historical context but may change with theoretical and/or technical developments. What is a black box to one generation of psychologists, or any other breed of scientists, may be an open book to the next. Cognitive psychologists as a group tend to the view that greater understanding of cognitive phenomena can at present be attained through abstract modeling, such as charting and quantifying the operations of a memory system in a flow diagram, than by trying to identify the brain sites that implement the operations. Few, if any of them, would deny the desirability in principle of being able eventually to slot their operations into known parts of the brain.

However, I believe that the cognitive psychologists' concentration on functional questions has less to do with our still quite primitive understanding of most brain functions than with the fact that we are still struggling to form what Marr (1982) has called a "computational theory" of psychology. That is to say, in less esoteric language, we are still struggling to understand with some exactness just what operations (computations in the computer scientists' vocabulary) need to be performed in order to achieve the ends that we, as human beings, do strive for, whether in thinking, memory, language, or perception.

Thus I believe that what Bunge sees as a serious weakness in modern cognitive psychology is actually the deliberate choice of a strategy which is thought to be more useful than a strategy based on neuropsychological principles. I know of no cognitive psychologist who believes that the brain is unimportant, just as I know of none who would be willing to state that all we know or need to know about language, memory, and thought can be studied only in the context of what we know about the physiological functions of the cerebral cortex. Gibson's point was not that the brain is irrelevant to psychology, but that studying the brain is not the best way to arrive at useful psychological explanations of the sort we in fact seek.

In taking the cognitive psychologists to task for treating the brain like a computer, I think Bunge does them a serious injustice. It is of course true that many have likened the brain to a computer, and mental functioning to the running of a program in a computer, but few have taken this to be a serious model of the actual facts. Rather, the computer–program analogy is found to be enlightening in thinking about functional properties of the brain and mind. Perhaps Bunge would be happier if one were to say functional properties of the brain–mind.

As an expository device, it is perhaps useful to contrast the cognitive psychologist with the neuropsychologist, but in truth I believe the boundaries between them cannot be drawn rigidly. Any reasonable cognitive psychologist is most likely to be willing to listen to what the neuropsychologist can tell him about the brain, just as the neuropsychologist would surely be out of business without the immense corpus of knowledge developed in the more traditional areas of psychology.

3. Neuropsychology

What, then, of neuropsychology? Bunge sings its praises, but his argument that it is the one and only true scientific psychology will, I believe, convince only those already converted to that point of view. Granted the scientific nature and status of neuropsychology as a branch of neuroscience, and granted the willingness of neuropsychologists to tackle every type of meaningful scientific question in psychology, one is still left with the fundamental question of what, in fact, constitutes a neuropsychological explanation? Perhaps more exactly one should say: What is it that the neuropsychologist seeks to explain that is different from other psychological explanations? As I have pointed out in another context (Dodwell, 1978), most of the questions for which psychologists seek answers in fact only make sense within an already understood and accepted matrix of concepts and processes. Thus it makes sense to study the question of whether chimpanzees can acquire language, or the effect of cortical damage or other brain lesions on language in humans, only if we already have a very good understanding of what language is. The understanding of what language is will never come about through the study of such experimental preparations. That is to say, the point of the behavioral or neuropsychological investigation assumes a *prior* conceptual knowledge of what a particular form of behavior is all about: in Marr's terms, its computational theory. That understanding is itself not explicable in neuropsychological terms exclusively. Show me a neuropsychologist who can explain language in terms of concepts which exclude the communicative function that language serves!

Of course a dyed-in-the-wool neuropsychologist might deny this point. As a matter of fact, I rather suspect that Bunge might be one such; consider, for example, the statement early in his paper: "We also wish to know what pain, love, and thought are, that is, what are the neural processes we *call* pain, love, and thought." This surely is begging the main question. Unless we already know what pain, love, and thought are from our personal experience and interpersonal observation and communication, there would be absolutely no way in which any neuropsychological correlate or theory of those states would make sense. If we choose to *redefine* pain, love, and thought in purely neuropsychological terms, that is another matter. We then must inquire whether or not that redefinition involves a misuse, not to say an abuse, of language.

Thus neuropsychology as such is too narrow—much too narrow—to yield *complete* explanations of psychological phenomena. It is no doubt the most promising way at present of furthering our knowledge of how brain and behavior are related. However, the exhortation to yield the field of psychology, and particularly of scientific explanation in psychology, to neuropsychology is a statement of faith, rather than one of proven validity. Viewed as one among several approaches to the problem of understanding human behavior, neuropsychology has many points in its favor, most of which Bunge mentions. One thing he did not emphasize, which to me is one of the more important points, is that within neuropsychology, and more specifically within the physiologically based theoretical models which are identified with what one might now call the McGill school, an accumulation of valid data and theory is much in evidence. One of the great weaknesses of most branches of scientific psychology over the last hundred years has been the comparative sparsity of such cumulative knowledge. This has been remarked on by many leaders in the field. In learning theory, mathematical psychology, cognitive psychology, to name a few, the theoretical predilection has been in the direction of following the latest developments in technology without a deep analysis and understanding of the nature of the sorts of explanation we should seek, and the sorts of answers which in principle it is possible to give. Thus the progress of psychology has been marked by the appearance of many fads and schools, which have held sway for relatively short periods of time, to be displaced by new schools of thought.

The building up of knowledge in one generation based on the theoretical models and research of previous workers, although not totally absent, constitutes a relatively small part of the progress of psychology. Neuropsychology has fared somewhat better than the average.

4. Explanation and Authority

A final point concerns the question of orthodoxy and authority. Bunge refers early in his paper to the fact that one of the important advances of scientific psychology was to outgrow the traditional authority of church, philosophy, and other established sources of wisdom and to institute its own scientific procedures. We should be aware of a subtle new form of appeal to authority, one which is evident in Bunge's own work. That is the appeal to the better established natural and life sciences as being "more scientific" than is our own behavioral science of psychology. Psychology has much to learn from the traditional sciences, but I believe it also has much to offer, including its own methodics and preferred explanatory schemata. A large part of what it has to offer is developing in the field of neuropsychology, but we should beware lest we come to think that this is the only valid facet of modern scientific psychology.

To attempt here a comprehensive analysis of the nature of psychological explanation and the extent to which the neuropsychological models can yield adequate explanations would lead too far afield. I will close by emphasizing the fact that explanation in psychology is not a well-understood and agreed upon matter. To the extent that Bunge has raised and advocated one particular explanatory model to the exclusion of others, he has issued a challenge worthy of serious response. I do not think that one can solve all the conceptual problems of psychological explanation by stating boldly that one and only one method of explanation is the right one. On the other hand the staking out of such a firm position will undoubtedly provoke from others a more careful analysis of their own points of view, and one hopes eventually a more adequate and sophisticated understanding of what it is we are trying to do when we propose scientific explanations of psychological phenomena.

5. References

Dodwell, P. C. (1978). Human pattern and object perception. In R. Held, H. Leibowitz, & H. L. Teuber (Eds.), *Handbook of sensory physiology* (vol. 8: Perception, pp. 523–548). New York: Springer.

Fodor, G., & Pylyshyn, Z. (1981). How direct is visual perception? Some reflections on Gibson's "Ecological Optics." *Cognition, 9,* 139–166.
Gibson, J. J. (1966). *The senses considered as perceptual systems.* Boston: Houghton-Mifflin.
Marr, D. (1982). *Vision.* San Francisco: Freeman.
Shaw, R. E., & Turvey, M. (1982). Coalitions as models of ecosystems. In M. Kubovy & J. Pomerantz (Eds.), *Perceptual organization* (pp. 343–415). Hillsdale, NJ: Erlbaum.

3

On Research Strategies in Psychology

Reply to Commentators

Mario Bunge

1. Notes on Corballis

Corballis's commentary on my paper contains the following controversial points.

1.1. Functionalism and Mentalism

It is true that a few functionalists, notably Pylyshyn, declare themselves to be materialists. However, others, Fodor, for instance, claim functionalism to be neutral between materialism and mentalism. And finally others, such as MacKay, are openly mentalists. Yet the interesting point is whether functionalism is *in fact* mentalistic regardless of the explicit professions of faith of its defenders. I claim that it is, for treating the mind separately from that which does the minding (i.e., the brain) and for ignoring developmental and evolutionary neurobiology and psychology.

1.2. Merits of Functionalism

Corballis sees more merit in functionalism than in Hebb's 1949 classic—and he does not even mention his former colleague Bindra, who expanded Hebb's hypotheses. In fact, he regards Hebb's book as having

Mario Bunge • Foundations and Philosophy of Science Unit, McGill University, Montreal, Quebec H3A 1W7, Canada.

been "more important as a morale booster than as a serious, verifiable psychological theory," (p. 137) and he shows great admiration for the "modern" (i.e., functionalist) theories of memory. These are extraordinary statements to make at a time when physiological psychology flourishes as never before, largely because of the work of Penfield, Hebb, Olds, the Milners, Hubel, and other former fellow-Montrealers of Corballis's. And it is doubly extraordinary since Corballis recognizes that the "weak" AI approach to cognition is metaphorical rather than literal.

1.3. Can Anything Think?

According to Corballis, "in Bunge's logic, strong AI implies that anything, even toilet paper and stones, can have a mind, and so mind is disembodied" (p. 138). I have never said this. I know that the advocates of strong AI state that only properly organized systems can have a mind—adding, of course, that the particular matter they are composed of does not matter.

1.4. Mind a Thing?

Corballis writes: "I suspect that the mistake that Bunge and Searle make is to treat mind as a thing rather than as a process" (p. 138). Had he taken the trouble to check my book *The Mind–Body Problem* (1980), he would have seen that I define mind as a collection of brain processes of a certain kind. Moreover I do so in exact set-theoretic terms.

1.5. Specifically Biological Phenomena

Toward the end of his commentary Corballis repeats his well-known thesis that "lateralization is fundamentally a biological phenomenon," (p. 141) suggesting that this is the one weak point of functionalism. However, any physical asymmetrical system, such as a chair, exhibits lateralization. On the other hand, only certain brains possess such properties as lateral inhibition, self-starting, and the capacity to create new ideas. A thesis of my paper and of my book is that if we wish to understand the formation of new ideas we had better study the brain rather than the computer. Computer design is an engineering field; cognition, being a biological phenomenon, is a matter for neuroscience and physiological psychology.

2. Comments on Dodwell

I am glad to learn that Dodwell is in broad agreement with my paper. This will allow me to deal only briefly with what I take to be the main questions he raises.

2.1. Should One Cease to Flog the Dead Horse of Radical Behaviorism?

It is questionable whether strict or radical behaviorism is as dead as Dodwell thinks it is. True, except for Skinner there seem to be no *famous* behaviorists around, at least in North America. But it is also true that many psychological journals still publish papers that approach psychological problems in terms of stimulus and response, reinforcement and conditioning without even mention of physiological mechanisms. The journal *Behaviorism* is committed to radical behaviorism. And, as Dodwell himself reminds us, there is the entire—and indeed fruitful—field of behavior therapy, which is based exclusively on conditioning models of neuroses, phobias, and like disorders. But even if behaviorism were really water under the bridge, it still would be appropriate to mention it when dealing with the various approaches to the problematics of behavior, perception, feeling, and ideation—which was exactly what I was doing in the paper.

2.2. Are There Any Pure Mentalists Left in Psychology?

Again, mentalism must be taken into account when reviewing the possible approaches to investigating the mind. And even if no scientific psychologist were to adopt it nowadays, there are plenty of linguists and philosophers who do. For example, the entire Chomsky school of linguistics is mentalistic and therefore totally disinterested in neurolinguistics and even in sociolinguistics (see, e.g., Chomsky, 1972). However, Dodwell is too optimistic when stating that there are no mentalists left in psychology. What may happen is that it takes a suspicious philosopher to ferret them out. One eminent recent example will suffice to prove my contention that mentalism is all but dead. Richard Gregory (1973) distinguishes two modes of generation of illusions: by the malfunction of physiological mechanisms (or physiological), and by the inappropriateness of cognitive strategies (or cognitive). This distinction implies that cognitive processes are not physiological but purely mental ones. Is this not a clear case of mentalism?

2.3. Is Cognitive Psychology the Same as Cognitivism?

In my paper I criticize *cognitivism*, or information-processing psychology, not cognitive psychology. I take the former to be the branch of cognitive psychology that denies cognitive processes to be brain processes, or at least not to be explainable in physiological terms; it asserts instead that cognitive processes are information-processing processes that are best understood by analogy with computers. Surely this approach to cognition is incompatible with the physiological one, not only because it is metaphorical but also because it bypasses the brain. Dodwell claims to know of no cognitive psychologist who believes that the brain is unimportant. Let me mention only three among the best known cognitive psychologists of the information-processing persuasion, who could not care less about the brain: Simon, Newell, and Pylyshyn. The latter states rightly that the computational models of cognitive processes "represent a move away from the study of material substance towards a more abstract study of form" and that "artificial intelligence systems are invariably directed toward cognitive rather than biological states" (Pylyshyn, 1978). The whole point of the cognitivist school of cognitive psychology is that matter does not matter: that only mind (which can be "embodied" now as a brain, now as a computer, and possibly also as an immaterial ghost) matters. This is why I claim that it is good old mentalism in fashionable garb.

2.4. Is It a Matter of Redefinition?

Dodwell states, correctly, that it would be impossible to explain mental states unless we were previously acquainted with them. Thus before attempting to explain problem-solving in neurophysiological terms we must have some experience with problem-solving and must even have studied it in the traditional fashion of pure psychology. Granted, but this does not imply that the only task left for neuropsychology is to "redefine" problem solving (or pain, love, imagination, etc.) in strictly neurophysiological terms, whence neuropsychology would be "too narrow." What Dodwell calls *redefinition* is no less than explanation proper. Think of a similar problem in the history of recent science, that of accounting for the gross properties of metallic bodies, such as the hardness, brilliance, and conductivity of a wedding ring. Man started by acquiring prescientific knowledge of these facts, and only recently he fashioned theories to account for them. The first theories were descriptive, not explanatory. Over the past half a century physicists have built an entire new discipline, namely solid-state physics, that has succeeded in *explaining* the macroproperties of metals in terms of the ions, electrons, and electric fields that compose them. Most emphatically, this has not been a mere redefinition of hardness, brilliance, and conductivity. We now understand what formerly we were able only to describe. I submit that neuropsychology is parallel, in that it attempts to explain what pure psychology was at most able to describe.

2.5. What Is Explanation?

Dodwell acknowledges that "explanation in psychology is not a well understood and agreed upon matter" (p. 149). I may add that philosophers of science are in no better a position since they abandoned Aristotle's notion of explanation as the search for causes. The dominant view on scientific explanation among philosophers is that it is identical with subsumption under laws. On this view Ohm's law would explain why, when a 1-volt battery is made to energize a circuit containing a 1-ohm lamp, an electric current of 1 ampere flows through it. But surely this is not the way physicists understand explanation; they would characterize that as a *description*. An explanation proper involves some *mechanism* or other, whether mechanical, electrical, chemical, or what have you. In our example the action of the electric field on the conduction electrons of the circuit explains the current (an even deeper explanation resorts to the particular metal structure unveiled by solid-state physics). No mechanism, no explanation proper, or at least no explanation of the kind which I used to call *interpretive* (Bunge, 1967, 1983). And, since every mechanism is material (though not necessarily physical), I submit that there are no psychological mechanisms that are not brain mechanisms. It follows that genuine psychological explanation is neurophysiological. To be sure, neurophysiology is insufficient in the case of language, mentioned by Dodwell, and other cognitive functions that make no sense except in a social setting. But this only shows that in addition to studying the individual brain we should study the interactions among brains through verbal utterances and actions.

2.6. A New Form of Argument from Authority?

Dodwell accuses me of promoting "a subtle new form of appeal to authority," namely, to the better established natural sciences. Perhaps, but it so happens that science started by attacking physical problems, not psychological ones, and all the other fields of knowledge became scientific to the extent that they adopted (and adapted) the scientific method that was born in physics, as well as the naturalistic world view created by the first modern physicists. Besides, there is the matter of overall consistency of human knowledge. One has no right to claim that one's favorite branch of learning is scientific unless it satisfies a number of requirements, among them its consistency with the more established branches of science. What would you think of a chemistry that ignored physics, or of a biology that ignored chemistry—or of a psychology that ignored biology? In other words, one of the criteria for establishing whether a given discipline is scientific, rather than pseudoscientific, is to make sure that it has intimate (and fruitful) contacts with established scientific fields (Bunge, 1983). This criterion suffices to decide that psychoanalysis and parapsychology—both of which are mentalistic, by the way—are nonscientific. If this be called appeal to authority, let it be—but with the proviso that it is the authority of the mode of knowledge that recognizes no *final* authority.

3. References

Bunge, M. (1967). *Scientific research*, 2 volumes. New York: Springer-Verlag.
Bunge, M. (1980). *The mind-body problem*. Oxford: Pergamon Press.
Bunge, M. (1983). *Understanding the world*. Dordrecht and Boston: Reidel.
Chomsky, N. (1972). *Language and the mind* (enlarged ed.). New York: Harcourt Brace Jovanovich.
Gregory, R. L. (1973). The confounded eye. In R. L. Gregory & E. H. Gombrich (Eds.), *Illusion in nature and art* (pp. 49–95). London: Duckworth.
Hebb, D. O. (1949). *The organization of behavior*. New York: Wiley.
Pylyshyn, Z. (1978). Computational models and empirical constraints. *Behavioral and Brain Sciences, 1*, 93–128.

4

Is Psychoanalysis Therapeutic Technique or Scientific Research?

A Metascientific Investigation

Carl Lesche

Abstract. The term *psychoanalysis* can refer to either a science or a therapy. We shall first deal with it as therapy and introduce step by step the explicata. As our first explicatum, therapy will be regarded as technique and we shall examine the conditions under which it may be justified. A rational technique ought to be based on a pure science, which in this case ought to be psychoanalytical science. Second, therapy will be explicated as the treatment of illness. A justified rational technique ought to be based on an empirical science which permits descriptions that are explainable universal sentences. It is only in the exact natural sciences that we can find such sentences, not in the behavioral and human sciences. A natural-scientific, pure psychoanalysis is impossible; it cannot exist, since psychoanalysis strives for self-reflection and emancipation. We shall then recommend a third explication of psychotherapy, namely, as a technique in which the goal is described in intentional and phenomenal terms and valued in terms of health values. It is impossible to explain phenomenal explananda with physicalistic explanantia. At most, psychoanalytical therapy, psychopharmacological therapy, behavioral therapy, and psychosomatic medicine can be considered unjustified empirical (Machiavellian) techniques. The fourth, and final, explicatum of psychotherapy will refer to a technique the goals of which (and even the antecedent conditions and technical interventions) are described in phenomenal languages and evaluated in terms of health values and which is justified by pure experiential psychology (and not behavioral, learning-theory, motivational, or pharmacological psychology). If we introduce this explicatum, the only reasonable one, it implies that psychotherapy cannot be justified as technique. *Thus the natural-scientific and therapeutic interpretations of psychoanalysis lead to an impasse.* The only alternative left is to pursue psychoanalysis as a form of human-scientific research. Therefore, it is necessary to acquaint oneself with what characterizes the methods of the humanities, namely, understanding and interpretation. These do not

Paper read at the Thirty-second International Psycho-Analytical Congress, July 26 to 31, 1981, Helsinki, Finland. This chapter was translated from the Swedish by Nigel Moore and improved by Leendert P. Mos. The "Reply to Commentators" was translated from the Swedish by Herman Tennessen.

Carl Lesche • Department of Psychology, University of Stockholm, Stockholm, Sweden.

inquire after explainable, universal descriptions of conscious phenomena; instead, the latter are systematized into intentional contexts of meaning. Of the various hermeneutics we shall consider only Apel's hermeneutic-dialectical model of knowledge acquisition, self-understanding, and emancipation in a pure psychoanalytic situation. The psychoanalytic process cruises between a "hermeneutic phase" and a "quasi-naturalistic phase." In the hermeneutic phase the analyst and analysand immediately understand each other. The function of the quasi-naturalistic psychoanalytical theories ("clinical theories") is that in the event of failure in the pure hermeneutic understanding they may be used in order to get at the unknown intention with quasi-explanations. Freud also introduced another type of theory, his metapsychology. I propose to explicate Freud's metapsychology as the metatheory of these quasi-explanations. It steers their construction by offering and accounting for concepts, rules, and patterns of those "naked models" borrowed from natural science. Thus the practice of psychoanalysis can either follow a pure psychoanalytical human-scientific method or a psychoanalytically based Machiavellian technique.

1. Introduction

Sigmund Freud wrote in 1922:

> Psycho-analysis is the name (1) of a procedure for the investigation of mental processes which are almost inaccessible in any other way, (2) of a method (based upon that investigation) for the treatment of neurotic disorders and (3) of a collection of psychological information obtained along those lines, which is gradually being accumulated into a new scientific discipline. (1955, p. 235)

Our question here will be whether psychoanalysis is therapeutic technique and/or scientific research. However, this cannot be answered by psychoanalysis itself, but by its metascience.

We shall first explicate psychoanalytic therapy as a technique and examine the conditions under which it may be justified. A rational technique ought to be based on a science, which in this case ought to be the psychoanalytical science. We shall determine whether the essence of this science is natural or human science. The outcome will show that psychoanalysis as a science cannot be natural science, from which it follows that psychoanalytical therapy cannot be justified. What remains then is to practice psychoanalysis as human-scientific research. Finally, we shall try to determine the location of metapsychology in the psychoanalytic interpretation.

2. Metascientific Assumptions

2.1. Object-Meta-Hierarchy

In order to clarify the relation between science and metascience (which is also used in the sense of philosophy of science), the concept of object-meta-hierarchy will first be introduced in Figure 1.

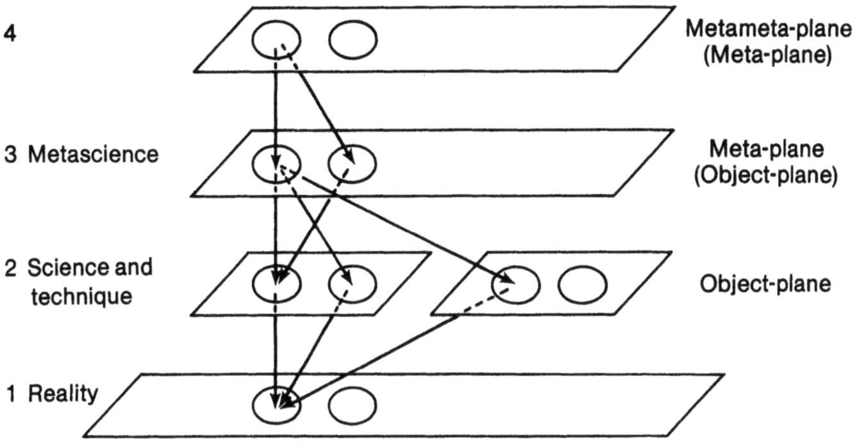

Figure 1. Object-meta-hierarchy.

Let us assume that the phenomena, which serve as objects for an empirical scientific study or technique, belong to plane 1, which symbolizes reality. At this level, for example, the real psychic phenomena with which the psychoanalyst deals are situated. What is considered to belong to a certain science's domain depends on how that science is defined, as well as its metascience, its meta-metascience and meta-meta-metascience, and so on. They all mutually assume each other.

Scientific research and its technical applications are positioned on plane 2 where, for example, both psychoanalytic research and the various forms of psychoanalytic therapy have their place. Also included here are descriptions of psychic phenomena, explanations, and confirmations of such descriptions, theories used in explanations, interpretations of the phenomena, and the technical prescriptions according to which phenomena may be changed, descriptions of these changes, and so forth. The reality located on plane 1 is therefore the object of the scientific and technical activity on plane 2.

This activity and its results on plane 2 are in their turn studied by metascientists. Thus, their object of research is scientific research and its technical applications. It is reasonable, therefore, to position metascience on plane 3, which can be called the metascientific plane in relation to object-plane 2. It is in this sense, then, that the theory of science is also called *metascience*. The metascience of psychoanalysis studies, for instance, how psychoanalysts conduct their research; how they formulate their descriptions, explanations, and interpretations; how they introduce their system of descriptions; how they justify their therapy; in addition to what constitutes the essence of these descriptions, explanations, interpretations, justifications, and so on. However, the metascientist of psychoanalysis is not directly interested in the psychic phenomena (plane 1) that primarily belong to the territory of the psychoanalyst.

There are several metascientific traditions and schools which, in various ways, have studied the scientific and technical activities of psychoanalysts and have given rise to many different metascientific conceptions about psychoanalysis. These different metascientific activities on plane 3 can be studied and compared on a higher, fourth plane, which in turn becomes a meta-level in relation to plane 3, which is now considered to be the object-level. When a metascientist discusses different metascientific conceptions of psychoanalysis, he practices metascience relating to plane 3, but meta-metascience in relation to the psychoanalytic science on plane 2. Object-meta-relations between different levels therefore arise. An object-meta-hierarchy is a contiguous one where one can shift between the various levels.

The advantage of calling attention to this object-meta-hierarchy is to become aware of what one is discussing in actual metascientific analysis and to find one's place in the "logical geography" so that confusion and misunderstanding can be avoided.

2.2. Sources of Information

Our analysis must begin somewhere, so what better place than a concrete source of information like Freud's *Gesammelte Werke*. This text may be called *DA* (discourse of analysis). Another text, more narrow, could be chosen, for example, Freud's *Introductory Lectures*, or, much broader, the various combinations of what is catalogued in Grinstein's *The Index of Psychoanalytic Writings*.

However, reports by themselves are not sufficient as sources of information. We also must acquaint ourselves with the actual research *activities*, namely, psychoanalytic work, which gives rise to the discourses. This activity cannot be fully described in writing, since one component of research training is personal knowledge (Polányi), which is imparted to the apprentice or trainee only under his master's guidance. Nevertheless, sources of information should be capable of enrichment by dialogues between researchers and metascientists.

2.3. Metascientific Analysis

There are, of course, several approaches open to the metascientific analysis of DA_1 (selected first discourse of analysis), concerning which only a few general considerations will be sketchily provided here (Lesche,

1979). In the first place, a *base metascience* (MS_1) is selected from among the different available metatheories which offer models and categories for the research of specific empirical sciences. Suppose that our first problem (P_1) were to study a certain metascientific category (C_1) in our specific science: for instance, how explanations are used and developed. The metascientist carries an initial preconception about C_1 and tries to steer his study so as to be able to confirm this notion.

Our first discourse of analysis DA_1 is hardly written for the metascientist and his investigation of a particular metascientific category C_1. Therefore, it is necessary for him to reconstruct the test of analysis by using only those concepts permitted by the base metascience. The reconstruction, DA_2, should naturally be governed and controlled by the metascientist's familiarity with the specific science at the level of the object-plane and this reconstruction should also be recognizable to other specialists in this scientific area. In a subsequent analysis of DA_2 in relation to C_1, we are able to sharpen our instrument of analysis MS_1, so that it may be especially suited to our specific empirical science. This improved base metascience may be called MS_2. (MS_1 can also be improved by borrowing a specific metatheory MT_1 about C_1; however, care should be taken to ensure that MT_1 is comparable with MS_1.)

If the DA_1 text presents the original specific discipline D_1 and DA_2 a reconstructed version of it, D_2, then there may exist several relations between D_1 and D_2, namely, D_2 and D_1 may coincide ($D_1 = D_2$), D_2 may be contained within D_1 or vice versa ($D_2 \subset D_1$ or $D_1 \subset D_2$), D_1 and D_2 may cut across each other ($D_1 \times D_2$), and D_2 and D_1 can exclude each other ($D_1 \perp D_2$). Of course these relations ought to be made more precise, since they can, for example, refer either to the descriptive content or the system of knowledge of the science.

It may even occur that the analysis of DA_1 and DA_2 leads to a new insight which can be generalized resulting in a more broadly based MS_3. In this way the study of the problem P_1 is continued. Even at the beginning or during the course of study, new problems may crop up, the solutions to which clarify the science itself, its metascience, and the relationships between them.

For the purpose of an analysis of Freud's texts, my point of departure was a general base metascience MS_1, namely, Törnebohm's "system-theory research theory." This originated with studies on physics but was later complemented by investigations in social and human sciences. This general metascientific theory will not be presented here; instead, the reader is referred to the relevant literature (for example, Törnebohm, 1957, 1973; Radnitzky, 1970).

3. Psychoanalysis as Therapeutic Technique

3.1. Prescription

Our first task is to explicate the terms *therapy, psychotherapy,* and *psychoanalytical psychotherapy,* since they are used in a vague and ambiguous sense. The explicata will be introduced gradually. With the aid of the metatheories of prescription and technique (MT_1), we shall (1) explicate the concept of technique, (2) with the help of this explicatum carry out a classification of techniques, (3) suggest an explication of psychotherapy, and finally (4) explicate psychoanalytical psychotherapy. These explicata may be used in metascientific discussions on the essence of psychoanalysis and also, we hope, in practical debates.

Both scientific research and technical intervention proceed according to prescriptions. A prescription lays down rules, given certain antecedent conditions (*AC*), on how certain actions (*A*) (or certain means) should be performed (or employed) in order to attain certain designated goals (*G*). One such norm could be expressed as follows: Given *AC*, perform *A* to attain G. The agent can pose the question why he should follow a prescribed rule and not some other one and, therefore, a prescription should be capable of being justified in order for it to become a binding imperative (see Figure 2; Lesche, 1962, 1973a, 1978, 1980, 1981*b*).

A prescription is justified, first, by taking into account empirical scientific descriptions. If the descriptive sentence "Assuming *AC*, if *A* then *G*" is not credible, then there would no be reason to obey the prescription. However, descriptions by themselves do not entail any imperatives or *ought*-sentences. So that in addition there must be, secondly, value judgments about the goals, as, for instance, "*G* (and *AC*, *A*) have the positive value *V*." One would not try to pursue a goal if it

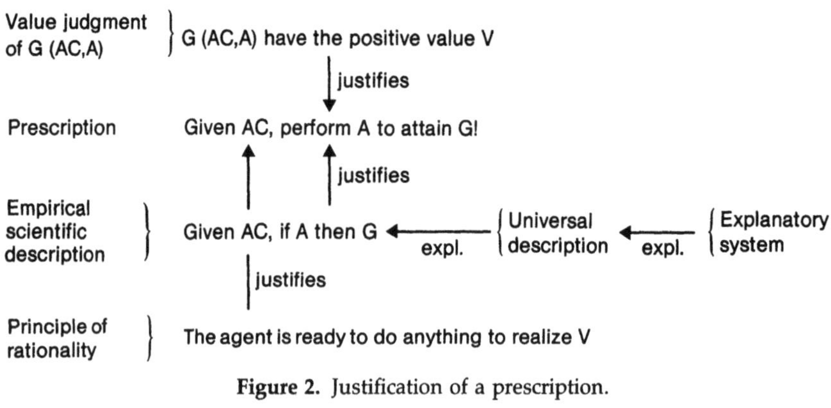

Figure 2. Justification of a prescription.

were not assessed as being worthwhile. Justification of prescriptions must also include value judgments of the goal, because the agent must share the same scale of values in order to convince himself to obey the prescriptions.

Third, justification ought to include principles of rationality which may be expressed as follows: "The agent is willing to try his utmost to realize *V*." The idea behind these principles of justication is that if the scientific descriptions are adequate, and the agent shares the value judgments about the goal and accepts the principles of rationality, then the resultant norm becomes a binding prescription. The relation between value judgments, descriptions, principles of rationality, and prescriptions is not one of logical deduction, however, but an intentional practical implication.

We shall further differentiate between *method* and *technique*. Within scientific research actions are taken in accordance with method, which obeys prescriptions wherein the goal is valued solely in terms of knowledge values. On the other hand, technique is ordinated according to technical prescriptions, in which the goal is assessed in terms of both knowledge values and other values.

3.2. Therapy as Technique: Explicatum 1

As our first explicatum, therapy will be regarded, as is indeed all medicine, as technique. It is possible that some other writer may mean something quite different by *therapy*, in which case he ought to give precise details. Our explication can then *mutatis mutandis* be repeated with this new explanation. In order to give a more accurate characterization of psychotherapy as technique, we shall see how techniques can be classified so as to include psychotherapy.

3.2.1. Classification of Techniques: Explicatum 2

In order to establish a classification of techniques, use can be made of the justification of technical prescriptions. By combining in different ways scientific descriptions and value judgments about goals, different kinds of technical prescriptions may be justified. A given technique is characterized by designating the science (or a part of it, or even several sciences) and the hierarchy of values that justify it (Figure 3). Furthermore, the classification can be refined by specifying the goals, antecedent conditions, and scientific theories. Thus, electronics can be characterized as that technique the prescriptions of which are justified by the laws of

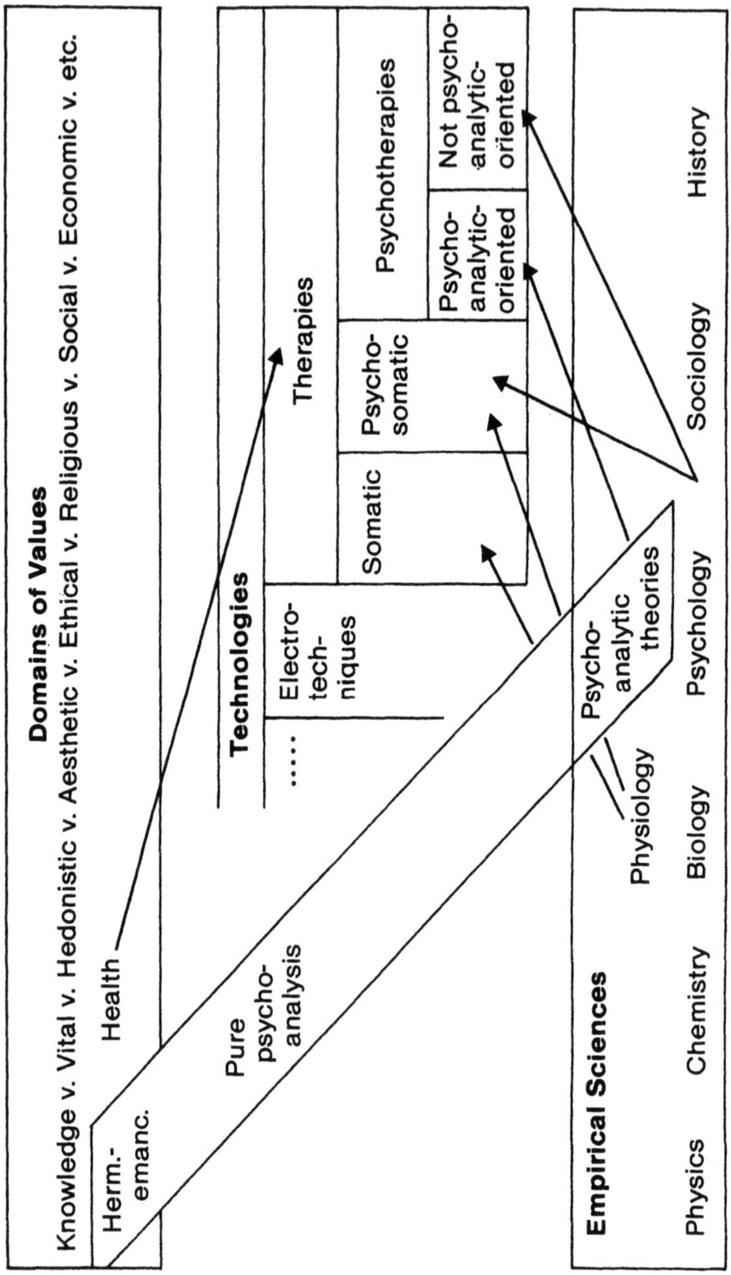

Figure 3. Division of techniques.

electricity and socioeconomic values. Medicine is that technique the prescriptions of which are justified by various sciences and a hierarchy of values in which the dominant values are those of health.

As our second explicatum, therapy will be regarded as a treatment of illness in keeping with its etymology, and therefore as those techniques wherein the values of health predominate. By combining physiology and health values, it is possible to derive medical somatic therapy, from psychology and health values psychotherapy, from physiology and psychology and health values psychosomatic or psychiatric therapy, and from the pure psychoanalytical science and health values psychoanalytical psychotherapy, and so on. In these cases the health values and the knowledge-constitutive interests result in the technical manipulation of human beings toward health (cf. Apel, 1968; Habermas, 1965 on knowledge values).

There is a distinction to be made between pure psychoanalysis as a science and psychoanalytical psychotherapy as technique. The science that justifies psychoanalytically oriented psychotherapies will be called pure psychoanalysis. The word *pure* in this context (see Husserl, 1900, §§ 15–16; Lesche, 1962, p. 58) has no other meaning than that this type of psychoanalysis is a science. And this science's goals are valued in terms of knowledge (especially in its hermeneutic and emancipatory sense): that is, psychoanalytic research contributes to an improvement of scientific knowledge and world view in addition to deepening the contemplation of existential themes (Radnitzky, 1970, part II, pp. 12–13). The relation between pure psychoanalysis and psychoanalytical therapy is identical to the relation, for example, between learning theory and behavioral therapy or between chemistry and chemical industrial technique.

The practice of psychoanalysis can then be pursued as scientific research, that is, as a method with knowledge values, and hermeneutic and emancipatory knowledge-constitutive interests. It is this to which pure psychoanalysis refers. It does not strive after health, social adaptation, or functional competence and the like, although this stipulation does not preclude the possibility that its results can be secondarily valued in this regard.

A clarification of the concept of therapy would demand an analysis of the concept of health. When health is defined in medicine, it is based on technical knowledge-constitutive interests as, for example, adaptation to society, increased functional competence, and so forth. However, health is understood in quite another sense among psychoanalysts, who are concerned with self-understanding and emancipation.

It is conceivable that the metascience MS_1 has developed to give rise to the subject-specific metasciences MS_{2s} (which applies to science) and MS_{2t} (which applies to technique). With their help DA_1 may be reconstructed to yield two purified discourses DA_{2s} and DA_{2t}. DA_{2s} may be exemplified by such common psychoanalytic textbooks as Fenichel's *The Psychoanalytic Theory of Neurosis* and DA_{2t} by such books on psychotherapy as Dewald's *Psychotherapy: A Dynamic Approach.* Suppose that the disciplines represented by DA_1, DA_{2s}, and DA_{2t} are designated ψ_1, ψ_{2s}, and ψ_{2t}. Then the following relations are conceivable: $\psi_{2s} \subset \psi_1$, $\psi_{2t} \times \psi_1$, $\psi_{2s} \perp \psi_{2t}$. If those parts of the discourse which contain the justification of pure psychoanalysis are not included in DA_{2t}, then it is possible that $\psi_{2s} \perp \psi_{2t}$.

For a closer characterization of a therapy which may well be suitable for the goals and values of psychoanalysis it is first necessary to state our position on justification.

3.2.2. Views on Justification

One may take various attitudes on the meta-metaplane toward justification as an adequacy criterion for technique. Let us use the designations *puristic technique* and *Machiavellian technique* where the former implies refraining from the action if it cannot be justified and the latter implies anything goes in the attainment of the goal: "The end sanctifies the means." Medicine has gained the character of the latter kind, since health and life are so highly valued that no effort is spared to attain them, even if it is not possible to justify the actions. Furthermore, within the medical context, one talks of "empirical therapies" and of "rational therapies." It would seem as though empirical therapy implied such therapy as is justifiable according to empirical scientific descriptions, although general laws are not necessary. Rational therapy, on the other hand, refers to techniques the prescriptions of which can be justified by empirical scientific descriptions of generalities which in turn can be explained with some sort of explanatory system (theory). Regularity belongs to the essence of technique: whenever identical actions are performed under identical conditions, the same results are always attained.

Even if desired goals are attained by applying a nonjustified technique, it is impossible to know why they were obtained, nor can the results confirm hypothetical descriptions and, even less, theories. One reason for this is that the results are not purely cognitive descriptions, for they also contain value judgments. The results, moreover, can be a consequence of other simultaneous, unintended circumstances. Non-justified techniques ought to give cause for risk-theoretical and decision-making considerations.

Our task then, will be to attempt a construction of justified therapies of neurosis (cf. the quotation from Freud at the beginning of the article). This may be regarded as an inverse task to the process of justification. For we have here a definite value, goal, prescription and principles of rationality, and we are to inquire after suitable scientific descriptions which allow the criteria of justification to be satisfied. This, then, is an attempt to define the scientific descriptions which are a solution of the system depicted in Figure 2.

3.2.3. Justificational Science

A condition for a justified rational technique is that the justificational scientific descriptions are explainable universal sentences. It is only in the exact natural sciences that we can find formulations and explanations of universal descriptive sentences; this is not true in the behavioral and human sciences. To demonstrate this consider the following.

3.2.3.1. Natural Science and Human Science. We shall continue our construction of justified therapies on the supposition that DA_{2s} serves as our source of information. Let us therefore assume that we have introduced metatheory MT_3 and that it has been our instrument to investigate the differences between the natural and human sciences.

What the researcher studies may be called the domain of research. The researcher himself is a subject and the domain and the activity within it are his objects. In the natural and behavioral sciences (the latter have been modeled on the natural sciences) these objects are considered as *external* objects in the sense that the researcher stands in isolation from them and objectifies them, as, for instance, when he studies the planets' movements, or man as a thermodynamic system or biological organism, or simply the behavior of a rat or man. In the natural sciences objective or objectified nature is *examined,* whereas in the human sciences the intending person's speech, modes of expression, actions (not behavior!) and their products are *elucidated.* Natural sciences exclude the possibility of an understanding dialogue between the researcher and the object, as it is the former's task to try to describe and explain the latter. In the human sciences, on the other hand, the researcher and his object are engaged in meaningful dialogue with each other. Here the object is not isolated and the researcher himself is an integral part of his or her own inquiry. Both the researcher and his research object must try to understand and interpret individual human actions and products. A human-scientific study implies that the researcher views his object as another subject. If human actions and speech are objectified, then the study results in behavioristic research of behavior (including verbal behavior) and, on this account, behavioral psychology and sociology belong to the natural and not the human sciences.

3.2.3.2. Pure Psychoanalysis as Natural Science. In MS_{2s}, the conception of psychoanalysis as a science, the views of logical empiricism on science, MT_{3ns} (*ns* stands for "natural science"), state that psychoanalysis should be constructed as a natural science. Suppose, therefore, that MS_{2s} is developed into MS_{3ns}, a point of view that has been advocated by Heinz Hartmann (1927), who has written:

> Wir sagten, dass Psychoanalyse *Naturwissenschaft* vom Seelischen ist; diesen Gedanken werden wir immer wieder in den Vordergrund rücken und es sei gleich hier betont, dass eine fruchtbare Weiterentwicklung dieser Wissenschaft *nur* von diesem Wege erwartet werden darf—nicht aber von einer mehr oder minder weitgehenden Annäherung an die sogenannte "geisteswissenschaftliche" Psychologie.[1] (pp. 6–7)

Hartmann applied the early view of logical empiricism to psychoanalysis, one to which many psychoanalysts still adhere. Some psychoanalysts hold this natural-scientific viewpoint even though they have corrected certain details in the light of logical empiricism's later development.

It has already been mentioned that according to logical empiricism's metascience, external and objectified nature together with man's behavior constitute the objects of the natural and behavioral sciences. The relationship between the researcher and the object of research is one of subject and object. For science the ideal is monistic: all accepted sciences form parts of a single unified science and, according to this reductionist idea, all sciences ought eventually to be reduced to physics. The principal tasks of science are to describe generalities, to explain causally these descriptions, and to predict deductively new descriptions which should be open to empirical confirmation. The ideal for this knowledge is an interpreted, axiomatic system and as such the knowledge should be expressed in an extensional language. Moreover, the body of knowledge is regarded as deductive. Induction is utilized to attain general laws from

[1] "We have asserted that psychoanalysis is *natural* science of the mind; we shall be continually throwing this idea into the forefront and it may be similarly stressed here that a fruitful furthering of this science may *only* be expected along these lines—and not by a more or less farreaching rapprochement with the so-called 'human-scientific' psychology." (Hartman, 1927, pp. 6–7)

descriptions of single cases. For all this to hang together, the ontology of science ought to be materialistic monism. Lastly, its conception of man is that of a biological organism, ideally well functioning and well adapted.

Hartmann actually sets out the following guidelines for the science of psychoanalysis: the formation of natural-scientific concepts, general laws of natural processes, causality, explanation, inductive and deductive reasoning, experiential science, and the reduction of qualities.

Hence, the discourse of analysis DA_{2s} has been reconstructed with the help of MS_{3ns} into a new DA_{3ns}. But is the corresponding natural-scientific psychoanalysis D_{3ns}, or ψ_{3ns}, practiced in reality? D_1, D_{2s}, and D_{2t} could exist, but their scientific concepts are so vague and ambiguous that it is difficult to give a definitive answer. On the other hand, the concepts contained in MT_{3ns} (and the corresponding MS_{3ns}) are so clearly formulated and unequivocal that it is possible to decide whether D_{3ns} actually exists. On this account, it is quite evident that D_{3ns} cannot exist, since psychoanalysis D_1 or ψ_1 strives for self-reflection and emancipation (and all that this implies), but MS_{3ns} is completely devoid of such concepts. Therefore, it is clear that what is meant by psychoanalysis in DA_1 was not intended to refer to anything like ψ_{3ns}.

3.2.3.3. Steering Factors. One leitmotif in the base metascience MS_1 is to clarify the way in which scientific research (in the natural as well as the human sciences) and technical interventions are steered. Conceptions on this point are contained in the metatheory MT_4 on steering factors. Those which are significant can be discovered in the weltanschauung, especially in the philosophical anthropology of knowledge. For reasons of brevity, only the most important steering factors will be mentioned here: the goal and its evaluation; conceptions and ideals regarding science, technique, knowledge, man, society, history, and nature; knowledge-constitutive interests and ontological presuppositions (cf. the relevant tables in Lesche, 1973b, 1978; Lesche & Stjernholm Madsen, 1976). A science or technique can be characterized by listing its steering factors, which ought to be selected prior to engaging in empirical activity. Of course, not just any combination of factors is appropriate to the task of steering research and technique, since a criterion of adequacy must be demanded, namely, that the steering factors ought to be consistent in the sense that no cognitive contradiction or conflict of values should arise among them.

This criterion renders the contemplated natural-scientific pure psychoanalysis impossible. The same applies to attempts to introduce natural-scientific elements into a human-scientific pure psychoanalysis or vice versa. On the basis of our earlier explications, we see that a justified technique and therapy can be founded only upon a pure natural science and not on a pure human science.

3.2.3.4. Explication of Psychotherapy: Explicatum 3. Our third explicatum of psychotherapy refers to a technique in which the goal is described in intentional and phenomenal terms and valued in terms of health values. In order to analyze this concept, we are obliged to take a closer look at the concept of *explanation* and with it the concept of *universal description*. A precondition for any justified, rational technique is that the justifying scientific descriptions are universal sentences. It is only possible in the exact natural sciences, and not in the behavioral and human sciences, to formulate and explain universal descriptive sentences.

3.2.3.5. Explanation in Natural Science. Scientific knowledge is systematized and in the natural sciences revolves around explanations, whereas in the human sciences it revolves around threads of meaning or intentional contexts. Unfortunately, the term *explanation* is used in a very vague and equivocal sense. We shall explicate the various concepts of explanation in order to be able to decide whether a metascientific operation is an explanation or not. In natural science several sorts of explanation occur, but they can be reduced to two basic types: subsumptive explanations and explanations which employ systems of explanation.

First, consider the concept of subsumptive explanations. Hempel and Oppenheim (1948; cf. also Stegmüller, 1969) presented a well-defined pattern of explanation, the so-called subsumptive explanation, in order to explain descriptions of single events. It is this type of explanation (or some imperfect copy of it) that is frequently made use of in the psychiatric, therapeutic, psychosomatic, and psychopharmacological contexts.

The explanation of a particular explanandum ed^f (formulated in a physicalistic language, hence the *f*) proceeds in such a way that ed^f is logically deduced from general laws $l^f_1, l^f_2, \ldots$ and descriptions of particular antecedent conditions $ac^f_1, ac^f_2, \ldots$: $l^f_1, l^f_2, \ldots, ac^f_1, ac^f_2, \ldots, s^f, \ldots, s^f \rightarrow ed^f \vdash ed^f$. This belongs to the essence of logical deduction, namely, starting from purely *f*-premises will always lead to a purely *f*-conclusion. It is impossible to derive a sentence with essentially phenomenal ϕ-terms, if these do not already appear in the premises. Therefore, to provide "cross-explanations" of descriptions of single ϕ-phenomena (or *f*-phenomena) only with the help of *f*-premises (or ϕ-premises) is obviously impossible. In order to complete such a cross-explanation, it is necessary to add a *f*-premise $s^f \rightarrow ed^\phi$. Such *f*-premises are usually included *ad hoc* as implicit premises, which are smuggled into the argument. These premises have not been confirmed as a description of generalities, nor have they been explained with the aid of any system of explanation. On the contrary, they are at best hypotheses belonging to a world view or an ontological assumption, for example, the psychological has a biochemical basis.

Hempel and Oppenheim (1948) held the view that even explanations of descriptions of generalities followed the subsumptive pattern. Apparently, this kind of explanation does not even hold in theoretical physics however, and Törnebohm has presented an alternative pattern of explanations by means of an explanatory system. It is briefly summarized in Figure 4 (cf. Törnebohm, 1952, 1957, 1973; for a résumé, cf. Lesche, 1962; Lesche & Stjernholm Madsen, 1976, for example).

In order to explain an observed universal description of a certain real, physical situation, the physicist chooses first a system of explanation (theory). With its help he constructs a cognitive or intellectual model by using the rules for schematization and model construction. At the same time, causal analyses of the schemata's characterization are conducted using theoretical concepts. Henceforth, the physicist no longer discusses the real physical system, but rather this intellectual model. The next step consists in applying new rules and principles to the model which prescribe ways to handle the model. Thus mathematical-logical sentences are formulated which, using calculus, are restated so that the results can be compared with the formulae in the original description. The resultant sentence must, however, first be reapplied to the real situation and the theoretical concepts operationally defined. In this way

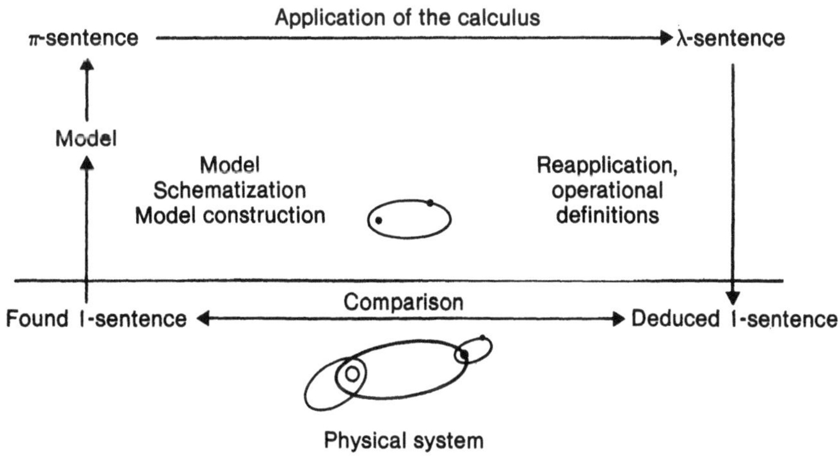

Figure 4. Explanation by means of an explanatory system.

a thought-out universal sentence is reached which can then be compared with the original observation. If the calculated sentence is acceptable when so compared, it may then be said that an explanation has been attained.

An explanatory system consists of sentences referring to schematization, including reapplication, model construction, and the principles. This is one way in which a theory may be explicated. These sentences are not descriptive of something in reality since they refer to the intellectual model. Therefore they have no empirical truth value, nor can they be empirically confirmed. Nevertheless, this entire construction is useful as an explanation.

It occurs frequently that models are borrowed by one science from another, although this is not a simple process. Assume that an explanatory system ES^1 is capable of explaining the descriptive sentence I^1, when the model M^1 in ES^1 has been infused with theoretical terms T^1. Assume further that ES^1 is not capable of explaining I^2. It is possible to borrow M^1 in order to explain I^2, but in that case it is not the model M^1, "dressed up" in terms T^2, that is borrowed, but rather the "naked" model M, which is the mathematical-logical structure of M^1, which is now attired with other theoretical terms T^2. Although there might be a chance that such an explanation will succeed, science S^2 has not been reduced to science S^1.

Some other types of explanations have been proposed: finalistic, teleogical, and motivational explanations of behavior (Apel, 1976). They will not be dealt with here, since it appears that they can all be traced back to the Hempel and Oppenheim formulation. If someone else claims to have introduced a new type of explanation, then he should be prepared to give a precise definition of it, just as Hempel-Oppenheim and Törnebohm have done, so that it may be possible to determine its status. However, there do not appear to have been other carefully formulated types of explanation. Occasionally, one hears the argument that developments in science will surely reach a point at which even the phenomenal will be explained by biochemistry. But desires for the future do not have any place in logical arguments.

We have seen above that if by explanation we mean the Hempel–Oppenheim explanation (or the Törnebohm explanation) then we cannot explain phenomenal explananda with physicalistic explanantia nor vice versa. Generally it is impossible to carry through an explanation if the explanans and explanandum are formulated in sentences which contain terms from different descriptive systems. The logical deduction poses an obstacle since it is defined in extensional language only. The truth value of an extensional sentence should be decided on the basis of its extension, without having recourse to understanding. Extensional sentences are, furthermore, truth-functional. The formulation of descriptions of regularities implies that the sentence is extensional and that it should be capable of being used in subsumptive explanations.

In this context, it may be opportune to mention the attempts to measure phenomenal attributes, like the attempts, for example, to work out scales of anxiety and depression. But such operationalizations hardly solve the problem, since it is just not plausible that metric descriptions of experiences are possible (Husserl, 1952, pp. 131–137; Lesche, 1971).

3.2.2.6. The Impossibility of a Justified Psychotherapy: Explicatum 4. From the above, it follows that the concept of justified rational technique can be based only on pure natural science. At most, psychoanalytical therapy, psychopharmacological therapy, behavioral therapy, and psychosomatic medicine can be considered Machiavellian empirical techniques.

It is now time to introduce the fourth and final explicatum of psychotherapy, namely, as a technique the goals of which (and even the antecedent conditions and technical interventions) are described in phenomenal language, evaluated in terms of health values, and justified by a pure experimental psychology (and not a behavioral, learning-theory, motivational, or pharmacological psychology).

Sentences describing experience cannot be formulated in an extensional language, but only in an intentional language of the genuine human sciences (and that excludes the behavioral sciences which imitate the natural sciences). It must be possible to understand the meaning of an intentional sentence in order to be able to determine its truth value. The extensional deductive logic cannot be used on intentional sentences, for *universal* intentional sentences cannot be formulated nor can intentional sentences be deduced from other intentional sentences. The general sentences of the human sciences cannot be formulated as universal sentences. Thus human science does not make any use of explanation. Finally, no justified rational technique can be based on pure human science.

There is no way in which any intentional concept can be connected with any natural-scientific one, since intentional and extensional sentences cannot be translated or explained by each other. No universal connections exist between external behavior and experience or intentional actions. Cross-explanations between the two are therefore logically impossible. Moreover, general laws relating physical stimuli and (behavioral) responses cannot be formulated either, because humans do not react with uniform regularity to external stimuli. Instead, they respond with intentional actions to understood, intentional stimuli.

If we were to introduce psychotherapy in the sense of our fourth explicatum, we would conclude with the following thesis, namely, that no psychotherapeutic prescriptions exist according to which it would be possible in a justifiable manner to treat (manipulate) a person. This explicatum is the only reasonable one and implies the impossibility of psychotherapy as a justified technique. Naturally, Machiavellian techniques can, and are, used, in which case risk-theoretical and decision-making considerations ought to be taken into account. Behavioral therapy, psychopharmacology, and the whole of psychosomatic medicine can at best only be Machiavellian empirical therapies. Yet it belongs to the essence of techniques that technique and natural science are interdependent. This has always been the case and only lately have attempts been made to manipulate man with psycho-, socio-, and political techniques. The view of man implied by these techniques is that humans can be manipulated.

The natural and human sciences have reached such a point of development and specialization today that they have almost approximated their respective ideals and hence diverged from and excluded one another. This differentiation may be regarded as a positive development as well as a crisis. There has always been a desire for a single unified science; however, this cannot be either naturalistic or humanistic or a combination of both, but should instead be constituted along quite a different line of thought as, for example, along phenomenological-philosophical lines (Husserl, 1954).

4. Psychoanalysis as Scientific Research

4.1. Psychoanalysis as a Hermeneutic-Dialectic Process

Thus the natural-scientific (which also includes behavioral-scientific) and therapeutic interpretations of psychoanalysis lead to an impasse. The only alternative is to pursue psychoanalysis as a form of human-scientific research. The next steps in our analysis is to reconstruct DA_{2s} with the help of the human sciences' metascience MS_{2hs} (*hs* for "human science") and especially the hermeneutic metatheory MT_{3h} (*h* for "hermeneutics"). This way of proceeding proves to be particularly fruitful since it permits the psychoanalysis D_{3h} or ψ_{3h}, both of which are not only possible but also exist.

Of the various hermeneutic schools, we shall consider only Apel's hermeneutic-dialectic model, MT_{3ha}, of knowledge acquisition, self-understanding, and emancipation (Apel, 1965, 1968; Habermas, 1968;

in this context mention should also be made of Ricoeur, 1965; Radnitzky, 1970; Lesche, 1973a, 1978; Lesche & Stjernholm Madsen, 1976). Its point of departure was orginally the psychoanalytical situation, but Apel realized that this pattern could be generalized to all the humanities, MS_{5ha}. In any case, it appears to be especially appropriate for the metascientific analysis of psychoanalysis. Psychoanalytical practice can indeed be pursued as scientific research, that is, as method, with knowledge values and the knowledge-constitutive interests of a hermeneutic and emancipatory kind. This is human-scientific pure psychoanalysis.

4.1.1. The Psychoanalytic Process

We shall now present the psychoanalytic process (at a metascientific level) as a hermeneutic-dialectic cognitive pattern for increasing self-understanding and emancipation. For the sake of brevity, this pattern is depicted in Figure 5.

4.1.1.1. The Hermeneutic Phase. The psychoanalytic process begins as a dialogue in which the analyst and analysand immediately understand each other, each other's speech, actions, and expressions. This phase of the process is the hermeneutic phase. In this the knowledge-constitutive interests are hermeneutical and emancipatory; also, the other steering factors are derived from the hermeneutic-dialectical metascience.

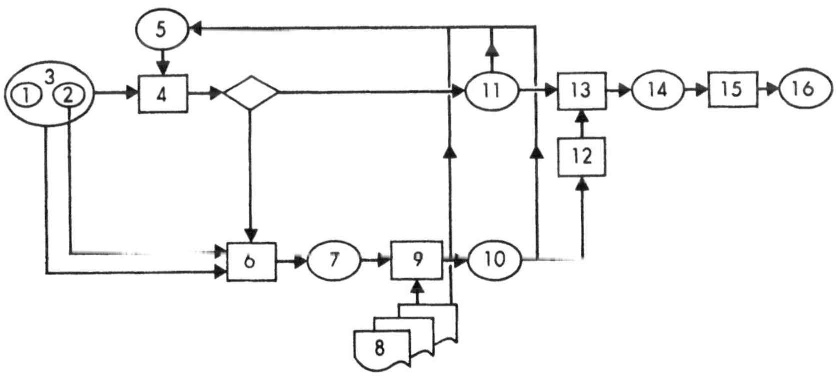

1. Analyst
2. Analysand
3. Dialogue
4. Understanding
5. Preunderstanding
6. Distance
7. Distanced dialogue
8. Explanatory system
9. Quasi-explanation
10. Hypothesis
11. Hypothesis
12. Historical explanation
13. Interpretation
14. Increased self-understanding
15. Position taking and choice
16. Emancipation

Figure 5. Apel's model.

4.1.1.2. Quasi-naturalistic Phase. If the pure understanding of the analysand's speech, actions, and expressions is successful, the analyst passes over into the quasi-naturalistic phase. The ideas about these quasi-naturalistic explanatory systems will be propounded in the three following steps.

1. The analysand's incomprehensible action, which *a priori* is intentional, is objectified and is considered tentatively as behavior or as an event. The analyst tries to explain this in a natural-scientific way with the help of a natural-scientific or behavioral-scientific explanatory system. The knowledge-constitutive interests are temporarily changed from a hermeneutic-emancipatory to a technical one, and the accompanying steering factors will be changed accordingly. A successful explanation will indicate the cause of the behavior. The behavior and its cause will be regarded by the analyst as the action and its good reason (intention), respectively; nevertheless, following the "historical explanation," the process is once again bound to the hermeneutic phase.

All this, however, is impossible—the idea is absurd! The intentional action is not objectifiable, and we cannot know for certain which behavior or cause would correspond to it. Similarly, to proceed in the opposite direction seems just as impossible (see Figure 6). Furthermore, sentences in the hermeneutic phase are expressed in an intentional language, whereas the sentences in the quasi-naturalistic phase are expressed in

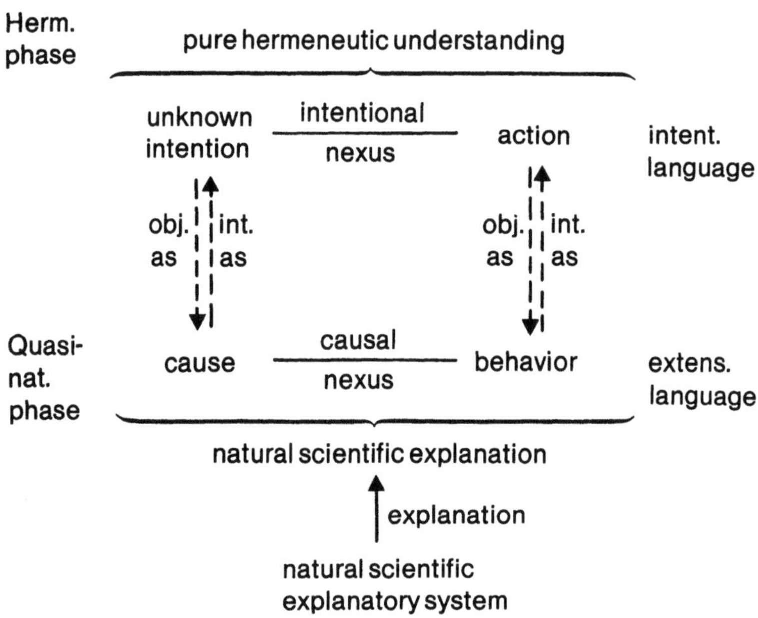

Figure 6. Relations between understanding and explanation.

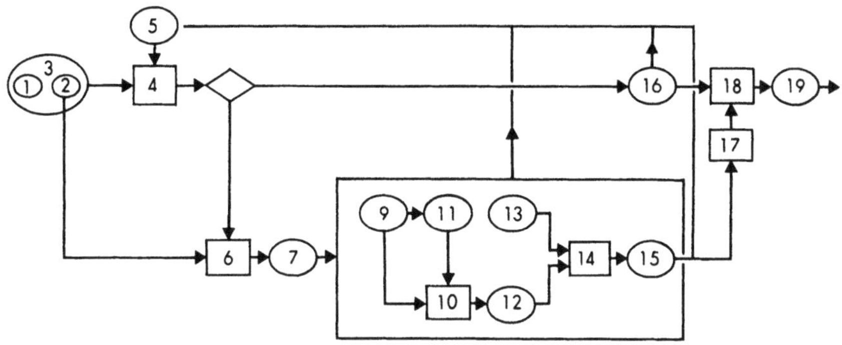

1. Analyst
2. Analysand
3. Dialogue
4. Understanding
5. Preunderstanding
6. Distance
7. Distanced analysand
8. Quasi-explanation
9. Preconscious disturbing intention
10. Repression
11. Affect, moral, rationality
12. Unconscious disturbing intention
13. Disturbed intention
14. Conflict
15. Hypothesis
16. Hypothesis
17. Historical explanation
18. Interpretation
19. Increased self-understanding

Figure 7. Freud's theory of parapraxes.

an extensional language. It is impossible to use terms in sentences expressed in an intentional language to explain, translate, or define terms in sentences expressed in an extensional language, and vice versa. Therefore, we must resign our efforts to explain when we are not able to understand. Explanations do not contribute anything to understanding.

2. Instead of natural-scientific explanations backed up by natural-scientific explanatory systems, the analyst introduces "quasi-explanations," based on psychoanalytic theories, for example, the theory of parapraxes, dream-theory, theory of neurosis, or theory of anxiety, which apply to one person only, the analysand. For instance the theory of parapraxes is depicted in Figure 7. Since the psychoanalytical quasi-naturalistic theory is expressed entirely in an intentional language and the human-scientific attitude, the hermeneutic-dialectical steering factors can be retained. The analysand is not objectified in the natural-scientific sense of object but is "distanced" as the human scientist's research partner.

3. Finally, instead of psychoanalytical theories about one person, the whole of the dialogue between the analyst and the analysand should be distanced and should include resistance, transference, and counter-transference as well as other intersubjective factors and the theories about them.

4.1.2. Quasi-naturalistic Theory

The function of quasi-naturalistic psychoanalytical theories ("clinical theories") is that in the event pure hermeneutic understanding fails they may be used in order to get at the unknown intention with quasi-explanations. The analytic theory offers certain conjectures about intentions which can be used later in the normal hermeneutic phase. That is, psychoanalytical theories are systematized preunderstandings (cf. Lorenzer, 1974). Freud crystallized these theories from out of the whole range of effective interpretations which he made following certain rules.

The structure of the psychoanalytical quasi-naturalistic theory and quasi-naturalistic explanation is reminiscent of the explanations and explanatory systems of natural sciences, without their being identical—hence the prefix *quasi-*. The major difference is that those of psychoanalysis are expressed in an intentional language, whereas those of natural science are extensionally expressed. From this there follow other differences.

Freud also introduced another type of theory, namely, his metapsychology, the essence of which we shall now try to define.

4.2. Metapsychology

4.2.1. Freud's Two Theories

Freud characterized metapsychology in *Das Unbewusste*: "I propose that when we have succeeded in describing a psychical process in its dynamic, topographical and economic aspects, we should speak of it as a metapsychological presentation" (1915, p. 181). Metapsychology has been analyzed by a number of modern analysts in the following manner. Metapsychology is a biological explanation, a psychology that goes beyond consciousness and a psychology of the unconscious (Gill, 1976, p. 74). Kernberg generalizes the definition of metapsychology so that it includes every proposition from other disciplines, be they from neurophysiology, biology, sociology, or others (Miller, 1976, pp. 146–150). Gill limits the term *metapsychology* solely to neuropsychological and biopsychological propositions (Gill, 1976, p. 71). Some psychoanalysts (for example, Klein, Gill, Kernberg, and Rubinstein) point out sharp differences between the two sorts of psychoanalytical theories: the one theory dealing with intentionality and meaning, the other with the physical substratum, neurology, and biology.

4.2.2. Freud's Two Weltanschauungen

In order to understand the reason why the above-mentioned researchers and possibly even Freud himself distinguished between two theories, I (Lesche, 1973b) have tried to interpret Freud's contradictory statements on the essence of psychoanalysis by referring to his two conflicting weltanschauungen. On the one hand, he wanted to represent it as natural science, which as some future point in time could be traced back into biology and neurology and, at least partly, to biochemistry. On the other hand, he looked upon it as psychology, or as an art of interpretation, which could not be reduced to biology.

We might call that weltanschauung to which a scientist adheres during a certain period as his steering factor, his current weltanschauung. A utopian weltanschauung expresses his humble wishes collateral with his current weltanschauung, although the former does not steer his current research project and neither is it consistent with the latter.

Critics of Freud's metapsychology appear to mean that Freud was working on two research assignments: (1) clinical psychoanalysis, which deals with intentionality and meaning and therefore is obviously steered by some phenomenological-hermeneutic conception of the world; and (2) metapsychology, which is concerned with the physical substratum and is steered by a natural-scientific weltanschauung. At best, the natural-scientific weltanschauung is utopian and carries no influence over clinical theory. At worst, implicit natural-scientific steering factors are jumbled up with the current hermeneutic factors so that the results become contradictory and corrupted (Gill, 1976, p. 76ff.; Schafer, 1976).

My thesis is that Freud's current weltanschauung is the hermeneutic-dialectical one and his utopian belief (or rather fear) natural-scientific reductionism. In order to determine the place of metapsychology in psychoanalysis, our point of departure will be psychoanalysis as a pure hermeneutic-dialectical process of knowledge acquisition in which we shall try to accommodate the supposed natural-scientific metapsychology (Lesche, 1981a).

4.2.3. Proposal on the Explication of Metapsychology

Klein held the position that metapsychology is the theory that explains the clinical concepts (Klein, 1976, p. 42). Gill (1976) mentions several other relations between metapsychological theory or propositions and clinical theory or propositions: deduction, abstraction, translation, metaphorical translation, neurological metaphors that parallel mental terms, transformation, transferring, restatement, consonance, legitimizing, relevance, and connection. Certainly everything has some connection with just anything, but the problem is just what kind of connection.

If we were to assume that the metapsychological explanation be of a subsumptive kind, the law sentences in the explanans would be regarded as the metapsychological theory, that is, the sentences would be extensional physicalistic sentences and the metapsychological concepts would be natural-scientific. That would, of course, be absurd! These metapsychological concepts cannot help to deduce logically or explain subsumptively anything other than natural-scientific sentences and certainly not intentional, phenomenal psychoanalytic theories (clinical theories), or even individual interpretations.

It would seem far from credible that even Freud would have viewed his metapsychological concepts as natural-scientific entities. If we were to conclude that Freud made strict use of natural-scientific concepts, then we would arrest the development of psychoanalysis, and in the end we would possibly end up with a general pure hermeneutics whereas Freud's own special brand of hermeneutics would have to be rejected.

So what did Freud actually mean by such concepts as power, energy, structure, drive, charge, binding, by such principles as psychic determinism, constancy, pleasure, and by such viewpoints as the topological, dynamic, economic, and structural?

I propose to explicate Freud's concept of metapsychology using the concepts which were introduced in the section on explanations and explanatory systems. This proposal is legitimized by the fact that a concept of metapsychology is obtained that can be developed and one that does not lead to paradoxes and stagnation. Taking as his model the explanatory system of the natural sciences, Freud appears to have constructed quasi-explanatory systems in psychoanalysis, which are the psychoanalytical clinical theories, used to make conjectures about intentions of purely hermeneutically nonunderstood actions. Therefore, metapsychology is the metatheory of these quasi-explanations. It steers their construction by offering and accounting for the concepts, rules, and pattern thought out for those models borrowed from natural science (and even from other sources, like sociology in the case of the structural viewpoint). These metapsychological terms and sentences are not descriptive, since they do not directly refer to physical reality nor to mental reality, but to models within the theories and only indirectly to psychic reality. Neither Freud nor his critics were clear about the ontological status of the metapsychological concepts as schemata. Some of the models are naked models borrowed from natural science, which have been dressed up in intentional psychoanalytical concepts. To put it concisely: metapsychology contains the concepts and rules by which psychoanalytic theories are constructed (see Figure 8).

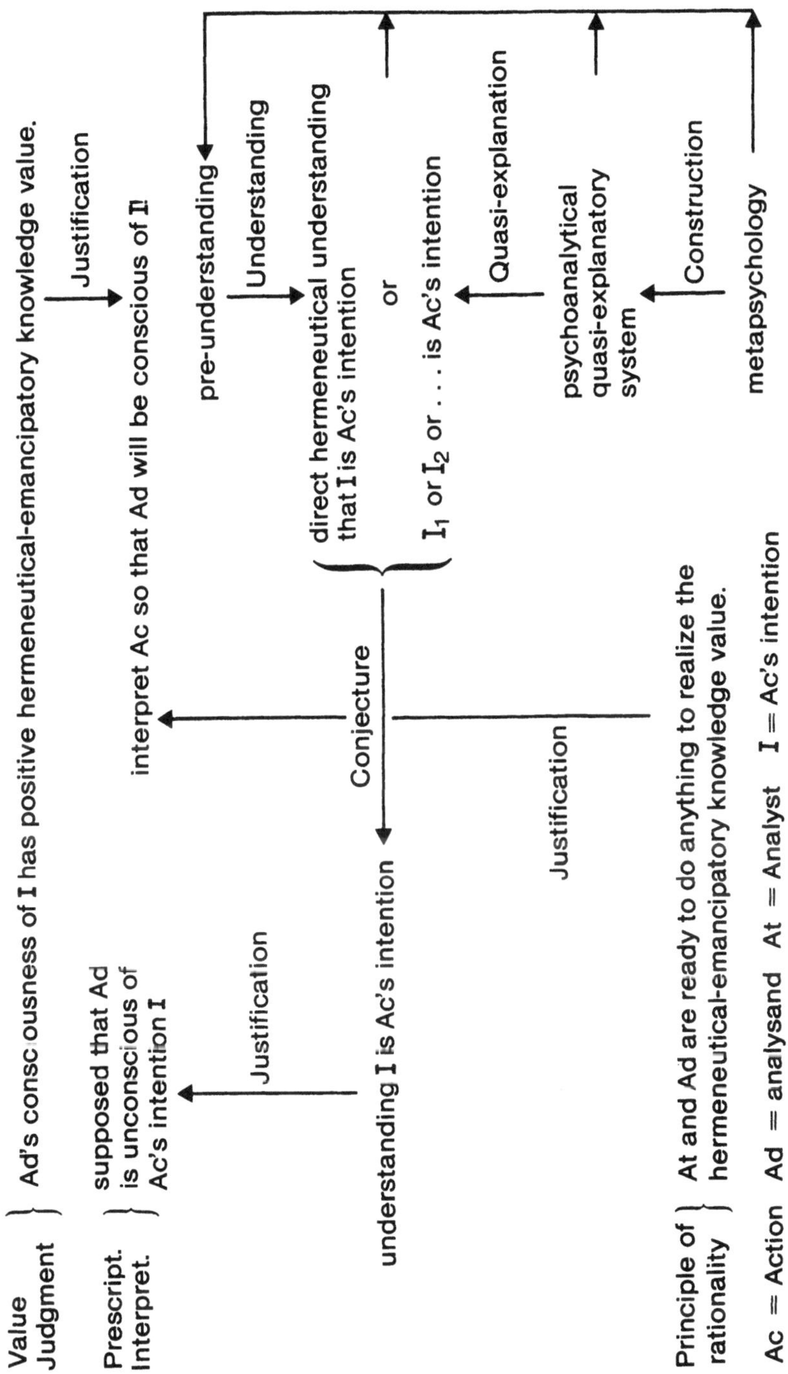

Figure 8. The psychoanalytical process.

Finally, a few words about the significance of metapsychological speculation. Freud (1937) wrote:

> If we are asked by what methods and means this result is achieved, it is not easy to find an answer. We can only say: "So muss denn doch die Hexe dran!"—the Witch Metapsychology. Without metapsychological speculation and theorizing—I had almost said "phantasying"—we shall not get another step forward. Unfortunately, here as elsewhere, what our Witch reveals is neither very clear nor very detailed. (p. 225)

The question whether metapsychology should be retained or rejected is not appropriate. For if it is abolished then Freudian psychoanalysis would have to be halted and rejected, which would mean that its buds would be cut off. Unfortunately, critics have not understood that psychoanalysis is a hermeneutic process wih quasi-naturalistic phases and hence have not comprehended the essence of metapsychology and its place in psychoanalysis.

Our arguments have also furnished an answer to the controversy of whether a purely psychological psychoanalysis is possible (for example, Klein, 1976; Rubinstein, 1976). If hermeneutic psychoanalysis is regarded as psychology, then this is the only possibility that exists. Any interference from natural-scientific elements makes psychoanalysis impossible.

5. Psychotherapy Research on Psychoanalysis

The practice of psychoanalysis can be either a pure psychoanalytical human-scientific method or a psychoanalytically based Machiavellian technique. The analyst acts in accordance with prescriptions when engaged in method or technique. The interpretations are justified by psychoanalytical understanding which he has hermeneutically gained by direct means or which he has surmised with the help of psychoanalytic quasi-explanations. The latter occurs by means of the psychoanalytic theories which have been constructed following the metapsychology.

The aforementioned considerations lead us to a couple of remarks about psychotherapy research on psychoanalysis (Lesche, 1976). This research ought to be metascientific and its domain the two areas already mentioned, namely, the decision-making processes of Machiavellian techniques and the relevant human-scientific methods, as well as the steering factors for each of them.

6. Summary

Both technical intervention and scientific research proceed in accordance with prescriptions. In order for a prescription to be a binding imperative, it ought to be justified, that is, by taking into account empirical scientific descriptions, value judgments of the goal, and principles of rationality. Therapy is first of all explicated as that sort of technique in which the values of health predominate. Puristic techniques imply a refrain from the action if it cannot be justified, whereas Machiavellian techniques imply that every means is tried to attain the goal. Even if the intended goal was attained by the use of a Machiavellian technique, it would not be possible to know why. By *rational technique* is meant those techniques the prescriptions of which can be justified with empirical, scientific descriptions, which in turn can be explained.

A requirement for justified rational technique is that the justifying scientific descriptions are explainable universal sentences which only occur in the exact natural sciences and not in the behavioral or human sciences. A pure natural-scientific psychoanalysis cannot possibly exist since psychoanalysis aims at self-reflection and emancipation and these concepts do not appear in natural science. Psychoanalytical therapy, psychopharmacological therapy, behavioral therapy, and psychosomatic medicine can at best be Machiavellian therapies. A meaningful explication of psychotherapy would now be: a technique the goal of which is described in phenomenal language and assessed in terms of health values and justified with a pure experiential psychology (for example, pure psychoanalysis). But this explicatum implies the impossibility of a justified psychotherapy.

What then remains is that psychoanalysis is practiced as a human-scientific, hermeneutic form of research. Of all the various hermeneuticians, Apel and his hermeneutic-dialectical model is singled out for discussion here since it deals with increased self-understanding and emancipation in a pure psychoanalytic process. This is understood as a dialectic between pure hermeneutic phases and quasi-naturalistic phases. Taking as his model the explanatory systems of natural science, Freud constructed psychoanalytical quasi-explanatory systems, clinical theories, which are used to surmise the intentions of purely hermeneutically nonunderstood actions. I interpret metapsychology as the metatheory of these quasi-explanations and not as natural-scientific theories. The metapsychological terms and sentences are not descriptive, for they neither directly refer to physical nor to psychic reality, but only to models and thus only indirectly to mental phenomena.

7. Glossary

Term	Definition	First Citation (page)
DA	discourse of analysis	160
DA_1	selected first discourse of analysis	160
MS	metascience	161
MS_1	base metascience	161
P_1	first problem	161
C_1	metascientific category	161
DA_2	reconstruction of selected first discourse of analysis	161
MS_2	improved base metascience	161
MT	metatheory	161
MT_1	a specific metatheory	161
D_1	original specific discipline	161
D_2	reconstructed discipline	161
MS_3	new base metascience	161
AC	antecedent condition	162
A	action	162
G	goal	162
V	value	162
MS_{2s}	metascience which applies to science	166
MS_{2t}	metascience which applies to technique	166
DA_{2s}	reconstructed discourse of psychoanalysis as science	166
DA_{2t}	reconstructed discourse of psychoanalysis as technique	166
ψ_1, D_1	discipline represented by DA_1	166
ψ_{2s}, D_{2s}	discipline represented by DA_{2s}	166
ψ_{2t}, D_{2t}	discipline represented by DA_{2t}	166
MT_3	a third metatheory, one about differences between the natural and human sciences	167
MT_{3ns}	a metatheory of natural science, logical empiricism	168
MS_{3ns}	metascience of natural science	168
DA_{3ns}	new reconstructed discourse of psychoanalysis as science	169
ψ_{3ns}, D_{3ns}	natural-scientific psychoanalysis represented by DA_{3ns}	169

Term	Definition	First Citation (page)
MT_4	a fourth metatheory in metascience, one on steering factors	169
ed	explanandum	170
ed^f	explandum formulated in a physicalistic language	170
f	physicalistic	170
$l_1, l_2,$	general laws	170
ac_1, ac_2	antecedent conditions	170
$\ldots \rightarrow \ldots$	. . . implies . . .	170
$\ldots \vdash \ldots$	. . . is provable from . . .	170
ES	explanatory system	172
M^1, M^2	model	172
M	naked model	172
S^1, S^2	science	172
MS_{2hs}	metascience of human science	174
MT_{3h}	hermeneutic metatheory	174
D_{3h}, ψ_{3h}	hermeneutic psychoanalysis	174
MT_{3ha}	Apel's hermeneutic-dialectic model	174
MS_{5ha}	Apel's model for all the humanities	175

8. References

Apel, K.–O. (1965). Die Entfaltung der 'sprachanalytischen' Philosophie und das Problem der 'Geisteswissenschaften'. *Philosophisches Jahrbuch, 72*, 239–289. (English translation: Analytic philosophy of language and the *Geisteswissenschaften*. Dordrecht: Reidel, 1967).

Apel. K.–O. (1968). Szientistik, Hermeneutik, Ideologiekritik: Entwurf einer Wissenschaftslehre in erkenntnisanthropologischer Sicht. *Man and World: An International Philosophical Review, 1*, 37–63.

Apel, K.–O. (1976). Causal explanation, motivational explanation, and hermeneutic understanding. In G. Ryle (Ed.), *Contemporary aspects of philosophy* (pp. 161–176). Stocksfield, England: Oriel Press.

Freud, S. (1915). The unconscious. In J. Strachey (Ed.), *The standard edition of the complete psychological works of Sigmund Freud* (vol. 14, pp. 166–215). London: The Hogarth Press, 1957.

Freud, S. (1923). The encyclopedia articles: "Psycho-analysis" and "The Libido theory." In J. Strachey (Ed.), *The standard edition of the complete psychological works of Sigmund Freud* (vol. 18, pp. 235–259). London: The Hogarth Press, 1955.

Freud, S. (1937). Analysis terminable and interminable. In J. Strachey (Ed.), *The standard edition of the complete psychological works of Sigmund Freud* (vol. 33, pp. 216–253). London: The Hogarth Press, 1964.

Gill, M. M. (1976). Metapsychology is not psychology. In M. M. Gill & P. S. Holzman (Eds.), *Psychology versus metapsychology: Psychoanalytic essays in memory of George S. Klein* (pp. 71–105). Psychological Issues, Monograph 36. New York: International Universities Press.

Habermas, J. (1965). Erkenntnis und Interesse. *Merkur, 19,* 1139–1153. (English translation: Knowledge and interest. *Inquiry,* 1966, *9,* 285–300).

Habermas, J. (1968). *Erkenntnis und Interesse.* Frankfurt: Suhrkamp Verlag. (English translation: Knowledge and human interest. Boston: Beacon Press, 1971).

Hartmann, H. (1927). *Die Grundlagen der Psychoanalyse.* Leipzig: Thieme.

Hempel, C. G., & Oppenheim, P. (1948). The logic of explanation. *Philosophy of Science, 15,* 135–175.

Husserl, E. (1900–1901). *Logische Untersuchungen* (2 vols.). Halle: Niemeyer. (*Husserliana, XVIII.* The Hague: Nijhoff, 1975).

Husserl, E. (1952). *Ideen zu einer reinen Phänomenologie und phänomenologischen Philosophie. Husserliana, V.* The Hague: Nijhoff.

Husserl, E. (1954). *Die Krisis der europäischen Wissenschaften und die transzendentale Phänomenologie. Husserliana, Vol. VI.* The Hague: Nijoff.

Klein, G. S. (1976). *Psychoanalytic theory.* New York: International Universities Press.

Lesche, C. (1962). *A metascientific study of psychosomatic theories and their application in medicine.* Copenhagen: Munksgaard and New York: Humanities Press.

Lesche, C. (1971). On psychophysical measurement. *Swedish Journal of Musicology, 53,* 91–106.

Lesche, C. (1973a). On the metascience of psychoanalysis. *The Human Context, 5,* 268–284.

Lesche, C. (1973b). Die Weltanschauung Freuds und der Psychoanalytiker. *Annales Universitatis Turkuensis, 126,* 85–100.

Lesche, C. (1976). *Über die Psychotherapieforschung.* Paper read at the Dixième Congrès International de Psychothérapie, Paris.

Lesche, C. (1978). Some metascientific reflections on the differences between psychoanalysis and psychotherapy. *Scandinavian Psychoanalytic Review, 1,* 147–181.

Lesche, C. (1979). The relation between psychoanalysis and its metascience. *Scandinavian Psychoanalytic Review, 2,* 17–33.

Lesche, C. (1980). Biochemical and mental depression: A meta-scientific analysis. In K. Achté, V. Aalberg, & J. Lönnqvist (Eds.), *Psychopathology of depression. Proceedings of the symposium by the Section of Clinical Psychpathology of the World Psychiatric Association 1979. Psychiatria Fennica Supplementum* (pp. 169–175).

Lesche, C. (1981a). The relation between metapsychology and clinical psychoanalysis. *Scandinavian Psychoanalytic Review, 4,* 59–74.

Lesche, C. (1981b). The management of anxiety: Therapeutic treatment or scientific research? Some metascientific considerations. In A. Okasha (Ed.), *Proceedings of symposium on psychopathology of anxiety and its management* (pp. 151–159). Cairo: Ciba-Geigy Scientific Office.

Lesche, C., & Stjernholm Madsen, E. (1976). *Psykoanalysens Videnskabsteori.* Copenhagen: Munksgaard.

Lorenzer, A. (1974). *Die Wahrheit der psychoanalytischen Erkenntnis.* Frankfurt: Suhrkamp.

Miller, J. (1975). A critical assessment of the future of psychoanalysis: A view from within, reported by Ira Miller. *Journal of the American Psychoanalytic Association, 23,* 139–153.

Radnitzky, G. (1970). *Contemporary schools of metascience* (2 vols.). Göteborg: Akademiforlaget.

Ricoeur, P. (1965). *De l'interprétation: Essai sur Freud.* Paris: Éditions du Seuil. (English translation: Freud and philosophy. New Haven and London: Yale University Press, 1970).

Rubinstein, B. B. (1976). On the possibility of a strictly clinical psychoanalytic theory: An essay in the philosophy of psychoanalysis. In M. Gill & P. S. Holzman (Eds.), *Psychology versus metapsychology* (pp. 229–264). Psychological Issues, Monograph 36. New York: International Universities Press.

Schafer, R. (1976). *A new language for psychoanalysis.* New Haven: Yale University Press.

Stegmüller, W. (1969). *Wissenschaftliche Erklärung und Begründung.* New York: Springer Verlag.

Törnebohm, H. (1952). *A logical analysis of the theory of relativity.* Stockholm: Almqvist & Wiksell.

Törnebohm, H. (1957). *Fysik och Filosofi.* (With an appendix: On explanation, predictions, and theories in physics: A case study). Göteborg: Acta Universitatis Gothoburgensis. Vol. 63.

Törnebohm, H. (1973). *Perspectives on inquiring systems.* (Rapport 53 from Avd. för Vetenskapsteori). Göteborg: Göteborg's Universitet.

4

On Lesche's Metascientific Investigations of Psychoanalysis

Bo Larsson

Lack of space makes it difficult for Carl Lesche to present his metascience of psychoanalysis in a comprehensible way. Also, it is difficult to do justice to his thinking in a still shorter commentary on his paper. This particular paper, it should be noticed, is only a small part of the penetrating and impressive work on the fundamental nature of psychoanalysis by this very erudite Swedish analyst and metascientist or, to use K. B. Madsen's term, metatheoretician. Maybe the best thing I can do is to try to translate some of Lesche's most central concepts into a less esoteric and less lapidary language than the one he has used, to point to some of his arguments which I find irrefutable, and finally to make some critical remarks on parts of his reasoning which I find difficult to follow.

Essentially, I believe that his argument contains a number of fundamental insights which psychoanalysis and, for that matter, philosophy cannot do without if these disciplines are to recover the intellectual position they held at the beginning of this century. It is a depressing fact that most thinkers who are seriously trying to explore the essence of psychoanalysis are likely to be rejected, isolated, and ignored by the majority of analysts, at least in terms of the impact they should have on psychoanalytic theorizing. This seems to be even more so as far as psychiatrists and psychologists are concerned. One reason is certainly the difficulty of the topic and the potentially revolutionary impact that a rethinking of the epistemological status of psychoanalysis might bring about. However, another reason may be that these authors often seem to be more preoccupied with formulating their own conceptions of

Bo Larsson • The Swedish Psychoanalytical Society, Västerlånggatan 60, S-111 29 Stockholm, Sweden.

psychoanalysis than with taking part in an integrated discussion which could become productive and fruitful and could in the long run influence the scientific community as a whole. I am referring to such profound thinkers as Apel, Gill, Holt, Klein, Lesche, Lorenzer, Ricoeur, Rubinstein, and Schafer. Most of these authors do discuss each other's work to some extent; nevertheless, one often gets the impression of listening to a number of monologues rather than dialogues that aim at mutual understanding in order to reach a higher level of discourse.

This impression of soliloquy rather than dialogue is perhaps even more conspicuous with Lesche than with other analysts and philosophers who have struggled with the tricky issue of the epistemological status of psychoanalysis. It is a regrettable fact that Lesche has had considerable difficulty in making his ideas understood outside a rather small circle that has had the privilege to discuss them with him personally. This may partly be due to the character of the subject matter (compare the complicated and personal terminology of certain phenomenologists), but it might also have something to do with Lesche's manner of presentation.

A colleague, basically sympathetic to Lesche's thinking, jokingly compared him to a tribal chief who, after having done much investigation leading to important insights, would stubbornly insist on presenting his ideas in a rather unknown idiom, expecting anyone who might be interested in the same issues to learn that special idiom, if a discussion is to get off the ground. Of course this is an exaggeration, but it contains a grain of truth. Lesche's major book on the metascience of psychoanalysis was published in Danish (1976), and some of his early work is not easily available to the general reader. However, in this respect the present paper represents a kind of breakthrough, as it was first read at an international congress of psychoanalysis in Helsinki in 1981. Still it seems to have little chance to exert any influence upon analysts interested in the same subject. Many of Lesche's seemingly simple concepts are certainly not known to the ordinary analyst and, to be comprehended, presuppose a familiarity with a wide field of knowledge in psychoanalysis, psychology, psychiatry, metascience, philosophy, phenomenology, formal logic, natural science, and onward.

Furthermore, Lesche seems to expect anyone who takes a serious interest in the problems he is dealing with to be no less versatile than he is himself. With such an attitude you cannot really hope to be understood, and it is no wonder that James T. McLaughlin, reporter at the Helsinki congress, did not grasp that Lesche's view of psychoanalysis as a hermeneutic-dialectical process is in fact based on detailed attempts to conceptualize specific psychoanalytic theories in that way. McLaughlin (1982) writes:

> [At] the same time, he acknowledged in discussion the awkwardness of this assignment—that psychoanalysis in *its* current state defies satisfactory categorization by metascience at *its* present levels of development, and urged us to seek our own ways and constructs. (p. 236)

However, neither is it true that Lesche finds such an assignment awkward nor that he thinks that psychoanalysis "defies categorization by metascience." He has made such a categorization in this paper and a somewhat more detailed one elsewhere (see his bibliography). However, this incident illustrates Lesche's rather evasive behavior in discussions, which sometimes obscures the fact that he does take his own view seriously, as it deserves to be taken.

Many of Lesche's fundamental insights are, in my view, much more similar to the views of such well-known authors as Gill, Klein, Ricoeur, and Schafer than he seems ready to admit. One is reminded of the widespread phenomenon of "the narcissism of minor differences" (Freud, 1929, p. 114), however abhorrent this "psychologizing" way of understanding is to Lesche. In any case, I do not mean to say that all of Lesche's ideas are merely variants of the ideas of other authors. On the contrary, some of his contributions are, as far as I can see, quite original and well worth a discussion as to their tenability. It is beyond my capacity to come to a final decision from a philosophical viewpoint, but I will discuss a few aspects of his theories which I find questionable. Among these are certainly *not* the opinion that psychoanalysis is basically a hermeneutic-dialectical science, even if most analysts disagree. Both Lesche and other authors who have come to the same conclusion are obviously ahead of their time. I shall try to elucidate why I think they are right, although I am aware that such statements can in no way spare anyone the arduous task of studying this difficult and prejudiced field for himself and draw his own conclusions.

First, however, I want to call attention to some of the formal peculiarities which make Lesche's writings so difficult to approach for anyone who does not know his thinking very well. His formalized language probably strikes most analysts as strange and unnecessarily difficult. It also seems a little curious that he uses the language generally found in formal logic, mathematics, and computer science, when, after all, his main critique is directed against logical empiricism and the "mathematicization" of the human sciences in general and psychoanalysis in particular. It is my impression that most of what he says could be said better without such symbols which do not add much, if anything, to the understanding of his main arguments.

Another difficulty for the reader is the fact that some of his central references are either out of print (Lesche, 1962, 1973a, 1973b, 1978; Radnitzky, 1970) or written in Swedish (Törnebohm, 1957) or Danish (Lesche & Stjernholm Madsen, 1976).

A third difficulty is that Lesche makes use of a terminology which is unfamiliar to most psychoanalytic readers, as it is borrowed from a philosophical rather than a psychoanalytical context. Furthermore, many concepts are taken from the European continental tradition of philosophy, not generally familiar to the English reader, and used without explanations or definitions. Such terms are, for example, *phenomenal, intentional, hermeneutic, emancipatory* and *pure* (in the sense of characterizing sciences and psychoanalysis).

Phenomenal refers to phenomena that would generally be characterized as psychological, or belonging to the sphere of the mind.

Intentional is used in the tradition of Brentano and Husserl as a defining characteristic of all mental phenomena. It expresses the fact that all mental acts (e.g., thoughts or affects) are directed toward something, or refer to something particular. This is a very important concept to Lesche, who points to the fact that in all natural sciences theories are ideally formulated in *extensional* language, that is, the symbolic language of the same type as that used in formal logic and mathematics. One of Lesche's ideas, which he shares with Schafer, for example, is that because all the natural sciences abstract from everything intentional or subjective they could not possibly have the subject or intentional phenomena as their subject matter. Strangely enough, as far as can be judged from the psychoanalytic literature and from my discussion with colleagues, very few analysts seem to realize that *intentionality* is a central concept of all psychology and philosophy of the mind (this seems to be true of most contributors to this volume on psychological metatheory).

Pure is also easily misunderstood by anybody not widely read in philosophical literature. It does not mean the opposite of *dirty*, as many of Lesche's critics seem to suppose. It refers rather to the general philosophical ambition to use pure concepts instead of concepts with an imprecise meaning borrowed from different and often incompatible fields of discourse. That it is possible to "purify" scientific concepts in ways other than by translating them into mathematical and formal logical ones Lesche has amply demonstrated by referring to phenomenology and hermeneutics.

Two of Lesche's central concepts are those of science and technique. These concepts may appear self-evident, but they are explicated by Lesche in a way that gives them a quite different meaning from what is customary in psychoanalytic texts. To Lesche, science comprises all those activities that aim at increasing knowledge in a systematic way. Thus, it will be observed that his concept of science is much broader than the general Anglo-Saxon one, which is usually restricted to that of natural science and does not apply to history, philology, or other human sciences, or humanities as they are often called. Schafer has expressed a similar view (e.g., 1970, 1973a, 1973b, 1976) in a more accessible way. The broad concept of science can be traced back to the beginning of philosophy, but much of the insight into its cogency was lost during the last centuries. However, it was rediscovered by some German philosophers of the last century (Brentano, Dilthey, Husserl, Heidegger, and others), all of whom were more or less contemporary with Freud. It has also been rediscovered by some present-day continental philosophers (Apel, Gadamer, Merleau-Ponty, Radnitzky, and Ricoeur), but scarcely by any significant number of psychologists or psychiatrists.

As mentioned above, the primary goal of science is the increase of knowledge. Following Jurgen Habermas, Lesche stresses the need to discern three types of "knowledge-constitutive interests," emancipatory, hermeneutic, and technical. Emancipatory and hermeneutic knowledge-constitutive interests are compatible with each other and are the knowledge-constitutive interests of human sciences. However, the technical knowledge-constitutive interest is not compatible with the other two. The technical knowledge-constitutive interest of natural science always aims to find invariances in *nature* in order to be able to make predictions and thus control nature. According to Lesche this is possible in a scientific way only when dealing with nonliving *things*. As a matter of fact, similar principles *are* applied in physiology, biology, and so on, but strictly speaking such invariances exist only if you disregard the *life* of living creatures, which is exactly what one has to do in natural science. Indeed, Lesche is of the opinion that it is restrictive to prescribe that only natural science should be regarded as science, as is usually done but not well argued. It appears irrefutable once you have understood his argument.

Natural science has a privileged position in relation to Lesche's concept of rational technique. Only natural sciences can justify rational technique and, consequently, no justified technique can be applied to human beings *qua* human beings. Lesche's idea seems to be that the only techniques that are applicable to human beings *qua* subjects are what he calls "Machiavellian techniques." There can be no other scientifically justified influence on human beings than human science, and thus even medicine must be at best a Machiavellian technique. Of course, this is an extreme and rather repugnant view to all physicians. If Lesche is right, we will have to console ourselves, as physicians, with the conviction that, after all, health and survival according to Lesche are such important values that Machiavellian techniques are morally justified in medicine, even if not scientifically so.

To distinguish between the activities of scientists and technicians, Lesche recommends that we use the word *method* to describe what the scientist is doing, and reserve the word *technique* for activities with goals other than the increase of knowledge. Hence we should not speak of psychoanalytic technique, as Lesche thinks that every single analysis can be looked upon only as either "a pure psychoanalytical human-scientific *method* or a psychoanalytically based Machiavellian *technique*" (p. 182). The pure science has its *method*, and what is generally referred to as (theories of) technique in psychoanalysis should, in Lesche's view, be called *methodology*.

Lesche's perspective has many advantages. All those fruitless discussions about what criteria you should use when judging what is psychoanalysis and what is not when defining its scope and its proper domain are brought to an end, and it becomes possible to solve many of the problems raised by analysts like Gill, Klein, Grossman and Simon (1969), and Shafer. It is also easy to understand why Heinz Hartmann's major attempt to reform psychoanalysis had to miscarry, as Lesche mentions in his paper, but regrettably he does not discuss this in a way that his metascience allows. Neither does Lesche refer to Schafer (1970) who, after all, came to the same conclusions.

Lesche's definition of clinical psychoanalysis as science is well founded but creates some new problems. Even if every single "pure psychoanalysis" aims at systematic knowledge of the workings of the mind of the analysand, there is nevertheless an important difference between this type of science or research, and scientific efforts that aim not only to increase the self-knowledge of a single analysand but also to formulate new general psychoanalytic theories. So if every single pure psychoanalysis (i.e., clinical analysis) should be regarded as science, a new concept ought to be invented to describe the activities of theoreticians, whose work is necessary for the science of psychoanalysis to develop.

Another important insight is Lesche's realization that if one borrows a concept from one science and uses it in another, this does not change or reduce that science to the one from which the concept was borrowed. Every science borrows concepts in this way, but it should be stressed that the concepts change their meaning, sometimes radically so, when they are used by another science. To realize that it is only a "naked" model that is borrowed from one science by another (e.g., concepts of energy, cathexis, and forces) is essential for anyone who must decide which concepts are acceptable and which are not. It is my conviction that most of the problems with which Klein, Gill, Schafer and others have grappled can be solved by taking Lesche's idea of the naked model into account. Thus, the many attempts to reformulate psychoanalytic theories according to the latest developments in neurophysiology, neurochemistry, and so on (e.g., Kernberg, 1976, Chapter 3), can easily be shown to be, at best, irrelevant for psychoanalysis.

When Lesche writes that the aims of psychoanalysis are self-reflection, self-understanding, and emancipation, one could easily get the impression that emotions are not involved in the psychoanalytic process. However, his idea is that not only thoughts but, even more so, feelings or, rather, emotional stereotypes can be changed only by way of increased understanding or insight. A purely intellectual understanding of one's emotions which does not result in a change of actions (not behavior) would not be regarded by Lesche as a genuine insight. Nevertheless, I do not feel quite convinced by Lesche's arguments in this particular area. To my mind, it is in fact possible to understand (to attain the hermeneutic goal) without acting accordingly (attaining the emancipatory goal).

In conclusion, I would like to restate that I find the corpus of Lesche's thoughts to be both correct and important. He gives us a clear indication where to look for the value of psychoanalysis and in what directions it could be developed. He has demonstrated that it is the *understanding* (not explanation) of the human psyche that can be and has been increased by psychoanalysis so much better than by other sciences and furthermore that there is no other science that is capable of "proving" whether its methods and findings are correct or not. We should be grateful to Carl Lesche for these accomplishments. One can only regret that he has presented them in such an uncommunicative manner.

1. References

Freud, S. (1961). Civilization and its discontents. In J. Strachey (Ed. and Trans.), *The standard edition of the complete psychological works of Sigmund Freud* (vol. 21, pp. 64–145). London: The Hogarth Press. (Original work published 1929).

Grossman, W. I., & Simon, B. (1969). Anthropomorphism: Motive, meaning, and causality in psychoanalytic theory. *Psychoanalytic Study of the Child, 24*, 78–114.

Kernberg, O. (1976). *Object relations theory and clinical psycho-analysis*. New York: Jason Aronson.

Lesche, C. (1962). *A metascientific study of psychosomatic theories and their application in medicine.* Copenhagen: Munksgaard & New York: Humanities Press.

Lesche, C. (1973a). On the metascience of psychoanalysis. *The Human Context, 5,* 268–284.

Lesche, C. (1973b). Die Weltanschauung Freuds und der Psychoanalytiker. *Annales Universitatis Turkuensis, 126,* 85–100.

Lesche, C. (1976). *Über die Psychotherapieforschung.* Paper read at the Dixième Congrès International de Psychothérapie, Paris.

Lesche, C. (1978). Some metascientific reflections on the differences between psychoanalysis and psychotherapy. *Scandinavian Psychoanalytic Review, 1,* 147–181.

Lesche, C., & Stjernholm Madsen, E. (1976). *Psykoanalysens Videnskabsteori.* Copenhagen: Munksgaard.

McLaughlin, J. T. (1982). Issues stimulated by the 32nd Congress. *International Journal of Psychoanalysis, 63,* 229–240.

Radnitzky, G. (1970). *Contemporary schools of metascience* (2 vols.). Göteborg: Akademiförlaget.

Schafer, R. (1970). An overview of Heinz Hartmann's contribution to psychoanalysis. *International Journal of Psychoanalysis, 51,* 425–446.

Schafer, R. (1973a). Internalization: Process or fantasy? *Psychoanalytic Study of the Child, 27,* 411–436.

Schafer, R. (1973b). Action: Its place in psychoanalytic interpretation and theory. *Annual of Psychoanalysis, 1,* 159–196.

Schafer, R. (1976). *A new language for psychoanalysis.* New Haven & London: Yale University Press.

Törnebohm, H. (1957). *Fysik och Filosofi.* (With an appendix: On explanation, predictions, and theories in physics: A case study). Göteborg: Acta Universitatis Gothoburgensis (vol. 63).

4

Metatheory and the Practice of Psychoanalysis

Lars B. Lofgren

I feel both honored and somewhat bashful trying to comment on this paper. I earn my living practicing psychoanalysis and psychotherapy and my interest in philosophy can only be regarded as amateurish. This paper has impressed me, however, and I think it may be important not only to me but to other practitioners in my field.

I started my medical career in the basic sciences, physiology, pharmacology, and mathematical biophysics. I never felt much at home among the rats and eventually, with a feeling of great relief, switched to psychiatry. My former colleagues used to chide me about the unscientific nature of psychiatry and especially psychoanalysis. Psychoanalytic findings were written off as fantasies, because they were not repeatable and generated no statistics. The interesting thing was that those persons who were certain that psychoanalysis was not a science could not define what science actually was. Science seemed to be good, though, because they approved of it. What was not scientific was, by the same uncertain criteria, bad.

I still do not think that anyone can define what science is. Maybe there is a cognitive kernel (however uncertain) containing such criteria as quantification, repeatability, probability, and other statistical manipulations, but this is surrounded by a halo of vague statements and value judgments.

Lesche's paper does away with a lot of this fruitless discussion. People will still dislike psychoanalysis—that is certainly their right. They can, however, no longer acuse psychoanalysis of not being a *natural* science. I find Lesche's argument that psychoanalysis is a human or

Lars B. Lofgren • School of Medicine, University of California at Los Angeles, Los Angeles, California 90024.

humanistic science (German: *Geisteswissenschaft*) entirely convincing. Psychoanalysis is a study of the evanescent moment, one following upon the other in the psychoanalytic situation. There are similarities enough so that one psychoanalyst can understand another, using the jargon of the trade they have developed. However, the differences among the glimpses of an inner life that we get through vicarious introspection will probably always be greater than the similarities. This is especially valid for the rather mature and highly developed person who would constitute an adequate client for psychoanalytic practice. Psychotic and other very primitive persons are much more alike—how many unfortunates have not been persecuted with ray guns by the Martians!

As a psychoanalyst in practice, I am of course eager to understand more about what I am doing in my daily work. Lesche's point that psychoanalytic treatment is actually *hermeneutic research* fits well with my own vaguely formulated ideas. For many years I have defined for myself—and for students in the field—my position as the patient's research assistant, putting my skills and my experience at his disposal but always letting him remain the boss of the enterprise. This not only describes what I am doing but protects me also from grandiose feelings. Because no matter how long and penetrating an analysis, the patient will always know infinitely more about his inner life, about subtleties of meaning and relevant connections than I can ever aspire to know.

With the understanding of a genius, Freud understood that every attempt to anchor psychological structures in anatomical substrates was doomed to failure. Yet all through his life he remained hopeful that his system would eventually be anchored in chemical-biological facts. No one can know if we will develop a true psychosomatic language, describing somatic and psychological events in the same terms. Nothing of the sort exists so far. After Lesche's investigations one can say with certainty that that will never be the case for psychoanalysis. Psychoanalysis will remain a humanistic science and psychoanalytic technique cannot be justified in a strict sense. This view has important consequences for psychoanalysis.

Every now and then papers are published that state something like "Dreams do not come from childhood wishes; they are generated by neurons in the pontine section of the brain, firing repeatedly." Such statements are nonsense and Lesche helps us to understand why. Actually statements such as the one quoted are of the general structure "This stone is not white but heavy" which immediately impresses us as meaningless. Psychoanalysis has to be judged from within, with criteria that are arbitrary or pragmatic at best. If psychoanalytic theory could be changed in such a way that this would lead to better treatment methods, this would be valuable for the psychoanalytic practitioner. Other persons using the theory might be better served by unrevised formulations. Truth criteria are simply of no value in dealing with the matter at hand.

The drive theory of psychoanalysis and the construction of a psychological apparatus obeying mainly hydrodynamic laws is an unexpected companion of a hermeneutic research practice. The absence of consideration of meaning except as a late, fine regulator is rather startling. Freud's speculations about sexual drives and the death instinct (sometimes modified as the aggressive instincts) seem outmoded and quaint in the light of modern biological understanding of behavior. Nowadays what Freud called fundamental drives would possibly be regarded as kinds of intervening variables. Is this a reason for psychoanalysis to revise itself? Not *per se* according to Lesche, and I think we have to agree. There are no strict criteria to decide what should and should not be part of psychoanalytic theory. Personally, I am not much in favor of the drive theory and the apparatus psychology—I think we could do better than that, especially in a humanistic science. I also believe that the psychoanalytic techniques that try to introduce drive concepts directly into the psychoanalytic situation lead to disastrous results for the patient. But my arguments against drive theory must remain essentially *persuasive* and cannot be defended by any strict theory. The importance of that is freedom. Lesche's paper liberates psychoanalysis. We can continue our work without any anxious glances at the brain chemists, the neurologists, and all the others.

I found the study of Lesche's difficult paper a rich and rewarding experience, leading to a deeper understanding of my practice and my theory. I hope that many other psychoanalysts will take the time and effort to penetrate this highly interesting work.

4

Psychoanalysis as Research, Therapy, and Theory

Gerard Radnitzky

1. Introduction

The main problems dealt with in Carl Lesche's paper are the epistemological-methodological status of psychotherapies based upon psychoanalytic theory, the ontological-methodological status of psychoanalytic theory, and the nature of the psychoanalytic process. The paper also addresses itself to the problem of the unity of the sciences and comes to the conclusion that psychoanalysis is not a natural science. It emphasizes that psychological theories (being theories about mental or, in general, phenomenal phenomena) are, in principle, not reducible to biological, physiological, neurological, or other theories, either in the sense of reduction through definitions or in the weaker sense of reduction through laws. There is an extensive literature on that topic. Let me just say that I agree with Lesche's thesis. I suppose that—to put it in a nutshell—the reductionists are confusing cause and motive. By claiming that my desire to do X is the "cause" of my doing X they disregard the fact that whereas physical causes necessitate, motives incline but do not necessitate. To assert that a person has a desire to do X may make it predictable or intelligible that he or she will in fact do X. But no contingent necessity or contingent impossibility is involved. *Man can, and cannot but, choose,* each and everyone choosing for themselves alone (Flew, 1983). Hence, the philosophical implications of behavioristic theorizing include an image of man that is not "half true," but completely wrong.

Gerard Radnitzky • Department of Philosophy of Science, University of Trier, Trier D-5500, West Germany.

Psychoanalytic theory views man, realistically, as a chooser; and the goal of psychological technology is to increase the capabilities of the person undergoing analysis to make his choices in a more conscious, more deliberate manner than before the analysis.

I agree with Lesche's view of the psychoanalytic procedure and admire his methodological analysis of Freud's theories of parapraxes, dreams, and neuroses, which are beautifully presented in Lesche and Stjernholm Madsen (1976). However, I consider the metascientific framework which he outlines in his paper to be inadequate.

2. On the Methodological Framework Sketched in Lesche's Paper

The metascientific framework outlined in Lesche's paper is a position in methodology that customarily is referred to as *"the received view": the logical-positivist position.* The two-language approach characteristic of logical positivism leads to an instrumentalist view of theories, which, in turn, leads to pseudo-problems such as the problem of "theoretical terms." Theory instrumentalism may be consistently combined with the view that singular statements—"basic" statements—are categorically and finally known, in some cases at least. Lesche limits instrumentalism in this way. He gives an *instrumentalist interpretation* not only of physical theories, but also of "metapsychological" theories (in Freud's sense, see §3.2.3). According to the instrumentalist view a scientific theory and the "model" of a real system which is constructed with the help of the theory are conceptual constructions. To the question, What makes the conceptual construction into a "model" of a real system? the instrumentalist points out that with its help he can make successful predictions. To the question Why? he has no answer. Why some theories are more successful than others in this respect becomes a mystery. Moreover, instrumentalism—be it in Törnebohm's version (to which Lesche refers) or in Wolfgang Stegmüller's "non-statement view of theories" (the "Sneedification of science" so to speak)—eventually leads to ontological idealism (as, for example, Alan Musgrave, 1979, has shown).

Why should theory instrumentalism be attractive? It offers a quick solution to something which many—wrongly—believe to be a dilemma. If one believes that science aims at true theories and has made some progress in meeting this demand and, like Thomas S. Kuhn, believes that scientific theories have all been falsified, then there is a simple and radical way to resolve the apparent conflict between these three statements: theory instrumentalism. It gives up all three in one sweep. It does so simply by declaring that the question whether or not a theory has descriptive functions is unanswerable, or by asserting that a theory does not describe, is but a fiction, a black box, that functions as *nothing but* an instrument for deducing testable consequences. Theory instrumentalism grows out of logical positivism's two-language approach, which, in turn, is rooted in the totalization of the distinction between observation statements and theoretical statements. Theoretical sentences or some theoretical terms are "empiricized" (given empirical content) by means of "bridge principles" (Hempel, 1966), or by means of operational definitions and re-applications (Lesche's approach), or by similar devices. One ignores that the issues to be faced are *problems of testing* and not *problems of meaning*. Also Stegmüller's "non-statement view of theories" (1979) is but a modern version of instrumentalism that combines it with Kuhn's view and presents it in a formalistic dress. The intellectual motive for theory instrumentalism appears to be the insight that a scientific theory cannot be justified because there are no infallible methods of ascertaining the truth value of its statements. This insight is then combined with the evaluation that, if justification—establishing the truth value with certainty—cannot be had, it is best to get rid of the descriptive function of theories altogether. In this way the failure of justification philosophy has first led to efforts to take *just* the theoretical stuffing out of theories and eventually to taking *all* stuffing out of them, reducing them to black boxes (Watkins, 1978; Radnitzky, 1979).

The logical positivist's position was decisively criticized by Karl Popper as early as 1934 (Popper, 1959). Popper showed that there is no epistemic difference between the so-called theoretical statements and the so-called basic statements. The methods of ascertaining the truth value of a particular statement are in each case fallible. The only difference is that basic statements, "observation statements," are easier to test than theoretical statements. Once the quest for certainty has been abandoned and the concepts of justification and criticism clearly separated, the problem situation is restructured. It is recognized that cognitive progress is possible in spite of pervasive fallibility: through the interplay of creativity and criticism. This interaction of "conjectures and refutations" may then be generalized into the schema: adaptive-(blind)-variation-and-selective-retention, a schema found to possess great heuristic value also for areas outside the methodology of research (Andersson, 1984; Radnitzky, 1982).

In my opinion, Lesche could drop the metascientific frame which he sketches without losing anything in the main part of his study, that is, his analysis of Freud's theories. Even more importantly, a *realist* interpretation fits his case studies perfectly. The core of his magistral analysis is a model that *describes* how the competition of two intentions, one of which is unconscious, leads to a conduct that can be seen as a compromise by means of which the conflict between the competing intensions is resolved, at least for the moment (Lesche & Stjernholm Madsen, 1976). What the model "maps" depicts, represents, or describes is a real situation in the life of a real person. The hypothesis that the person a at time t_0 had a certain intention—a phenomenal or mental entity (Popper's world-2 entity)—attempts to describe something in reality in exactly the same way in which the hypothesis that at a later moment a particular slip of the tongue or some other parapraxis occurred describes. Of course, when speaking of a slip of the tongue we already make use of a psychoanalytical conceptual framework, whether we recognize it or not. However, this does not conflict with the view that the statements constituting the model or the theory have descriptive functions and that some of them may be closer to the truth than others.

In sum, the explanations and predictions that occur in what Lesche, following K.-O. Apel, labels "quasi-naturalistic phases" make use of psychoanalytic theories. At least to this commentator it appears reasonable to give these theories a realist interpretation. Lesche's analysis could only profit from getting rid of the received view, a view in which I once believed myself, but which I later on recognized to be but a blind alley. Much the same holds for some of the details such as, for example, the question of extensional language. It, too, is a problem induced by logical positivism. Why should one wish to formalize here at all? What advantages could be gained by introducing an "ideal language" *à la* logical positivism into the methodological study of psychoanalysis? Since ordinary logic is extensional, it cannot, for many contexts, provide reconstructions of intentional sentences or of causal sentences that are adequate in these contexts. But, unless there are concrete advantages to be gained by formalizing, this fact is irrelevant.

Lesche's reconstructions of Freudian theories stand on their own. They would not gain anything by formalization. What makes them so convincing is that Lesche is able to show that the elementary schema describing the process and mechanism of parapraxis can be developed into a schema that covers the basic mechanism described by dream theory and eventually into a schema that reconstructs Freud's theory of neurosis. This is what makes Chapter 2 of Lesche and Stjernholm Madsen (1976) so illuminating. It is hoped that an English translation will soon be available.

3. Psychoanalysis as Research

3.1. Research to Improve Psychoanalytic Theory

Lesche clearly distinguishes between pure psychoanalysis (§§3.2.1 and 4.1) and technologies based upon psychoanalytic theories. He points out that *psychoanalytic research,* or *pure psychoanalysis,* belongs to the humanities and not to the (natural) sciences. Although this appears to me to be correct, two comments may be in place. (1) Insofar as the activity called *psychoanalysis* is research it has to use scientific method. There is only *one* scientific method: the continuous interplay of creative imagination and criticism, testing. In this sense there is a unity of science: the unity of global method. The *goal* of this activity is *cognitive progess,* whereby the concept of (absolute) *truth functions as a regulative principle.* If a research enterprise is successful, its result is a new theory or an improvement of some old theory, that is, it strives to produce *general* hypotheses. In the case of psychoanalysis, for instance, there are hypotheses about how transfer and counter-transfer function, hypotheses with the help of which the phenomenon of parapraxis may be explained, and so forth. In these essential respects psychoanalytic research is on a par with any other (nomothetic) scientific research. (2) Every discipline and every subject matter requires, in addition to the above-mentioned general scientific method, also special methods. Hence the fact that psychoanalytic research also needs special methods is nothing peculiar to psychoanalysis. The thesis that psychoanalysis is one of the humanities or, better perhaps, one of the *sciences humaines*—which include not only the humanities, but also the social sciences and historiography—is certainly correct. The knowledge (hypothetical knowledge as is all scientific knowledge) psychoanalytic research produces helps us to improve the image of man as a *species.* This is possible because the hypotheses or theories which are the result of its endeavors are general hypotheses. Hence, I would place psychoanalysis together with what is sometimes called "general psychology" (*allgemeine Psychologie*). Psychoanalytic research has ingredients of historical research and may use an evolutionary perspective, but this does not turn it into historiographic research. Psychoanalysis as a discipline is a nomothetic science—*science* used in the wide sense in which the corresponding terms are used in German, French, Italian, and so on.

3.2. Research in Which Psychoanalytic Theories Are Applied

To be distinguished from research conducted with a view to improving psychoanalytic theory is the application of already existing psychoanalytic theories in *a special kind of historical investigation.* In history what is to be explained are events or sequences of events. They are explained with the help of general laws taken from various disciplines together with hypotheses about the initial conditions. When psychoanalytical theories are used to explain (or to predict) a particular individual's conduct, experience, emotions, and so forth, the goal of the research enterprise is better to understand why a particular, individual event took place and eventually to understand an individual life-style, biography, and so on. That means that also in this case *the aim is cognitive progress*: to explain and better to understand individual events, an individual's life, and so on. The regulative principle is, here too, *getting closer to the truth,* to interesting truths. The result of such a research enterprise, if it is successful, will help the individual whose conduct, feelings, attitudes, and so forth are being explained or predicted to *improve his self-understanding as a person.* He will do so by trying to fit this information, this new knowledge, in with the information he gets from other sources such as biographical studies, the history of his family, of his region, his nation, and so forth. The knowledge that the humanities together with history produce may enable him to improve his self-conception as an individual. This sort of psychoanalytic investigation is on a par with the normal work of the historian.

4. Psychoanalysis as Technology

To be clearly distinguished from both the above-mentioned historical research enterprises is a technology that is based upon psychoanalytic theory. This technology has something in common with the historical research, namely, already existing *psychoanalytic theories are applied*; but now the goal is not cognitive progress—a goal and a criterion internal to science ("science" in the wide sense)—rather, *the goal is utility* of some sort, namely, "health, social adaptation, or functional competence" (§3.2.1).

Here is the place where one, if one wishes to do so, may speak of *emancipation,* namely in those cases in which the aim of the psychotherapeutic treatment is to free the patient from certain things that he evaluates negatively, for example, from a certain type of fear, anxiety, compulsion, aversion, or what have you, to "emancipate him from" something that he does not like, or that somebody else whose advice he follows does not like. Emancipation from something is clearly a practical goal, a goal or a criterion *external* to science (again, in the wide sense). Hence, I would not claim, as Lesche does (see §4.1.1), that emancipation is the aim or one of the aims of pure psychoanalysis, of analysis as a research activity. Habermas's and Apel's concepts of the so-called technical, the hermeneutic, and the emancipatory interests—as *transcendental* interests, that is, as conditions for the possibility of a certain sort of activity—are in my opinion, mistaken. (I too once believed that they offered interesting perspectives but later recognized that this was a mistake.) This classification of transcendental interests is but a replication of Max Scheler's three types of knowledge: *Herrschaftswissen, Verständigungswissen,* and *Erlösungswissen.* One of the consequences of Habermas's theory is that everything becomes governed by the technical interest, namely, hermeneutic becomes a technique for understanding, understanding becomes a technique for emancipation, and so forth. His thesis that the so-called technical interest is the only transcendental (or quasi-empirical) interest underlying natural sciene emanates from confusing the testing of theories with the appliation of theories.

4.1. The Explication of the Concept of Technology and the Classification of Technologies

A technology is sometimes defined as law hypotheses in the context of application (e.g., Albert & Stapf, 1979). With Lesche, I use *technology* for a system of hypothetical prescriptions based upon law hypotheses. I would prefer a somewhat different wording than the one used by Lesche (see §3.1). A typical technological rule would have the form: "If you want to realize G and you find yourself in situation *S*, then procedure *P*—a certain technique—is recommended, because *P* is an effective and efficient means for realizing G. This is so because the law hypotheses underlying *P* are highly corroborated or at least sufficiently well tested and corroborated for the case at hand." The level of aspiration will depend upon an evaluation of the risks involved if one should be mistaken and, in general, on the outcome of a cost-benefit analysis.

I agree with Lesche's classification of technologies into those wherein the law hypotheses can be explained ("puristic techniques"; see §3.2.2) and those wherein this is not the case ("Machiavellian techniques"). However, the terminology chosen does not appear very suitable. In particular, it appears to me to be misleading to call the former type of technology "rational" and to speak of "rational therapy." Let me first make a terminological proposal: the distinction between handicraft technology and technology based on science. A *handicraft technology* is based upon a presumed connection "If initial conditions *J* are brought about then the result is *P*"; but one cannot explain why this is so and thus, also, not be certain that a causal law is involved and not a correlation which might be brought about by a third factor. If one cannot explain the if–then sentence, then one does not know its realm of application either and one can suddenly find oneself in situations in which the technology does not function (Radnitzky, 1980, 1983). It may nonetheless be perfectly rational to use such a technology, even to use a technology based upon a falsified law hypothesis. All depends upon whether or not the falsification applies to the sector of phenomena one has to deal with, that is, whether or not the predictions made with the help of the theory are sufficiently good for the practical purpose at hand. To navigate through a dangerous passage one does not need to know where all the underwater rocks are; it suffices to know where there are no rocks. How much accuracy it is rational to request of the predictions derived with the help of the theory will depend upon the outcome of a cost–benefit analysis which will have to take into account the decision-maker's subjective evaluation of the disutility of failure of the practical action under consideration and to a lesser degree the utility of success.

4.2. On the Role of the Problem of Reductionism

The problem whether or not psychoanalytic law hypotheses can be reduced to physiological or neurological hypotheses or theories appears to me to be irrelevant for the issue of the rational use of a particular technology. If the law hypothesis upon which a particular psychoanalytic technology is based can be explained, that is, be deduced from a more general psychoanalytic law hypothesis, then the technology in question would qualify as a technology based upon science (in the wide sense), in Lesche's terminology, as "rational therapy." If this is not the case, it would, according to the terminology here proposed, be classified as "handicraft technology." This classification has nothing to do with the problem of the reducibility of psychology to biology or the problem of the unity of science in the sense of a reductionist (materialist) program. The question of whether or not it is rational to use one or the other technology cannot be decided merely by ascertaining whether or not the law hypothesis upon which the technology is based is explainable by a more general law hypothesis that also explains why the realm of application of the first-mentioned, low-level law hypothesis is limited in the way it is.

4.3. Psychoanalysis and the Demarcation Problem

Is psychoanalysis—as an umbrella word for research enterprises conducted with the aim of cognitive progress—scientific? Obviously, the answer will depend in part, upon what demarcation criterion is used. As is well known, Popper mentions psychoanalysis as an example of a theory or an activity that is outside of science. However, it does not appear fruitful to speak of psychoanalysis in the singular. The question is rather whether or not a particular psychoanalytic theory fulfills certain criteria that appear to be a *sine qua non* for qualifying as scientific, that is, scientific in the sense of systematic procedure of rational problem solving. These criteria are of two kinds. Firstly, that the theory in question is not placed in a context that prevents it from being criticized and, secondly, if the theory is placed in a criticist context, a context in principle permitting criticism, it does not itself contain criticism-deflecting devices. Thus, for example, Freud's theory of parapraxis as it occurs in a particular context—for instance, in Lesche's reconstruction—should be investigated with a view to finding out whether or not that context and the theory itself meet the above-mentioned criteria.

This leads to the question of how psychoanalytic theories are tested and what in that context constitutes a "severe test." It is a characteristic of basic science (science in the wide sense) that with the help of the theory, testable consequences are deduced in order to test that theory. If such consequences are deduced for some other purpose, for example, to make a prediction that will help to make a decision, that is, if what is at issue is the prediction, then the research activity in question belongs *eo ipso* to applied research or constitutes an application of the theory. It is not possible to test a theory in the context of an application. This holds good not only for physical theories—where it is obvious—but for theory testing in general. It follows that a psychoanalytic law hypothesis can, in principle, not be tested in the context of psychoanalytic therapy. Hence, only if it can be shown that it is possible to subject a particular psychoanalytic *law hypothesis* to severe testing in the context of psychoanalytic research—an activity under the regulative principle of truth—does that hypothesis qualify as science, as science in the wide sense which includes the humanities as well as the social sciences. Only if it can be shown that for a particular *singular* hypothesis that has been deduced by using one or more psychoanalytic law hypotheses in a historical explanation (or in the prediction) of individual events there exists a severe test, does that hypothesis qualify as scientific in the sense in which hypotheses of historiography may qualify as scientific. It may be worthwhile to note that being "testable" or being "falsifiable" is a logical relation. A hypothesis H in the context of a theory T is falsifiable if and only if there exists a singular statement S such that S describes a situation that is logically possible, and which is logically to be observed, and S is incompatible with H. Falsifiability has nothing to do with the empirical or with the technical possibility of falsifying. The problem of falsifiability (of having empirical content) and the problem of being corroborated—having survived tests of a certain degree of severity—are distinct problems. The reconstruction of Freud's basic theories by Lesche and Stjernholm-Madsen may open the way to investigating the old question of whether or not psychoanalysis is a science.

5. References

Albert, H. & Stapf, K. (Eds.). (1979). *Theorie und Erfahrung*. Stuttgart: Klett-Cotta.

Andersson, G. (Ed.). (1984). *Rationality in science and politics. (Boston Studies in the Philosophy of Science*, vol. 79). Dordrecht: Reidel.

Bean, P. H. (Ed.). (1983). *Mental illness: Changes and trends*. London: John Wiley & Sons.

Flew, A. (1983). Mental health, mental disease, mental illness: "The medical model." In P. H. Bean (Ed.), *Mental illness: Changes and trends* (pp. 115–136). London: John Wiley & Sons.

Hempel, C. G. (1966). *Philosophy of natural science*. Englewood Cliffs, NJ: Prentice-Hall.

Lesche, C. (1981). The relation between metapsychology and psychoanalytic practice. *Scandinavian Psychoanalytic Review, 4*, 59–74.

Lesche, C. & Stjernholm Madsen, E. (1976). *Psykoanalysens videnskabsteorie*. Copenhagen: Munksgaard.

Levinson, P. (Ed.). (1982). *In pursuit of truth. Essays on the philosophy of Karl Popper on the occasion of his 80th birthday*. New York: Humanities Press, and Brighton: Harvester Press.

Musgrave, A. (1979). Theorie, Erfahrung und wissenschaftlicher fortschritt. In H. Albert & K. Stapf (Eds.), *Theorie und Erfahrung* (pp. 21–54). Stuttgart: Klett-Cotta.

Popper, K. (1959). *The logic of scientific discovery*. London: Hutchinson (German original 1934), 9th rev. ed. 1977.

Radnitzky, G. (1980). What limits do technology and science have? *Crítica: revista Hispanoamericana de filosofía 12*, 15–54.

Radnitzky, G. (1982). Popper as a turning point in the philosophy of science. In P. Levinson (Ed.), *In pursuit of truth: Essays on the philosophy of Karl Popper on the occasion of his 80th birthday* (pp. 64–82). Atlantic Highlands, NJ: Humanities Press.

Radnitzky, G. (1983, Summer-Autumn). Science, technology, and political responsibility. *Minerva, 21,* 234–264.

Radnitzky, G. (1985). Cost–benefit analysis in the methodology of research. In G. Radnitzky & P. Bernholz (Eds.), *General Economy: The economic approach applied outside the traditional areas of economics.* New York: Paragon.

Radnitzky, G. & Andersson, G. (Eds.). (1978). *Progress and rationality in science.* (*Boston Studies in the Philosophy of Science,* vol. 58). Dordrecht: Reidel.

Radnitzky, G., & Andersson, G. (Eds.). (1979). *The structure and development of science.* (*Boston Studies in the Philosophy of Science,* vol. 59), Dordrecht: Reidel.

Schilpp, P. A. (Ed.). (1974). *The Philosophy of Karl Popper.* (*Library of Living Philosophers*) 2 vols. LaSalle, IL: Open Court.

Stegmüller, W. (1979). A combined approach to the dynamics of theories: How to improve historical interpretations of theory change by applying set theoretical structures. In G. Radnitzky & G. Andersson (Eds.), *Progress and rationality on science* (pp. 151–186). (*Boston Studies in the Philosophy of Science,* vol. 58). Dordrecht: Reidel.

Watkins, J. (1978). The Popperian approach to scientific knowledge. In G. Radnitzky & G. Andersson (Eds.), *Progress and rationality in science* (pp. 23–43). (*Boston Studies in the Philosophy of Science,* vol. 59), Dordrecht: Reidel.

4

Is Psychoanalysis Therapeutic Technique or Scientific Research?

A Metascientific Investigation: Reply to Commentators

Carl Lesche

1. Larsson accuses me of being imperspicuous and presents himself as a spokesman for the great majority of ordinary, average psychoanalysts. But my paper was not at all specifically geared to such an audience (even if the current version was indeed presented before a psychoanalytic congress). Most psychoanalysts are inclined to reject, isolate, and ignore the work of theoreticians, partly because the topics are difficult and partly because they prefer the performance of monologues over participating in well-integrated discussions. According to Larsson, what makes the content of my paper difficult to comprehend is primarily that my terminology is borrowed from Continental rather than from Anglo-Saxon philosophy and that I presuppose concepts and viewpoints drawn from a multitude of different areas. Presumably, this presents difficulties for the average psychoanalyst; however, in my ears this sounds more like *laus stultitiae* (praise of stupidity).

My paper, as clearly stated in the subtitle, is intended as a metatheoretical examination of psychoanalysis. Needless to say, mention is made of various psychoanalytic concepts and viewpoints, but they are dealt with in a metatheoretical language. The psychoanalytic language will not suffice for talking *about* psychoanalysis. I attempt to demonstrate that the logical empiricist theory of science is inappropriate for an analysis, and possible reconstruction, of psychoanalysis and consequently

Translated from the Swedish by Professor Herman Tennessen.

Carl Lesche • Department of Psychology, University of Stockholm, Stockholm, Sweden.

we are forced to employ a metatheory from the sciences of the humanities, which do indeed have the Continental imprint of phenomenological and hermeneutical thinking.

Larsson believes that I am contradicting myself when I employ a formal and mathematical language. My strategy here was to demonstrate that by taking one's point of departure in the conception of psychoanalysis as natural science and make use of the logical empiricist's science-theoretical method, then one is indeed bound to end up with contradictions. However, I only employ a single such formalism (the deduction in the Hempel-Oppenheim explanation), plus a few ordinal numbers from 1 to 5. One could, of course, omit the symbols, but then one would be forced to replace them with long transcriptions. I admit, however, that my presentation is an extremely concentrated one, but I did try to demonstrate how my science-theoretical analysis of a psychoanalytic text is carried out (see Lesche, 1979).

The difficulties here touched on give rise to some thoughts concerning psychoanalytic education. Psychoanalysts have been very much involved in discussions about their own identity. This is a most intricate problem which, needless to say, has very much to do with one's conception of psychoanalysis. In my paper I tried to show that the explication of psychoanalysis presupposes various points of view. The prevailing state of affairs in this area, however, is one of ignorance about science-theoretical presuppositions. To bring about a conscious awareness and possibly even a change of world-view factors *is* extremely difficult and it actually arouses resistance in the psychoanalytic sense of that word. After all, it shakes the student's total spiritual foundation. This is certainly not surprising though, since the natural scientific world-view is so deeply rooted in our culture. The prospective student is required to have a good grounding in (natural scientific) medicine or (most often, behavioristic) psychology. We are already indoctrinated with this world-view in kindergarten and the primary grades; in fact, it is absorbed with the mother's milk. The domain of psychoanalysis has been expanded from analyses of sexual and aggression instincts to elucidations of the Ego, Superego, and evaluations, religious and political ideas. What I recommend is that the prospective analyst's self-analysis include his view of science and more particularly of psychoanalysis. These same views should be considered in the guidance and instruction of the candidate.

In this context, I should like to note that I have never made any claims as to the originality or priority of my ideas.

Larsson finds that my explications of science and technique are quite original. But, in retort, my question is, what then are the usual approaches? My point of departure is that action is rule-governed. *Method* and *technique* differ in that one acts *within science* only according to method, which follows certain directives and where the merits of the goal are appraised *solely* in terms of the *value* of the *growth of knowledge*. On the other hand, technique is a question of proceeding according to technical instructions where the merits of the goal may be appraised according to *other* values as well. Furthermore, I distinguish between puristic and Machiavellian technique. The former is to be considered justified, whereas Machiavellian technique is by definition unjustified, that is, one acts according to the motto "the end sanctifies the means." Larsson concludes from these distinctions and concepts a point of view which is bound to appear extremely repulsive to physicians, namely, that medicine in general and psychotherapy in particular can at best be said to be employing Machiavellian and, hence, unjustified techniques. These conclusions in turn allegedly contribute to the incomprehensibility and unacceptability of my views. However, I should like to add that, after all, it is not always, as it may seem, a question of *disagreement* in scientific matters, but rather a question of competition between different occupational groups.

I have not attempted to define clinical psychoanalysis (Klein, 1976); rather, I have spoken of pure psychoanalysis as a science. Scientific research can embrace even the activity of theoreticians (wherein may be included meta- and meta-metatheoretical aspects) without necessarily bestowing on this activity a new name.

Finally, permit me a few words—from the viewpoint of pure psychoanalysis—concerning the growth of knowledge in terms of the values of both hermeneutic self-understanding as well as emancipation. By *emancipation* I mean a "liberation from misconceptions" (Lesche, 1973). I have attempted to explicate the concept of *insight* in terms of self-understanding and emancipation, but emancipation does not necessarily manifest itself in external behavior, and hence the two concepts must be clearly distinguished.

2. Dr. Lofgren's comments on my paper warmed my heart. I have here nothing in the direction of a riposte. Had his comments been negative, then I should of course have been forced to defend myself. But Lofgren has clearly understood what I intended to convey, namely, a liberation, an emancipation of the psychoanalyst, whether physician or psychologist, from the ballast of misconceptions with which he has already become encumbered by his preanalytic, basic education. I have in mind such misconceptions as apply with regard to science and technique (in general), and to natural science and the human science. As a training analyst I have frequently experienced the liberation felt by a student who, during his psychoanalytic training period, begins to understand

that psychoanalysis in essence can be neither a natural science nor a therapeutic technique.

Although Lofgren mentions his doubts concerning Freud's theory of instincts and indeed the total psychic apparatus, he finds no reason to revise psychoanalysis. Modern biological knowledge cannot steer psychoanalysis. But we need to reach a more thorough understanding of what Freud meant with his metapsychology. General hermeneutics without metapsychology does *not* become a Freudian psychoanalysis.

3. I am very grateful to Gerard Radnitzky for his comments. These form an importantly relevant and systematic discussion of my paper. They would be of invaluable help to anyone engaging in a venture to reformulate psychoanalysis. But *I* never aimed at such a reconstruction.

My original problem was the following. To Freud *psychoanalysis* signified both scientific research and therapeutic health treatment, a distinction often made among psychoanalysts, psychiatrists, philosophers with an interest in psychoanalysis, and others. My paper was read before an audience of psychoanalysts, some with a medical (natural scientific, technological) background, others with a psychological (most often behavioristic) education. Among them it is quite common to regard psychoanalysis as a natural science or as though its techniques were rooted in the natural sciences. I attempted to explicate psychoanalysis as scientific research and therapeutic technique in a way that I believe they were intended by Freud.

In the argumentation for explicata I chose an instrumentalist interpretation of theories demonstrating that if one were to permit theoretical terms to refer to physicalistic entities, then one would never come close to an understanding of Freudian psychoanalysis. I arrived at the conclusion that psychoanalysis may be conducted as an unjustified technique (psychotherapy), or as a *geisteswissenschaftliche* (human-scientific) method within the humanities.

Are psychoanalytic theories to be given an instrumentalist or realist interpretation? I am inclined to interpret realistically the "quasi-naturalistic" or "clinical" theories which enter into the interpretations of parapraxes, dreams, neurotic symptoms, and anxiety. (Credit for the reconstruction of these theories as well as the piecemeal development of the conflict model belongs to Freud, as he organized his *Vorlesungen* [*Introductory Lectures on Psycho-analysis*] in exactly this way.)

But these theories are not theories within a natural science. I find it hard to imagine any general theory which is either realist or instrumentalist. I believe, on the other hand, that Freud originally developed his metapsychological theories in an instrumentalist atmosphere. He borrowed his models or analogies from contemporary physicalistic, biological, and even sociological theories, and other psychoanalysts have tried to interpret his metapsychological terms as concepts extracted from these disciplines. However, I view them as psychoanalytic concepts that refer to consciousness. Freud's metapsychology steers and determines the construction of his quasi-naturalistic theories which are employed in psychoanalytic interpretations. The mistake made here was that the psychoanalytic interpretations were systematized according to the natural-scientific metatheory *of that time*. But before one rejects or reformulates Freudian metatheory, a good deal of work is required in order to arrive at purely psychoanalytic, experiential interpretations of metapsychology.

Radnitzky chooses to view psychoanalysis as nomothetic, scientific research aimed at establishing general hypotheses, whereas I read it as a typical idiographic science. In this connection I want to make it clear that I have not tried to formalize my reconstructions. When formalized sentences do actually occur in my paper, they are invariably quotations from others' attempts at employing the Hempel-Oppenheim or Törnebohm explanation in an extensional language. I have—as should by now be obvious—always emphasized that this type of explanation is bound to fail when put to use within intentional psychoanalytical contexts. Nor am I capable of conceiving of psychoanalysis as a "general psychology" (*allgemeine Psychologie*), which is an obscure concept.

I shall persist in using the term *emancipation* to designate the aim of psychoanalytic science, which is the emancipation from fallacious apprehensions, for example, infantile conceptions. Emancipation from displeasure, anxiety, neurotic symptoms, and so on belongs entirely to the objective of psychotherapy described in terms of health values. (Emancipation from ugliness should then belong to the art of cosmetology and emancipation from poverty to some sort of economical technique, etc.)

I am grateful to Radnitzky for his ideas about [the] testing of psychoanalytic theories, but I shall not use this opportunity to take issue with his points of view, as it would lead us too far astray from my original article.

1. References

Klein, G. S. (1976). *Psychoanalytic theory*. New York: International Universities Press.
Lesche, C. (1973). On the metascience of psychoanalysis. *The Human Context*, *5*, 268–284.
Lesche, C. (1979). The relation between psychoanalysis and its metascience. *Scandinavian Psychoanalytic Review*, *2*, 17–33.

5

Psychology and Philosophy of Science

C. Sanders and H. V. Rappard

Abstract. In this paper the authors take a functionalist epistemological starting point. A fundamental idea is the concept of the *structuring of reality* as it takes place in science. The scientific community gives prospective scientists a dual training: a perceptual training which makes them see reality (the object of science) in a special way (structuring) designed to make it invariant, and a methodological training which makes them formulate knowledge in a way that is acceptable to the scientific community.

From this point of view we deal with the role that human values play in science, the difference between the various kinds of science, and the special character of psychology as science. Finally, from within this framework we also address differences between the formal structures of scientific languages.

In conclusion, we consider some methodological consequences for psychology, the suitability of the hypothetical-deductive method in psychology, and the possibility of other forms of verification in psychology. We reject both the "received view" and the idealistic conception of the incommensurability of theories.

1. Scientific Knowledge

1.1. The Problem of Knowledge

In this paper we will understand every form of knowledge to be ultimately derived from *action*, from *interaction* between the knowing organism and reality. This approach to knowledge is variously reflected in the history of philosophy and psychology and can also be found in

This paper is an abbreviated and reworked version of the authors' *Tussen ontwerp en werkelijkheid* (*Between conception and reality*) published by Boom, Meppel, The Netherlands, 1982. Translated from the Dutch by Jane Debrot.

C. Sanders and H. V. Rappard • Department of Theoretical Psychology, Free University, De Boelelaan 1081, 1007 MC Amsterdam, The Netherlands.

contemporary thought, for example, American functionalism, the cultural historical school represented by Vygotsky, Leont'ev and others, Piaget, and the Dutch philosopher C. A. van Peursen (1958). Van Peursen points out that "knowledge is first and foremost a relation, an interconnection" (Van Peursen, 1958, p. 41). According to him, the pragmatists were correct in emphasizing the relational character of knowledge, but they took a wrong turn when they interpreted thought and the search for truth as focussed solely on utility. He argues:

> [K]nowledge should be functional. Often either reality is defined as conditional upon knowledge (especially in idealistic schools of thought), or knowledge is defined as conditional upon reality (especially in realistic schools of thought). But in fact both knowledge and reality are equally familiar or unfamiliar. Everyday life experience and ordinary language do not suggest that either one of them is familiar or unfamiliar; but subsequent philosophical reflection makes it so. Knowledge and reality are inseparably interwoven. If it is realized that knowledge is always incomplete, that further disclosure is always possible, then reality gives scope to the process of knowing. Isolated from every possibility of knowledge the word *reality* loses its meaning. (Van Peursen, 1958, p. 34)

Reality channels the process of acquiring knowledge. As such reality is a *normative* principle.

1.1.1. Implications

From this point of view the acquisition of knowledge may be described as a *historical* process of interaction between man and reality. The way in which we as adults perceive reality as a structured whole is highly influenced by the store of knowledge that we have accumulated in the past. Cognitive structures which have proven adequate in past interactions ("systems of transformations," Piaget calls them) function as perceptual schemata through which sense impressions are formed into meaningful gestalts. However, the order introduced in this way does not constitute a *necessary* structure. In principle it is possible to find other structures that might offer a more favorable point of departure for the acquisition of knowledge. Moreover, in concrete situations our perceptual schemata leave room for several interpretations of what has been perceived. The way in which the world appears to us, our world view, depends on the culture and society in which we live and on the specific store of knowledge that we have acquired during our personal history. From our point of view ahistorical perception is not possible. Animal "knowledge" is also functional but even so there are large differences between humans and animals with respect to knowledge. Man's uniqueness lies in the fact that he is an *animal symbolicum* (Cassirer), that is to say, he uses symbols to represent reality.

It is true that higher animals also attach symbol values to sounds and objects, but in human reality symbols play a role of a different order. Says Von Bertalanffy (1967),

> Whatever else the psychology of a monkey, a rat or a sea urchin may be . . . their universe is one of physical things, food, obstacles, enemies and so forth. Man lives in a symbolic world of language, thought, social entities, money, science, religion, art—and the objective world around him, from trivial surroundings to books, cars, cities and bombs, is *materialization* of symbolic activities. (p. 22)

Among the symbols that are found in human reality, *language symbols* are the most important. Piaget especially—but Soviet psychology as well—has emphasized the fact that language opens the way to conceptual representation of reality, which in its turn leads to new possibilities in the interaction with the environment. Concrete action can be postponed and preceded by symbolic or internal action or, in other words, by thinking. The advantage of symbolic action is its flexibility and the fact that it may be more easily corrected than actual behavior. Moreover, to be in possession of a conceptual apparatus means that it is possible to escape from the limitations of the actual moment and situation, from the limitations of the here and now in which animals live; and although we also depend primarily on the here and now because of our corporeality, our thinking enables us to live in the past or future or in any other conceivable situation since our thinking is limited neither temporally nor spatially. The extent of reality with which we can interact is thus much larger than that of the animals.

Through the use of language symbols humans are capable of a high degree of abstraction. They can unite individual things, individual characteristics, and individual relations into single concepts, thus recognizing interrelations within reality. They can use abstract knowledge, developing constantly, in their interaction with reality, and thus their adaptive behavior, which is already quite flexible in comparison with that of animals, becomes yet more adequate. In complex situations man is able to solve problems whereas animals find themselves without resources.

A further implication of our point of view is that knowledge is *meaningful* and hence the acquisition of knowledge cannot be seen without considering the question of *norms* and *values*. The concepts of meaning, value, and norm are closely interrelated. It could be said that something has value for someone only if it is meaningful for that person and that we can talk about what someone ought to do only if he is confronted with meaningful alternatives.

We will understand the concept of meaning to be a functional one, namely, to understand the meaning of something is to know how it functions in the interaction between man and reality. Things functioning only marginally in a person's life will also have only marginal meaning to him. He or she hardly notices them, hardly *knows* them. Meaning and knowledge are two sides of the same coin. In *meaning* the subjective aspect is emphasized, in *knowledge* the objective aspect, but both are inseparably connected. Thus values and norms always assume cognition.

Values constitute *goals* to be pursued; *norms*—a concept that is often mentioned in the same breath with values—constitute *criteria* for human behavior and knowledge. They prescribe what should be done in certain situations; they determine whether certain insights are true or false, whether they should be considered good or objectionable, and they determine which values (goals) one should try to realize.

Not only are values and norms implied in cognitively regulated behavior; they are also known in themselves, even though one cannot always formulate them explicitly. They constitute a peculiar store of knowledge which precedes the acquisition of other knowledge.

In human life norms and values play an important part. Because we can transcend the *hic et nunc* in our symbolic, internalized (cognitive) behavior, we are not limited to instinctive or habitual behavior patterns. Decisions concerning intentional behavior are largely controlled by more or less explicit values. In the psychology of Eduard Spranger (1929) we even find an attempt to understand (*verstehen*) man on the basis of six basic values or *Lebensformen.* The Dutch psychologist Duijker (1972) distinguishes four systems of behavior determinants: motivational determinants, operational determinants, limitative determinants, and *normative determinants.* By motivational determinants he has in mind those determinants that cause an optimal stimulus level; operational determinants are based on acquired habits; and limitative determinants are behavior-determining factors connected with limits imposed on the behavior of an organism. These three kinds of behavior determinants may be found in the human as well as in the animal realm. Normative determinants, however, are specifically human. According to Duijker, they play the major part in determining human behavior. "Perhaps a fundamental human possibility (or necessity) is at work here, that is to say, the possibility of structuring our own world, a world which is not preformed on the basis of instinct, as in the case of animals" (Duijker, 1972, p. 134).

In this paper we will also take the point of view that cognition is inherent in norms and values. This implies that it is possible to speak of norms and values only if meaning is involved; moreover, it implies that norms and values may be constituting and directional factors in the acquisition of knowledge.

1.2. The Acquisition of Knowledge in Science

The acquisition of scientific knowledge is a human activity which may be characterized as a special case of the way in which knowledge is acquired generally.

Science approaches reality (1) from a selective perspective and (2) in a systematic way. *Perspective* means first of all that in the process of acquiring scientific knowledge certain phenomena are selected which will constitute the subject of study. But in the empirical sciences, besides selecting specific categories or sets of objects, a second kind of selection is also found. In physics, for example, we do not study specific sets of objects, but reality as a whole; however, a certain *aspect* of reality is always selected: the material, spatio-temporal aspect. In so doing we disregard other aspects of meaning, which we do attribute to reality in our everyday life. In this sense choosing a perspective means "reducing" reality, abstracting a certain aspect of it.

In those sciences in which (unlike classical physics) we focus our attention on a certain part of reality (on certain sets of objects) we are reducing reality too. The classical behaviorist chooses as his subject matter human and animal behavior, not as intentional behavior, but as physical movement. Not only do we select our subject matter and reduce reality to some extent (*what* do we want to investigate), but we also *regulate* the process of acquiring knowledge: we set down rules regulating *how* to investigate this abstracted reality.

1.2.1. Perceptual and Methodological Training

The pursuit of science requires *perceptual* (Koch, 1964) and *methodological* training. On the basis of this training we develop perceptual and methodological schemata which allow us to select a subject of investigation and to study this subject according to the appropriate rules.

For the most part these schemata function implicitly. They constitute the conditions for the process of acquiring scientific knowledge and, as such, they are hidden from view while we are engaged in scientific activity. The implicit character of these schemata derives in part from the fact that in the course of time they are taken for granted, but also because they have never been explicit. We could say that they have their roots in paradigms, in the sense of weltanschauung (Masterman, 1970).

By considering the significance of the results of scientific investigation, and because problems and controversies with other scientists arise in the course of this investigation, we begin to reflect on these schemata. This is a process of bringing them into consciousness, and in so doing implicit schemata become partially explicit so that they may be corrected.

Reflection leads to a sharpening of our perspective. This reflection concerns first of all the subject on which we focus our attention (the *what*). The more subtle the perceptual scheme, the more structured the cognitive scheme will be through which we perceive reality, and the more *invariant* (Rappard, 1979) the subject of investigation will prove to be. At the same time this sharpening offers the possibility of extending our methodology (the *how*).

1.3. The Acquisition of Knowledge as a Functional Structuring

1.3.1. Functional Structuring

Elsewhere we have given the name *functional structuring* to the process of the scientific acquisition of knowledge (in what follows we shall speak in short of *structuring*). We described that concept as a framework "*of which* the scientist expects that it will harmonize with reality, *in which* he organizes reality on the basis of an at least partially explicit point of view, *through which* he sets aside other possibilities of giving meaning (reduction), and *in the context of which* he studies action and reaction of the objects under investigation constituted in this structure" (Sanders, Eisenga, & Rappard, 1976, p. 375).

If we analyze this formulation the similarity with what has been discussed before will be evident. The concept of a framework is the (objective) pendant of the (subjective) perceptual scheme. We have already indicated that a perceptual scheme, functioning as a point of departure, is but partly explicit; and the reduction mentioned in the above description is similar to what earlier was called *selection*—particularly in the second sense of the word (aspect of reality).

The definition distinguishes between the action and the reaction of objects. This distinction will prove to be related to the nature of the subject of investigation. Physicists, for example, study reacting bodies, whereas the organisms studied by biologists, especially the higher organisms, do more than react; they manifest independent activity, which is described as action.

The above definition has two features which demand further consideration. It concerns the adjective *functional* in the concept of functional structuring in the part of the definition which states that the scientist expects a particular way of structuring reality to *harmonize* with that reality. Accordingly, these two features of the definition are interdependent. The adjective *functional* is a further qualification of structure. Structuring may be described as a cognitive ordering of a certain part of reality. As such, this structure is a *realization* (*realizing* in the sense of charting or mapping out) of that area of reality. Seen in this way, structure belongs to the domain of the object. The adjective *functional,* on the other hand, does not point to an area of reality, but to the investigator, to the person who structures. He has not introduced an arbitrary structure; he (or his scientific school of thought) has certain intentions in mind. The structure in question is functional, that is to say the structure is in harmony with assumptions which the investigator takes for granted concerning the possible cognitive development and practical use of the chosen structure in realizing his objectives. The concept of function and its derived adjective *functional* are used in this paper in a way that is rooted in our views on the process of acquiring knowledge. As we have mentioned earlier, this view is closely related to the approach of those thinkers who emphasize the connection between knowing and acting and at the same time emphasize the importance for the acquisition of knowledge of the process of interaction between man and reality in the course of history.

Again, the notion of a framework as harmonizing to a certain extent with reality is in keeping with this view. The investigator relies on the fact that a certain way of structuring will not strain reality but will prove to be adequate, that is to say, it will be a good "fit." To a certain extent he has an *a priori* view of the nature of that part of reality that was chosen for investigation, and this view is of decisive importance for the way in which scientific investigation is conducted. Thus there is here a close connection with the problem of values and norms.

1.3.2. Levels of Structuring

When the types of functional structuring in the various empirical sciences are compared, we become aware of an overall hierarchy. As we climb this hierarchical ladder, the structural level will become higher, which is to say that the possibilities of attributing meaning will be relatively more numerous. The exact sciences appear to be able to dispense with this *meaning variance* rather easily, but in psychology, history, sociology, and so forth this is, generally speaking, not the case. Here it proves to be more difficult, if not impossible, to arrive at a tolerable degree of meaning invariance of the subject of investigation. A high level of structuring corresponds to a relatively high degree of meaning variance, whereas at the lower levels a relatively high degree of meaning invariance is found. In other words, we suggest that there is a hierarchical order of sciences along a continuum of relative (meaning) invariance to relative variance.

At this point some remarks should be made about the concept of reduction. We have already introduced this concept as the rendering invariant of the subject of research. We are now in a position to make a further observation.

It was mentioned that as one draws nearer to the level of spatiotemporal materiality, it becomes easier to render the subject of investigation invariant. As a result sciences at a higher structural level often use cognitive schemata borrowed from lower levels. Thus in psychology physicalistic reductionism and physiological reductionism can be found. So there is a second, additional sense of reductionism, namely, to revert to lower structural levels. Clark Hull's behaviorism offers an example of physicalistic reductionism in psychology.

The levels of structuring may not be seen as isolated from each other. Because of their common basis in concrete reality they show a hierarchical interconnection. This is also the case *within* psychology. In psychology various ways of structuring may be distinguished at different hierarchical levels. For example, in experimental psychology, perception is at a lower level than in social psychology.

No psychological subject can be pursued creatively in isolation from its neighbors. Royce (1976, p. 2) calls psychology, among other things: "multi-methodological, multi-paradigmatic, multi-theoretical, multi-disciplinary." Duijker (1980) speaks of a "pluricontextual" psychology. And, in fact, in every kind of psychological investigation a number of other sciences are involved. First of all, within the cluster of psychological disciplines we shall have to take into account the results of those disciplines situated on a hierarchically lower level. The lower-level disciplines limit the disciplines situated at a higher level: physiological processes limit psychological processes; but the converse does not hold true. The same relation applies between physical and physiological phenomena. On the other hand, sciences of a higher level may be of importance for investigation in sciences of a lower level (see Fortmann, 1971, p. 164ff., Leont'ev, 1979, p. 219ff.) For example, the Russian scientist, Gregory Razran (1972, p. 310ff.) assumes a hierarchy of behavioral possibilities which are related to the development of the neural network. The higher behavioral possibilities are not without relevance to the lower ones; they color them, so to speak. Also, Van Olst (1971) showed that basic processes like habituation and dishabituation (lower) may be influenced by cognition (higher).

1.4. Knowledge as Human Activity and the Normativity of Science

1.4.1. Internal Normativity

As was described above, knowledge, and also scientific knowledge, is acquired in and by human action. In science, regarded as a body of formulated knowledge, one is dealing with symbolic or internalized action. In the formal sciences this internalized action takes place in a strictly formal manner, which is to say that in those sciences "empty" operations are applied in which one abstracts from the semantic and pragmatic functions of the symbols used, namely, one abstracts from their meaning and the part these symbols play in the concrete process of communication. The symbolic operations are of a purely syntactical nature, and one operates by means of rules that refer exclusively to the graphical form of the symbols.

In the other sciences both syntax and semantics play a part. Here the laws and interrelations within a certain discipline are described. Insofar as the different disciplines approach reality from one basic perspective, they should be distinguished from philosophy (i.e., from metaphysics), which attempts to describe reality in its complete interrelatedness. But in neither of the cases mentioned does the pragmatic function play any part. Science, in the sense of knowledge formulated in language, is as such not interested in "origins, uses and effects of signs within the total behavior of the interpreters of signs" (Morris, 1955, p. 219).

This is not to say that scientists are not interested in the effects of their work and do not take those effects into account in their publications. Science *as such*, however, abstracts from the concrete researcher and his views and intentions; science is only interested in the *knowledge* which he formulates. As we move away from formal operations action becomes more concrete. In view of this we may wonder what this means for the question of norms and values: could we call symbolic action, as contrasted with practical action, norm and value free?

We cannot. Indeed, even logic is a normative science. The science of logic offers rules or criteria enabling us to judge whether our reasoning is correct. It is worth mentioning that these norms concern operations as such and disregard the changes which we might want to bring about in reality and the goals (values) which we might want to pursue. These norms are not concerned with values *external* to the subject of investigation, but only with *internal* values. They concern correct scientific activity as such.

In many other sciences the norms are *internal* as well insofar as they are concerned with correct syntactical and semantical activity. Of course, it could be called common knowledge that the sciences are not value-free. The choice of subject matter and especially of the perspective from which the subject is to be studied are rooted in values. Max Weber called this the *Wertbezogenheit* of science. But just as it proved to be the case in the formal sciences, so here it is a question of internal normativity concerned with scientific activity *after* a perspective has been chosen. Here the meaning of norms is synonymous with methodologically correct action *within* a functionally structured part of reality.

1.4.2. External Normativity

When in the foregoing we spoke of science, we had *pure* science in mind, that is to say, we were concerned with sciences which are stimulated in their development by questions which arise internally: questions which are derived from the system of the discipline itself, or as stated by de Boer (1980, p. 155), questions which are derived from *theory*. External factors of a more practical nature, like applicability, social relevance, and externally imposed norms, do not apply, or at least they do not apply directly in the case of pure sciences. That is to say, we understand the concept of pure science to be that form of scientific activity that has been purified as much as possible of external-normative and practical aspects. Besides pure science, forms of science are conceivable where, although the topics are internal (that is to say, they are primarily related to the system and theory of a particular discipline), still the scientist deliberately and explicitly takes a point of view which falls outside of the scope of science. This point of view then constitutes the "ground value" (Rappard, 1979, p. 140) of that particular science. For example, in the past one could find psychologies that took their point of departure in the Christian world view (Brennan, 1946; Waterink, 1941). The psychology of the Soviet Union offers recent examples of a science with a Marxist ground value (Gal'perin, 1980; Leont'ev, 1979).

It is in the nature of the case that such external-normative sciences (in what follows for short: *normative sciences*) run the risk of giving greater weight to their point of departure than to empirical fact or, in other words, they run the risk of becoming dogmatic. Although the relationship between weltanschauung and theory on the one hand and empirical fact on the other cannot be said to be without problems, we feel that normative sciences have the tendency to stress unduly the importance of the chosen ground value at the expense of empirical fact. Thus, writers on Soviet psychology such as Payne (1968) and McLeish (1975) find here many positive aspects, at the same time regretting the dogmatic character of Soviet psychology. In spite of this objection, which according to us is an inherent difficulty in normative science, we want in principle to admit the possibility of such a science. However, the question as to where a topic of investigation comes from appears to us to be of less importance than the question whether the investigation meets scientific criteria. "How somebody comes by his ideas is in the context of justification six of one and half a dozen of the other; credentials are not called for" (Hofstee, 1980, p. 17).

We shall now unite pure science and normative science into the one concept of *systematic science* because the topics of investigation are derived from the system or theory of the discipline, in contrast to the practical field.

Systematic science should be distinguished from metatheoretical science (i.e., metascience, metapsychology). By the latter term we refer to sciences that have as their subject the empirical sciences. In metapsychology, for example, one reflects on psychology and is specifically concerned with the foundations of that discipline.

Besides systematic psychology there is psychology that is primarily focussed on certain practical fields such as labor and organization, education, psychotherapy, and so on. It is characteristic for field psychology that the topics of investigation are derived from real-life settings and hence are external and derived from a field outside of systematic psychology. They are *field-induced*. However, this does not imply that field psychology is less scientific than systematic psychology, and this has two reasons. For one, because, as was just stated, the degree to which the acquisition of knowledge is scientific is not determined by the origin of the problem but by the manner in which the investigation is conducted. And, second, it should be remembered that field psychology is *not* applied psychology. Applied psychology refers to the application of knowledge acquired by (systematic or field) psychology. In field psychology as well as in systematic psychology one is concerned with acquiring knowledge. In both cases it can be said with Drenth (1980, p. 12) "that the criterion whether psychological investigation is scientific or not is found in the question whether it is meant to lead and in fact does lead to an increase in our knowledge of behavior."

Within field psychology the same distinction can be made as before within systematic psychology, that is to say, between *pure field psychology* and *normative field psychology*. We shall refer to the latter as *praxiological psychology*. "Empirical sciences," wrote Wolman and Nagel (1965), "concern themselves with propositions that are true or false, while praxiological sciences . . . use sentences describing what ought to be done and how to do it" (p. 4).

The objections which were raised against normative science evidently also apply to this form of science. It should be noted that in praxiological sciences, as in Holzkamp's (1973) critical psychology, and even more so in the many psychological "gospels of salvation" under the guise of psychotherapeutic theory, the realization of certain ground values of an ideological nature has gained the upper hand at the expense of the scientific acquisition of knowledge. In such cases one does not even find an application of scientific knowledge anymore, but here personal value and/or social utopia have become the determining factors for praising certain behavior as "psychological." Such activities bear little resemblance to science, praxiological or otherwise.

In our opinion there is reason to be even more reticent toward praxiological sciences than toward normative sciences. But here again, and for the same reasons, we shall not preclude the possibility of such a science.

2. Science and Reality

2.1. Natural Sciences and the Humanities

About the turn of the century it was thought possible to separate the various sciences of reality into two categories: *the humanities* and the *natural sciences*. The natural sciences were supposed to investigate a system independent of man called *nature*; the humanities were supposed to direct their attention toward social and cultural reality, in other words, the reality formed by human beings in the course of history.

Closely connected with this distinction, which is considered by many to be of a fundamental nature, is the problem whether science is or is not value-free. As above, no form of knowledge is value-free and that is why this distinction must be rejected. On the other hand, insofar as a science chooses as its subject a product of the disclosure of reality as manifest in cultural development, we do think that norms and values will play a more dominating role. Subjects of investigation which are themselves the products of human culture cannot be understood (*verstehen*) if the human values and norms that are expressed in them are not recognized.

Moreover, it becomes increasingly difficult for the scientist to have a neutral outlook as the phenomena investigated are a more outspoken product of culture. The role of an impartial judge, recommended at some time by many philosophers of science, becomes impossible. One can quite safely distinguish between nature and culture as long as it is a distinction of degree. The concept of nature is a boundary concept denoting that which is basically given and as such has only a marginal meaning. But this does not imply that we are not in a position to say anything about nature in a scientific way. On the contrary! The knowledge in question, however, is not knowledge about nature-in-itself but is a product of the human disclosure of nature and as such cannot be called value-free.

2.2. Nature as the Given

In developing the above ideas, we shall take a nonidealistic, commonsense point of view, taking the concept of nature to denote something concrete as the foundation for the disclosure of reality.

In this connection we would like to draw attention to the work of P. F. Strawson (1958, 1959). Strawson makes an attempt at writing a *descriptive metaphysics.* He does not want to make existential judgments about nonempirical subjects like the existence of reality as such; he simply wants to give a description of the way in which we speak and think *about* reality. His intention is "to lay bare the most general features of our conceptual structure" (1959, p. 9). Contrary to what he calls revisionistic metaphysics, which tries to improve our thinking about the world, his point of departure is concrete talking and thinking about our world of everyday life. In doing so he follows a tradition in British philosophy which often finds a point of departure in commonsense notions. In his investigation Strawson (1959) turns his attention to

> [the] massive central core of human thinking which has no history—or none recorded in histories of thought; there are categories and concepts which, in their most fundamental character, change not at all. Obviously these are not the specialities of the most refined thinking. They are the commonplaces of the least refined thinking; and are yet the indispensable core of the conceptual equipment of the most sophisticated human being. (p. 10)

2.2.1. Particulars

Strawson (1959) starts out with the simple fact that we talk and think about the world

> [as] containing particular things some of which are independent of ourselves, and we think of these particular things . . . as included in the topics of our common discourse, as things about which we can talk to each other. These are remarks about the way in which we think of the world, about our conceptual scheme. (p. 15)

Particulars, according to Strawson, are concrete things and persons, "people and their shadows," as well as events, states of affairs, processes, and so on. However, properties and qualities, numbers, kinds, in short anything that can be predicated of something else (predicates) are not particulars. Each of us possesses a specific conceptual scheme with the help of which we can identify particulars, think about them, and communicate about them with others. The conceptual schemata of people appear to be largely similar. This overlap of conceptual schemata is not unaccountable from the point of view of the ideas which we developed earlier. Concepts that allow us to identify particulars are, as argued above, the result of the activity of a human community disclosing reality. In relating to reality particulars have arisen and the intersubjective identification of particulars has led to the development of names and concepts allowing us to reidentify them. According to the philosopher Quine (1960), these kinds of terms possess "built-in modes, however arbitrary, of dividing their reference" (p. 91). These modes are ways of delimiting and sorting those states of affairs which are to be understood as things.

The overlap of conceptual schemata does not mean that every person has an identical conceptual scheme. Firstly, it will be found that constant growth takes place in individual conceptual schemata. That is why one person may have a more clearly differentiated conceptual scheme than another. Moreover, the differentiation does not concern the same domain for each person. It is also possible that in the course of development a shift takes place. What all this boils down to is that there are differences between people concerning their ability to identify particulars.

2.2.2. Identification of Particulars

Generally speaking, it is a sufficient but not a necessary condition for common identification that the listener be able "to pick out by sight or hearing or touch, or otherwise sensibly discriminate" the particular in question. This most simple case of identification occurs when the particular in question can be directly located: *demonstrative identification.*

However, as a rule this is not possible. More often than not someone refers to a person, object, or event that is not present, without the listener having any problem in identifying the particular in question. Apparently the listener is acquainted with a particular that fits the description given by the speaker. But such a *nondemonstrative identification* has its foundation in the aforenamed demonstrative identification. However indirect the identification of particulars may sometimes be, basically, according to Strawson, "all identifying description of particulars may include, ultimately, a demonstrative element" (1959, p. 22).

In this sentence one of Strawson's crucial points is to be found (and this is also the reason why his line of thought is of interest for the section on logic, which is still to follow, where demonstrative reference, *denotation*, is a central issue). According to Strawson, for each concretely demonstrable particular there is a possible description which connects this particular in a *unique* way to all the other particulars which are relevant in a certain universe of discourse. But such a unique, that is to say, individuating description is unthinkable without a structure, a system of relations which embraces all the directly demonstrable particulars. Indeed "for all particulars in space and time, it is not only plausible to claim, it is necessary to admit, that there is just such a system: the system of spatial and temporal relations, in which every particular is uniquely related to every other" (Strawson, 1959, p. 22). The existence of this system of relations has the advantage that it is hardly ever necessary to demonstrate explicitly the particulars one is talking about. It is this framework that plays an essential role in each identification and that is increasingly extended as new particulars enter into our field of view, even if we have never ourselves had any sense perception of them. But for all these extensions and systematic structures the system of spatio-temporal relations is the foundation upon which this extension takes place.

2.2.3. Basic Particulars

According to Strawson the framework that is essential for the identification of particulars cannot be thought of independently of the concrete objects within our universe of discourse. "If we ask what constitutes the framework, we must look to those objects themselves, or some among them" (Strawson, 1959, p. 39). From this it follows that, because we are concerned with a spatio-temporal framework, only those particulars can constitute such a framework the properties of which coincide with this fundamental characteristic, that is to say, they have to be publicly observable, three-dimensional objects with at least some duration in time. This means that "things which are, or possess, material bodies must be the basic particulars." The addition "or possess" is important because it means that persons, too, belong to the class of basic particulars and, according to Strawson, "persons and material objects are what primarily exist."

It is necessary to take another look at the first category of particulars (persons) because it is especially important for psychology. Strawson attacks the old but still persistent Cartesian dualism by stating that the concept of a person is a primitive (irreducible) concept. States of consciousness and corporeal characteristics may be ascribed to a person, but one cannot split a person up into these states of consciousness and characteristics (taken as substantialized, that is to say, particular states and characteristics):

> The concept of a person is to be understood as the concept of a type of entity such that *both* predicates ascribing states of consciousness *and* predicates ascribing corporeal characteristics, a physical situation, etc., are equally applicable to an individual entity of that type. (Strawson, 1959, p. 104)

In this connection Strawson speaks of *P*(erson)-predicates and *M*(atter)-predicates. Thus we ascribe *mental acts* and *intentions, sensations, thoughts, feelings, perceptions,* and *memories* to ourselves. These are *P*-predicates. Although not every *P*-predicate describes a state of consciousness, it may safely be said that what all *P*-predicates have in common is the fact that they imply the possession of consciousness.

An example of a *P*-predicate not directly ascribing a state of consciousness to a person is "going for a walk" (Strawson, 1958, p. 343).

> [I]t is essential to the character of these predicates that they have both first and third person ascriptive uses, that they are both self-ascribable otherwise than on the basis of observation of the behavior of the subject of them, and other-ascribable on the basis of behavior criteria. (Strawson, 1958, p. 346)

Strawson argues that these *P*-predicates are primitive, that is to say, irreducible to *M*-predicates.

As far as persons are considered material bodies, one can ascribe *M*-predicates to them, for instance, relatively permanent corporeal characteristics, such as height, form, weight, and the like. We do so on account of public observations; no private experience is involved in these cases.

In general it can be said that *M*-predicates are "properly applied to material bodies to which we would not dream of applying predicates ascribing states of consciousness" (Strawson, 1958, p. 342).

2.2.4. Person and Organism

The question arises how one should think of organisms which cannot be called persons. May only *M*-predicates be ascribed to them? Strawson does not enter into this matter explicitly. Concerning the identification of "private particulars," he says that it depends upon the identification of the set of persons. But he adds something to this: "Perhaps we should add 'or animals'; for perhaps we sometimes refer identifyingly to the particular experiences of animals. But this is a complication I shall neglect" (Strawson, 1959, p. 41). From this quotation it appears that Strawson wants to maintain the difference between animal and person, but without denying mental events to animals. One would be inclined to conclude from this that *P*-predicates are applicable to animals as well. But does this not conflict with the specific logical character of the person-predicates?

The logical difference between *M*- and *P*-predicates is that the latter cannot only *be ascribed* to the appropriate "material bodies" but that these material bodies can in principle *ascribe these predicates to themselves* other than on the basis of behavioral criteria. And may one expect animals to be able to do so? At best animals possess a "shadow-like inner life" (Buytendijk, 1961).

In Strawson's work points of departure for a possible solution may be found. Talking about the specific logical character of *P*-predicates (Strawson, 1958, p. 345) he adds within brackets: *"or at least of a crucial set of P-predicates"* (italics ours). One could say that the noncrucial *P*-predicates may be ascribed to the higher species of animals. In the case of animals one cannot speak of an explicit self-ascription of these *P*-predicates, but one does find with them some form of knowing about what is designated by the *P*-predicates "other than on the basis of behavioral criteria."

In a similar vein Harré and Secord (1972) attempt to ward off the danger of a rigid discontinuity between person and nonperson. They suggest thinking in terms of "a spectrum or ordered series reflecting more and more stringent conditions (culminating in full self-consciousness and self-awareness) for calling a thing a person" (pp. 7 and 110). The crucial *P*-predicates are ascribed to those material bodies which satisfy the more stringent conditions.

This means that according to Harré and Secord "sufficiently advanced robots might be considered people for some purposes; even a talking chimpanzee might be similarly regarded" (p. 8), so that under certain conditions *P*-predicates may also be ascribed to nonpersons.

For Harré and Secord the central condition for being a person lies in the condition that Hampshire (1965) has formulated, namely, that an individual should be equipped with the possibility to say what he is doing. This criterion has less emphasis on the functioning of attention and consciousness than is the case in Strawson's work, while at the same time the typically human possibility of self-reflection remains decisive.

2.3. Natural Sciences and the Given

The world we live in could be seen as a disclosure of concrete, basic reality. In relating with the world of basic particulars these particulars have increasingly acquired a human functional meaning in the course of the development of culture and personal life-history. But this is not to say that the reality of the basically given itself does not have a functional meaning, but only that it is marginal.

Where material bodies are concerned, it is a question of meanings which relate to the fact that the human being himself possesses a material body. That which is basically given is thus far away or near to him, it is an obstacle to him, it is sensible and impenetrable, and so on.

Where persons are concerned, the marginal functional meaning is also connected with the fact that the subject is a person too; within the concretely given he distinguishes between particulars which actively address themselves to him, as he does to other particulars, and particulars that do not manifest intentional activity. Thus the concept *person* does not coincide with human persons, even though they will be preeminently persons.

One might suppose that the natural sciences attempt to describe basic reality as such, insofar as basic reality concerns material bodies. But this is not the case. Even in the natural sciences one is concerned with a product of culture, with a specifically scientific and thus human way of disclosing reality. The difference between the natural sciences and the humanities lies in the fact that the regulated, methodological, and scientific way of disclosing reality in physics starts with basic reality, whereas in the humanities one starts with products of culture that have been formed in the course of history. In physics one does not end up with a true-to-nature likeness of a real or absolute reality, but with a blueprint in abstract and hypothetical terms of a way in which man can interact with concrete, tangible reality. It is thus incorrect to identify the *physical* reality with the *concrete* basic reality of Strawson. This is all the more important to remember as concrete reality includes not only material bodies but also persons.

2.4. The Sciences and the Distinction Person Versus Material Body

2.4.1 General Observations

In the definition which was given of functional structuring the following passage is found: "in the context of which (the investigator) studies action and reaction of the objects under investigation constituted in this structuring" (see §1.3.1). One may wonder whether this definition is applicable to all *Realwissenschaften* (empirical sciences). For not only are the natural sciences comprised under this heading, but also such typical human sciences as linguistics, law, aesthetics, and so forth. One can hardly object to the reaction of physical bodies or to the action of humans, but what sense does it make to speak in terms of reaction or action when studying icons? If an icon is regarded as independent of its maker, then words like reaction have no meaning. But the question remains whether the objects of the humanities can be studied without understanding them as *products* of human activity. In the humanities, therefore, psychology can be taken as a propaedeutic science. We do subscribe to this view, however, without falling into psychologism. We are of the opinion that to reduce the humanities to psychology is to disregard the limitations of psychological structurings, but that is not to say that the humanities can ever be regarded as independent from human activity, which transcends the framework of psychology because man is the creator of that framework. Or, in other words, because man himself is the *condition* for scientific psychology.

But setting this aside, even though the objects of the humanities cannot be studied without paying attention to the human being which creates them, still the above passage from the definition of structuring is not literally applicable to the humanities. In this connection we shall add the following: "in the context of which he studies action and reaction of the objects under investigation constituted in this structuring *or in the context of which he studies these objects in their interrelatedness with human activity*."

Thus the concepts of action and reaction remain the crucial concepts in our argument. One may wonder whether the difference between reaction and action should be regarded as gradual or absolute. Do the activities of (higher) animals also have the characteristics of action? We would tend to answer this question in the affirmative and to defend a gradual distinction. At the same time it should be observed that even so there is descriptively speaking a large difference between typically human and animal behavior. Just think that man alone is a constructor of culture. The impressive organization of bee colonies cannot be called culture because it shows a rigid structure in which no development guided by reflection is inherent.

The distinction between action and reaction of course does not mean that human beings always act and never react. Human beings quite evidently do react, for they are also material bodies. The possibility to act on the part of the object under investigation, as is the case in psychology, poses special difficulties for scientific investigation.

In psychology one deals foremost with acting persons, with *objects* that on the basis of this action structure their reality. Often this active character of the objects of psychology is emphasized by remarking that the factor of *situation* is specific for the social sciences. But one may wonder whether this is correct. For physical bodies, too, react according to the situation. Scientific laws offer insight into the ways of reacting (consequent variables), given certain conditions (antecedent or situational variables). So it can be said that even in physics the situation plays a part, although there the concept has a *minimalized meaning* which can be described as a "state of affairs in the spatio-temporal material reality." In psychological theory, however, when we speak of situation we are primarily concerned with a reality *perceived* or categorically apprehended by the object under investigation, a reality in which the concretely given is integrated in a world full of human, functional meaning.

Here, as a consequence, in experiencing reality it is not primarily a matter of states of affairs which one may fairly *record* in a material reality, but rather a question of activity in which the present is *interpreted* from the point of view of the past with an eye to the future. This apperception lies at the basis of all intentional activity. In those psychologies which are to some extent based on the phenomenology of Husserl, the concept of intentionality came to have a broad connotation. We would like to use the concept in this broad sense too, namely as an essential feature of human activity—a feature that is lacking in the reaction of physical bodies. Intentionality and human action/apperception presuppose each other.

At this point we should like to introduce a concept originating with Leibniz, namely that of "possible worlds." In the case of physical descriptions it is sufficient to review only successive actual situations. Physical, extensional language designates the events in the successive actual worlds and places them in a causal relation. Could the same thing be done in psychology? In psychology it is also a question of successive actual worlds. Contrary to events in physical reality however, this succession cannot be clarified according to causal laws. Real causal laws are not to be found in cultural-historical reality. Here it is a question of activity guided by rules. The succession of situations in which a human being finds himself can only be clarified if attention is paid not only to the actual situation but to a whole scala of *alternative situations* as well, that is to say, to those situations which are compatible with the actual situation. In other words, we have to take into account that human beings not only live immediately and factually but also have expectations and live in uncertainty with respect to future and actual situations. The requirement that the actual situation be seen against a background of possible situations does not refer solely to future developments but also to the actual situation itself insofar as this situation allows alternative interpretations. *What someone perceives is often compatible with different "states of affairs" and different "courses of events."* Both are forms of possible worlds. All this makes a strictly causal description inadequate since a causal explanation is possible only within the succession of unambiguous actual worlds. But as soon as one has to view the actual in light of the possible, intentionality is characteristic for the behavior of a human being and there the causality of the natural sciences will not be sufficient.

2.5. The Logical Structure of the Sciences

2.5.1. Extension–Intension

Scientific knowledge is knowledge that is formulated in propositions and thereby rendered objective. Does the logical structure of scientific languages change as one reaches higher structural levels? Before answering this question we shall first treat the concepts *extension* and *intension.*

The concept of extension could be translated with *extent* or *range.* When one speaks of the extension of a *concept* one is dealing with a reference to a set of given individuals. Thus the concept *star* refers to each of the members of the set of all stars. If one is dealing with propositions, then the extension is a factual state of affairs. In the case of extension one abstracts from the intension or *connotation.* Intension is understood to be the active aspect of a concept or proposition, the content, and the meaning consonant within language is said to be independent of any reference to empirical fact. Natural science, being empirical, has always had a preference for extensional languages in which the truth of propositions is decided purely by the state of affairs in empirical reality. Intensions are then said to be vague, mostly affective consonant meanings, which should be eliminated as much as possible.

As we move nearer to the basically given, which has only marginal interest for the investigator *personally,* it becomes easier to make use of an extensional language. But it remains questionable whether on the higher structural levels intensions can be eliminated.

2.5.2. The Logic of Propositions and Modal Propositional Logics

Two-valued propositional logic forms the skeleton of the language of classical natural science. Within this language it is especially the one-dimensional theory of reference (denotation) which is of importance. Two-valued logic is *extensional.* Two-valued propositional logic can be extensional because it uses *assertoric* propositions. It is a matter of a language which records, in which actual facts are described.

Modal logic is different in this respect. Besides assertoric propositions it makes use of, for example, *problematic* and *apodictic* propositions. Here we have gone beyond the limits of actual facticity. Other dimensions arise. *Thus in modal logic we take into account not only the actual facts of a course of events or state of affairs (the actual world), but other possibilities (possible worlds) as well.*

If propositions are provided with modal operators in which the truth or falsity of the propositions in question is modified in one way or another—for example as being apodictic or problematic—then one leaves the sphere of the factual course of events and one regards the truth of a proposition within the context of the totality of *all possible worlds* (De Wit, 1977, p. 45ff.). According to De Wit, a psychology which has its own dimensional level as a basis and does not, as does for example Hull's behaviorism, propagate a physicalistic science uses a logical language that has a modal structure. He illustrates this by showing (De Wit, 1977, Chapter IV), that the logical structure of the languages of humanistic psychology can be clarified with the help of the concept of "propositional attitude" (Hintikka, 1969, 87ff.), which may be found in formal logic.

2.5.3. Propositional Attitudes

Propositional attitudes or acts are concerned with states of consciousness or mental acts, which are intentionally directed toward propositions. These acts or attitudes are linguistically symbolized by verbs such as *seeing, thinking, believing, supposing, fearing, judging*. The characteristic feature of linguistic expressions in which propositional acts are described is that the verb is often followed by *that*: "I believe that. . . ." And according to Hintikka (1969, p. 174), where this is not so, as in the case of the verbs *to see* and *to understand*, verbs which also designate mental acts, the applied direct-object construction can be reduced to a that-construction. Edwards observes:

> The propositional acts or attitudes signified by those verbs are mental acts involving a dyadic relation between a subjective constituent of the subject's mind and an objective constituent consisting of the complex entity named by the noun "that" clause. (Edwards, 1972, Vol. 6, p. 495)

In propositional acts or attitudes human engagement in reality plays an essential role. That is why psychology which studies human behavior needs propositions which inform on this kind of behavior.

Physics, however, tries to avoid propositions containing propositional verbs. Physics attempts to abstract from the human activity that is realized in attitudes and regards as a scientific datum only the proposition to which a human being is directed in his attitude.

One cannot say that people who talk in their everyday lives of seeing, hearing, feeling, and so forth or use other nonperceptual *propositional verbs* "are ever interested in anything as fancy as 'possible worlds.' . . . Surely what they are interested in is just the unique world of our that happens to be actualized" (Hintikka, 1969, p. 154). Insofar as they are prepared for more than one course of events, however, or insofar as they choose in their perception for a certain interpretation (state of affairs), they are quite obviously dealing with different possible worlds. In general, reference to possible or thinkable worlds takes place tacitly. That is, what we say about our actualized world implicitly presupposes other possible worlds. In this connection Hintikka remarks, "[A] logician might say that we often succeed in saying something about the actual world only by locating it, as it were, on the map of all the different possible worlds" (1969, p. 154).

Now the question is whether this applies only to modal propositions. Does this not equally apply to many propositions within the two-valued language of classical physics? Hintikka thinks that this is indeed the case. In this connection he points to scientific predictions. These are possible if we can make a distinction between several different courses of events which can be subdivided into two sets, namely, those in which the predictions turn out to be true and those in which they are false. Therefore the specific characteristic of modal predictive propositions lies not so much in the fact that in them we take account of more than one possibility, but in the fact that for the verification or falsification of such propositions we have to take into consideration not just one possible world—as in the case of nonmodal propositions—but *"several possible worlds in their relation to each other."* When an investigator in classical physics makes a prediction, he is uncertain about the results. To obtain certainty however, it is sufficient to take into consideration the invariant (monointerpretable) spatio-temporal state of affairs at the point in time to which the prediction relates.

So even though possible worlds are found in modal as well as in nonmodal propositions, there is an important difference in the fact that in the last case one refers to only one possible world, whereas in modal propositions one always finds multireference. To illustrate this and so demonstrate that propositional attitudes have a modal character, we would like to make a few remarks on the *logic of perception*.

We could say that we know what someone perceives "at a given moment in time insofar as he can distinguish between states of affairs (at that moment in time) which are compatible with what he perceives, and states of affairs which are incompatible with his visual perceptions at the time" (Hintikka, 1969, p. 153). In passing we should like to add that every ascription of a propositional attitude (such as, for example, belief, wish, knowledge, hope) may be paraphrased in the same way in terms of possible worlds that are compatible with what someone believes, wishes, knows, hopes. The example of the logic of perception is certainly not unique in this respect.

The shape of a propositional attitude with respect to perception is as follows:

> *a* perceives that *p* =
> in all possible states of affairs compatible with what *a* perceives, it is the case that *p*. (Hintikka, 1969, p. 155)

A concrete example would be: *a* perceives a running man.

The compatible possible worlds which determine the content (intension) of the proposition are here all the possible interpretations of the situation at the moment of perceiving "running man" that do not contradict the proposition in question. In analyzing this construction one is concerned with possible answers to questions such as: who is it that is running? what is the cause of his running? what does he think to achieve by running? Thus the proposition in our example is multireferential in the sense that its contents can only be understood in relation to several possible worlds.

Propositional attitudes are clearly distinct from observational propositions in physics. These could be formulated as follows:

> *a* observes that *p* =
> *p* is the case in a concrete monointerpretable situation (only one possible world).

In the observational propositions of physics one also deals with perception, albeit not with the perception of an object where this object is a perceiving being as in psychology, but with the perceiving subject or investigator.

2.5.4. Closing Remarks

After the foregoing it is not difficult to explain what Hintikka means when he interprets the meaning of concepts (meaning, connotation, intension) as having *multiple reference*, that is to say, reference to (all) possible worlds. Following the ideas that were developed in this chapter, the lower-higher structural levels continuum that was earlier connected with invariance of meaning may now be described as a mono-multireferential continuum. According to De Wit (1977), *intensionality* and intentionality are related: that which is intentional in behavior can be described intensionally.

We subscribe to this view and would even like to go a bit further. It is well known that concepts and propositions may have more or less different meanings for various individuals (Osgood, Suci, & Tannenbaum, 1957). We should not interpret this as an argument in favor of the opinion that intensional meanings are nonreferential. We would like indeed to defend the opposite *by connecting intension or connotation and the (partly tacit) knowledge about the functional possibilities of the given.* The intension of an expression lies in its multireferentiality. We cannot escape from this multireferentiality when we are dealing with meaningful behavior (intentionality). A description of human behavior is meaningful only in the context of the realization of *possibilities,* which can only be called possibilities in the context of alternatives.

As the structural level is further reduced, that is to say, as it shows a relatively higher degree of *invariance*—or reference restriction—the intension (multireference) of the concepts will become poorer. On the other hand, the higher the structural level, the more connotations will be integrated in the concept. The more real to life a proposition is, the more "indwelling" will have to take place (Polanyi) to understand the meaning correctly.

3. Methodological Consequences

3.1. Functional Structuring as the Product of Object and Subject

In the foregoing we have argued that the term *psychology* denotes a cluster of sciences differing in kind and structural level. That is to say that in this chapter we cannot possibly give one simple answer to the question of the consequences of the foregoing. In order not to complicate our argument needlessly we shall limit ourselves for the time being to systematic psychology. We shall deal with the consequences for field and normative psychology in §3.4.2.

Human behavior can be studied as it is basically given. It is also possible to turn our attention to human activity as culturally and historically determined. In both cases one is dealing with the acquisition of knowledge each of its own kind.

In the first case one could speak of basic psychology or psychology in the vein of natural science, in the second case of a psychology in the vein of the humanities. In psychology, in which the basically given could be described as "things that *possess* material bodies," it proves to be more difficult to isolate the basic phenomena than in physics, where "things that *are* material bodies" are investigated. Although it could be defended that each of the fundamental disciplines that have been distinguished by Duijker (experimental psychology, theory of personality, developmental psychology, and social psychology) has its foundation in basic reality, it is in fact the case that only in a few branches of experimental psychology and of developmental psychology may we speak of a basic scientific psychology. As soon as it is a question of investigating higher and complex mental functions which go through a period of development, it is difficult to abstract the cultural and historical dimensions. Think, for example, of semantic memory, the functioning of which is partly determined by culture. Generally speaking, it can be said that three of the four fundamental disciplines (Duijker, see above) show an ascending structural level and along with this an increasing degree of cultural determination of their results of investigation. Developmental psychology is rather different in nature from the other three for psychological functions as well as persons, and social behavior may be investigated from the point of view of their development. Developmental psychology should be defined not so much in relation to its objects as in relation to its perspective. That is why parts of it may differ strongly with respect to their structural level.

From the above one should not conclude that the nature of the subject (psychological functions, personal or social behavior) wholly determines the structural level. The subject chosen may be looked at in several different ways, and only after allowance has been made for that factor can we get a complete picture of the kind of psychology we are dealing with. Object and subject together determine the kind of structuring that will take place.

3.2. Structural Levels and the Acquisition of Knowledge

3.2.1. Explanation and Understanding

The acquisition of knowledge, it has been argued, takes place through interaction between man and reality, which is to say that reality becomes structured reality, not only in a predominantly static sense (finding clearly distinguishable entities such as objects, organisms, persons, states of affairs, and events), but in a dynamic sense as well. Patterns of development can be distinguished that give us something to go by. To a certain extent reality is calculable and predictable. Scientific acquisition of knowledge attempts to increase our insight into this order and in so doing to increase our behavioral possibilities. As has been explained, this happens systematically and consistently upon a process of disclosing reality. This process gains a specific character according to the structural level used.

At the lower structural levels, which are distinguished by a high degree of invariance of the object under study (see §1.3.2.), the problem of interpretation plays a relatively minor role; without much difficulty intersubjective agreement can be reached about what shall be considered to be factual. We are here dealing with reacting objects which must be studied in the context of their actual situation. The emphasis is on *reaction*. The *entities* we deal with are as individual facts not important. They can be regarded as representative instances of the object under investigation, as they do not act on the basis of their individual nature and history, but react in a uniform way. At the lower structural levels the acquisition of knowledge can be narrowed down to searching for the external conditions that form a sufficient cause for certain reactions, and at this level causal laws can be looked for.

At the higher structural levels we find those sciences the objects of which do not belong to the sphere of basic reality but are selected from social and cultural reality. These objects are formed by aspects of human action or by the products of that action, like the objects of cultural sciences such as linguistics and law. In the following we shall leave them out of consideration. When we speak of sciences on higher structural levels, we shall mean behavioral sciences. These disciplines are characterized by a higher degree of variance of their data than sciences at lower levels and hence a reasonable response consistency is possible only among adherents of the same school of thought.

As has been said, the cause of this may be found in the fact that human beings act intentionally, and in such a case it is not sufficient to take the actually given as datum. Here data cannot be understood adequately within the limits of the moment. Sciences studying human behavior at higher structural levels will therefore be concerned with data which are different from those of the lower levels.

In the first place the data are never ahistorical, that is to say, nonintentional phenomena. Furthermore, this means that as one arrives at higher structural levels it becomes less and less possible to regard the data as representative of the phenomena to be investigated. The higher the structural level, the more a science becomes oriented toward what is individual. A scientist who wants to investigate the behavior of iron under heating and his colleague who wants to investigate the phenomenon of fear empirically are each working in a completely different context. About the turn of the century this used to be expressed by adding the adjective *idiographic* (uniquely describing) to sciences which are oriented toward what is individual.

In the third place, as one arrives at a higher structural level, it becomes increasingly difficult to distinguish between entities and reactions of entities. Here recognizing an object entails recognizing its function, its meaning in human intercourse. At this level the activity of objects is not a mere reaction, as is the case at the lower levels, but it is determined by a subject on the basis of his or her own nature, former experience, interpretation of the reality of the moment plus the expectation about the way in which that reality will develop. That is why psychology at higher structural levels cannot be reduced to the search for causal relations, for the final aim of psychology at these levels is the description of a datum as intentional behavior.

Around the turn of the century psychology at higher structural levels was called a science of understanding (*verstehen*). It was thought that understanding phenomena entails seeing them in their intrinsic interconnections or structure.

One could say that in the case of understanding the role of the subject is more pronounced than in the case of explaining (where one minimizes the influence of the subject). In this connection we would like to point to the role of intuition that cannot be eliminated from the process of understanding. In intuitive knowledge it is a matter of acquiring insights by inspiration. In the explanatory sciences this plays an important role as well. However, although here the product of intuitive knowledge can be separated from the process of knowing that goes on in the mind of the investigator, this is not so easy in the case of sciences which use understanding as a source of knowledge. In the explanatory sciences intuition brings forth conjectures about connections in basic reality. Moreover, they are suppositions of a predominantly analytical nature (it is a matter of linear connections between intentionally isolated quantities). In the cultural sciences a more complicated situation is found, not only because those sciences investigate aspects of a reality that have been disclosed by man himself, but also because the emphasis is on investigation of totalities (structures). When what we have "understood" is put into words, it will have a provisional, incomplete character. Understanding takes place in a relation with the process of tacit or implicit knowing of the person. The investigator as a person and his or her implicit cognitive world must be intuited if one wants to understand what he or she understands. At the lower structural levels this will be less of a necessity. There reality has methodologically been reduced in such a way that it is possible to have an invariant perception of it and the role of the investigator can be mostly disregarded.

3.2.2. The Unity of the Acquisition of Knowledge

The foregoing could suggest that we are propagating a methodological dualism. But this is not the case. Just as we have rejected the dualism of material bodies and persons, just so we reject methodological dualism.

As the object of investigation becomes gradually more historical and intentional as we climb the hierarchical ladder, the acquisition of knowledge does change its character, but the change is not essential. In all cases it is a matter of actively disclosing an *order* in reality, whatever the structural level of this order might be. Explaining and understanding are interconnected variants of one and the same process of knowledge acquisition. In this connection we should point out that the term *idiographic* which was given to psychologies at higher structural levels could create some confusion. It might be true to say that at those levels it is more difficult to regard a concrete datum as a member of a set of identical (re-)acting instances of the object under investigation, but that does not alter the fact that at those levels we are all in the same pursuit of *general* knowledge. In theory formation at higher levels, however, one does not look for general *laws* as much as for general *structures*, types, syndromes. The datum is recognized as a specific, or even a unique variant of such a general structure. Conversely, studying unique phenomena may contribute to refinement, modification, or recognition of general phenomena. In this connection one could think of phenomenological reduction (Husserl).

The unique as such cannot be described (De Groot, 1961, Popper, 1934). It can only be formulated by approximation with the help of the structures, types, and so forth that we mentioned.

Our next remark relates to the manner in which explaining and understanding occur in an integrated way. In this connection we would like to refer to the work of Polanyi (Polanyi, 1958; Polanyi & Prosek, 1975). Polanyi explains that at a certain structural level understanding *includes* the explanatory knowledge which has been acquired at lower levels. As he expresses it, this explanatory knowledge is known "tacitly." In understanding we rely on this knowledge. We could say, then, that explaining and understanding are complementary aspects of the acquisition of knowledge. At the higher structural levels understanding dominates whereas at the lower levels explaining dominates. But even at the lowest levels understanding is never completely absent, just as at the higher levels explanation will never be completely absent. Explaining and understanding are complementary activities. The more that exact explanation proves possible, the less the role of understanding will be and vice versa.

3.3. Verification and Structural Level

3.3.1. Introduction

The requirement of verification of psychology's scientific propositions was already stated in the last century—Mill and Helmholz—but it could be defended that this was only concretely realized in neobehaviorism. In saying this we have in mind the hypothetico-deductive method propagated by Clark Hull. Hull developed an inductive psychology. It is true that he took theoretical conjectures to be the starting point, but these had to be verified by physical fact. The protocol sentences in which these are described are considered to be theory-independent criteria. That the inductive acquisition of knowledge has to proceed by way of testing conjectures is due to human limitations. It happens to be the case that one cannot inductively envisage the totality of interrelations in a specific area of reality at one glance (Hahn, 1933) and that is why we need the indirect way of testing hypotheses. Related to the hypothetico-deductive method is the *empirical cycle* (De Groot, 1961). De Groot distinguishes five phases in the empirical investigation: *observation, induction, deduction, testing,* and *evaluation.*

In principle we take a positive view of the concept of an empirical cycle, because it is placed in the framework of Popper's thinking, where testing does not have an absolute character as in the strictly hypothetico-deductive method of Hull.

Testing is a fundamental feature of the acquisition of knowledge as it happens in everyday life. We are referring to factual reality all the time; we support or refute insights. So it is no more than natural to record this procedure in our methodology and give clear rules for it. Still, in doing so a problem arises. It is an open question whether the empirical cycle is relevant for *every* area of science and thus whether it may always be applied profitably in exactly the way formulated by De Groot. We claim that this is not the case. The formulation of the method of testing becomes more precise as we approach lower structural levels. The hypothetico-deductive method is strictly focussed on sciences at lower structural levels and the empirical cycle can only be applied satisfactorily in psychology insofar as that psychology has an explanatory character.

According to the principle of falsification as stipulated in the empirical cycle, hypotheses should be formulated in the form of a material implication. A higher level of knowledge acquired by understanding can be tested by means of the empirical cycle only insofar as explanatory hypotheses can be deduced from that knowledge. Now we have already argued that this possibility is not to be excluded, so that even in psychology at higher structural levels the empirical cycle could in principle be applied. But the question remains whether its application at these levels is sufficiently decisive.

3.3.2. Mechanistic, Organismic, and Humanistic Theories

3.3.2.1. Mechanistic Theories. Mechanistic theories assume models of human activity that have the character of a mechanical object that can be taken apart into separate units. In large part these models are inspired by existing industrial products, for example the model of a telephone switchboard by Hull, the locomotive analogy in the psychology of Freud (Russelman, 1982), and the computer as a basic model in cognitive psychology. Overton and Reese have discussed the problem of models in psychology at length (Reese & Overton, 1970; Overton & Reese, 1973). They characterize the mechanical model as "the *reactive,* passive, robot or empty organism *model of man.*"

> A most significant characteristic of this model . . . is that change in the products of the machine or behaviors of the organism is not seen as resulting from change in the structure of the organism itself. The appearance of qualitative changes is considered either as epiphenomenal or as reducible to quantitative change, since the organism, like the elementary particles of classical physics, does not exhibit basic qualitative changes. (Reese & Overton, 1970, pp. 132–133)

Mechanical models in psychology are presupposed and thus form limiting frameworks within which the experience of the object under investigation has to be pressed. In this way the behavior which is to be studied is canalized and controlled and, moreover, cast in such a form that it is accessible to precise description. Mechanistic theories are focussed on explanation and prediction. In both cases it is a matter of recognizing the sufficient conditions for behavior. Because of the central role of the pursuit of causal interconnections, the empirical cycle is here the appropriate method of investigation. In this category of theories one is looking for formalisms. The ideal of constructing theories that are in fact axiomatic calculi can be pursued in this area.

3.3.2.2. Organismic Theories. These theories do not work with mechanistic models. Insofar as it is proper to speak of models here, it is a matter of a selection of viewpoints or principles of description that as a rule are borrowed from organisms. In organismic theories human behavior is interpreted as an *open system* (Von Bertalanffy) and accordingly the holistic character of the behaving organism is emphasized. The structural totality comes first; moreover, one is no longer dealing with reacting mechanisms, but with active systems interrelating with their surroundings. Behavior is regarded as related to an active "functional center" (Piaget). Accordingly, "the organism (is) inherently and spontaneously active . . . (it is) the source of acts, rather than . . . the collection of acts initiated by external (peripheral) force" (Reese & Overton, 1970, p. 133). "The individual who accepts this model will tend to emphasize the significance of processes over products, and qualitative change over quantitative change" (Reese & Overton, 1970, p. 134). Qualitative change involves development of "emergent systemic properties and levels of organization while quantitative change is 'strictly additive' " (Overton, unpublished). Because phenomena are studied as they present themselves and are not cast in preformed frameworks, there is less emphasis on the method of testing hypotheses than on naturalistic observation and description as exemplified in Piaget's (1970) clinical method.

3.3.2.3. Humanistic or Personalistic Theories. The characteristic feature of these theories is that the self-conscious human subject or the human person is emphasized. To understand this one should think of the experiential, phenomenologically given self. Here knowledge is assumed of human reality that is rooted in primordial intuitions experienced as evident. In consequence, "conscious goals (are) a necessary feature of the explanatory process," while organismic "teleological explanation consists in the postulation of regulative principles and does not necessarily involve individual awareness of needs, motives, or desires . . ." (Overton & Reese, 1973, p. 83).

The psychologies in question have a holistic character as well. The uniqueness of every individual human being is emphasized. This is expressed in taking personal responsibility for choices and decisions made with personal freedom. Here neither naturalistic observation and description nor testing of hypotheses is crucial. We try to interpret human behavior, to describe what is essential about it. Empathy and encounter are the central methodological concepts.

In phenomenological descriptions we also take certain viewpoints or principles of description as our starting point, the viewpoints being in this case mostly of an anthropological-psychological nature. Sometimes in these theories an attempt is found to make models of schematic representations of the mental apparatus, for example in the psychology of Freud. In this case we would rather speak of "metaphoric models" because of their vague and rather dubious character.

Natura non facit saltus (Nature makes no leaps)—Leibniz. There appears to be a gap between the humanistic and organismic as opposed to mechanistic thinking. But the question is whether that gap exists in reality. From our point of view, the human being himself creates the structures and hence the gaps. The fact that spatio-temporal, material reality has tolerated the mechanistic-causal structuring in the past—as witnessed by the history of classical physics—does not mean that this structuring is true to the essence of reality. In this connection it is useful to point out that according to the results of modern physics the mechanistic-causal structuring is not tolerated without limits by material reality (Harré & Secord, 1972). In the case of human reality this shows itself sooner and more clearly.

3.3.3. Types of Theory and Structural Level

One could say that the division into mechanistic-organismic-humanistic is roughly parallel to a division according to structural level wherein the mechanistic theories would be situated on a low level and the humanistic theories on a high level. In mechanistic theories one abstracts as much as possible from historical and cultural reality, whereas this is impossible when dealing with humanistic psychologies.

3.3.4. Testing in Organismic and Humanistic Psychology

The importance of testing hypotheses as explicated in the empirical cycle is not undisputed by organismic psychology. Of course the organismic psychologist looks for order as well. He too has conjectures that, as we shall see, may in a reduced form be tested by the method of the empirical cycle, but there are limitations to this kind of testing. The procedure used in acquiring knowledge looks more like that of *Anschauung* (unprejudiced observation) than like that of *Begrifflichkeit* (abstraction), to use a distinction made by W. Wundt. The organismic psychologist looks for identification in concrete situations, whereas testing is repeated concrete identification. As much as possible he pursues an *experimental* psychology in his method, in which the conditions are described under which a certain phenomenon, process, or relationship will take place in such a way that it can be identified and described as clearly and ideal-typically as possible. In organismic psychology one cannot always work in an experimental way. This applies especially to organismic field psychology where the conditions cannot be controlled. In this case one has to resort to descriptive research (Drenth, 1980; see also §3.4.2.).

In the experimental form of organismic psychology testing is *repeating*. It differs from the hypothetico-deductive method in that we do not deduce a hypothesis in order to test it, but to reproduce the phenomenon as such. An example of this *anschauliche* method may be found in Piaget's clinical method. As we remarked before, we are of the opinion that organismic and humanistic theories present differences of degree only. The humanistic psychologist also uses a method of *Anschauung*, but his descriptions often do not bear the stamp so much of being *true to nature* as of being a *disclosure of reality*.

That is why humanistic theories are more often disputed, that is, describe less the intersubjectively given fact of common sense, but bear more a stamp of an *aprioristic* view of man. In other words, they are more theoretical or philosophical. In this sense Soviet psychology, as a psychology that postulates Marxism as its point of departure, may also be called humanistic.

The humanistic psychologist, too, uses a method of testing. Aside from the hypotheses that have been gained by means of a reduction to other structural levels (see below), a procedure which is not rejected by humanists (see Rogers, 1964), the testing of hypotheses at the higher structural levels is more like a rejection or a confirmation of a personal feeling of evidence. When one is confronted with new unique cases these may or may not fit into the structuring used. In the case of a misfit this usually results in further differentiation or correction. Of course this kind of testing does not bear an exact character and does not offer the possibility of complete explication. We look in vain in methodological handbooks for detailed instructions on this method of testing. Feelings of (dis)harmony between structuring and phenomena play an important part in this method of testing; it is a matter of empirical fit. It should be noted, however, that the investigator has to stay within the semantic domain of discourse.

This kind of verification is very much dependent on the person who is employing it. It is not easy to detach the process from the investigator as something that goes on outside of him, something impersonal. The effectiveness of this form of testing can be heightened by training in correct thinking and by being open to the ways in which it can interact with reality. The testing becomes more *intersubjective* as the process of acquiring knowledge is determined more strongly by teamwork and discussion about the tentative results of the investigation. At the higher structural levels testing is not a one-shot deal, strictly regulated, but usually a historical process.

3.3.5. The Empirical Cycle and Organismic and Humanistic Theories

Psychological knowledge at the higher structural levels does not necessarily have to be tested at its own level. For from this knowledge conclusions can be drawn about interconnections at lower levels. So there exists a possibility of indirect testing according to the empirical cycle. If this method is used, one deliberately simplifies the functional structuring and so reaches a lower level. The manner in which this takes place and the structural level that is reached is rather arbitrary. One creates *ad hoc* a new structure, so to speak, and in that context a clearly delineated methodology can be used.

In this way of thinking we can interpret understanding as a source of scientific hypotheses formation. This process takes place within the *context of discovery* which, according to many methodologists, is important for the *psychology* of science but is of no interest for the *methodology* of science. The latter deals only with the *context of justification*. In view of the foregoing, it is clear that we do not share this opinion.

If we want to use the empirical cycle for the investigation of hypotheses at higher structural levels, we substitute explanatory knowledge of a lower level for knowledge by understanding. The latter is of another order than the explanatory knowledge, just as the black and white projection on the wall of a colorful event is different from that event itself. The event can never be reconstructed from the projection, although it is possible to falsify certain descriptions of the event by means of it. Testing hypotheses at the lower structural levels by means of the empirical cycle may offer support or correction for knowledge by understanding. That is why it remains an invaluable instrument of investigation, even for theories at a higher structural level; and it should not be brushed aside too easily, even though at the same time it should be remembered that this instrument becomes less relevant as we are dealing with more specifically human phenomena, that is, at higher structural levels. One could also say that as the logical distance between theory and hypothesis becomes larger, the importance of testing decreases (see also §3.5.2). The decrease in relevance of this method is often expressed in complaints about the slight "ecological validity" of the investigation (Brunswik, 1955). In §3.3.3. we argued that as a result of the nonextensional character of the language of organismic and humanistic theories it is not possible to deduce propositions that are susceptible to just one interpretation. As Overton and Reese (1973) put it:

> Within (the mechanistic) view there is a *unidirectionality* of causal application or a one-way causality, in which effect is strictly dependent upon cause. . . . [The logical form of this type of causality is the material implication in *extensional* symbolic logic (Carnap, 1966).] The organismic model leads to a quite different type of representation. The view that the organism is inherently and spontaneously active means that an external condition can never in itself be the sole determinant of an effect. Rather, cause and effect or environmental event and organism stand in a relationship of reciprocal action in which each member affects and changes the other. (Overton & Reese, 1973, pp. 78–79)

Again, this whole question is not rendered easier by the fact that in general the organismic and especially the humanistic theories are logically speaking still stated so much in intuitive terms that one can speak of deduction only in a metaphorical sense. In most cases there is no question of a logically compelling deduction. The knowledge by understanding which is offered in theories at a higher structural level is too implicit for compelling deductions and so it is not easy to test it for logical consistency. It is only an occasion to formulate hypothetical causal relations.

Developing hypotheses is a matter of explication, a matter of becoming conscious of subsidiary knowledge that cannot be deduced from the knowledge by understanding through explicit logical operations. The falsification of these hypotheses may therefore argue against a theory, but there is no question at the higher structural levels of crucial falsification. Think in this connection how arbitrary it would be to deduce unambiguous causal implications from existential theories of personality and how slight the significance would be of the results of the test for these theories.

But we have not yet summed up all the factors that render investigation by the method of testing hypotheses, as it takes place according to the empirical cycle, less significant. We should like to point out one more factor mitigating the importance of such investigation, a factor not of a theoretical nature this time, but taken from the concrete process of research.

The panacea that has been recommended in past decennia to make room for the testing of hypotheses in organismic and humanistic psychologies is that of the operationalization of concepts. An analysis of the process of operationalization in psychology, brings to light some further limitations of testing. This time we are talking about the dubiousness of attempts to render invariant concepts that function in theories at a higher structural level by way of operational reductions to concepts at lower levels.

When concepts are operationalized, it is a matter of a *partial* analysis of meaning. Operationalizations other than the one chosen are always possible. Once a concept has been operationalized, it is not only a limited but also a *frozen* concept, it has become a "fossil" (Van Peursen, 1965).

Limitation and dehistorization by way of operationalizing concepts has methodological advantages: concepts become unambiguous. The purpose of operationalization is *reference restriction;* it is an attempt to limit the multireferentiality, that is to say, the connotations of concepts. All possible worlds, all the situations to which the concept is thought to be applicable, have to be limited or restricted to just one or to a few worlds: concepts should be rendered as monoreferential as possible so that we can limit ourselves to human activity in an unambiguous situation in which we look at the factual results only. How this restriction takes place is arbitrary. As the operationalization takes place at a higher structural level, it is more evidently *ad hoc*. However, the results of the investigation are strictly limited by this *ad hoc* operationalization. Often it remains unclear how investigations in which the operationalizations are different may be related to each other, and this applies even more so as the number of differently operationalized concepts increases.

The difficulties which have been pointed out in connection with deduction are mirrored in the problems concerning the *ad hoc* character of the operationalization. At the higher structural levels we are dealing with such arbitrary and also limited explications that investigation into the conditional relations between these operationalized concepts are of only slight significance for insight at the original level.

3.4. Testing and the Various Psychologies

The discipline of psychology is characterized by a conglomerate of psychologies which have different structural levels and may be distinguished into normative or pure field theory and normative or pure systematic theory.

Testing in the sense of the empirical cycle can be better applied as the structural level is lower. In the foregoing we have given our arguments for this view. But in doing so we have abstracted from the distinction between normative and pure and from the difference between systematic and field psychology. That is why we shall now have to determine to what extent a more finely grained classification of the discipline of psychology necessitates a further differentiation in our discussion about testing.

3.4.1. Normative versus Pure Psychology

First we should like to make a few remarks about the difference between normative and pure psychology and the problem of testing. When in the foregoing we developed our view about testing, we were mainly thinking of pure systematic psychology. If normative psychologies are taken into account, whether of the systematic or of the field variation, then it should be noted that these are in principle out of the reach of testing in the sense of the empirical cycle. It might be objected that the distinction between normative and pure psychology is of little importance because all acts of disclosing basic reality are of a normative nature—and, given the foregoing, we would not want to deny this. In psychotherapeutical theories, for example, one is often dealing with a sort of theory that, although it does not offer itself as a world view, still possesses all the elements of such a view. Similarly, Freud presented himself as an investigator who held the tradition of natural science in high respect, but his publications are soaked in dogmatism. The same could be said of the publications of many positivists.

Therefore, even if it is correct to say that all science is based on norms and values—or in other words on a weltanschauung—we still have to keep the distinction in mind between the inevitable *implicit* normativity of science (and all human activity for that matter) on the one hand and, on the other, those norms that have been consciously *chosen*. In the latter case we speak of normative psychology because norms are then taken as an explicit point of departure, as a ground value.

It is clear that normative psychology attaches great value to theory. Especially when it is a question of theories which are strongly true to life, they may have the character of a world view. In philosophy of science this is not usually referred to as science any more. At most, such theories will be described as prototheories or as potential, prescientific sources of hypotheses. In the case of normative science it is difficult or completely impossible to come to intersubjective agreement outside of the norm community in question. For that matter this phenomenon occurs in all schools. The boundary between pure and normative science is in no way absolute. In the latter case, however, the walls have often built to such a height that any contact with other schools or norm communities becomes extremely difficult.

If we argue within the context of the usual methodology, normative theories cannot be tested. But if we interpret the concept of testing in a broader sense, then testing proves to be not wholly absent even in this category of theories. Though we may cast human reality into a certain preformed mold, reality will protest sooner or later if the mold strains reality too much. The theory will then be falsified in the course of history. "Truth is a daughter of time" (Bacon). The theory is rendered out of date by the development of science. Sometimes this happens because inconsistencies and contradictions occur or by the fact that there is no further progression. Alternative theories prove to be more applicable and offer more perspective.

One cannot work without assumptions. Completely pure psychology cannot exist. Even so we can formulate a criterion which a theory must satisfy in order to be called pure. Pure science is prepared *to put her assumptions up for discussion.* Of course within normative psychology discussion will occur as well; the discussion is often livelier and is taken more to heart than in pure science, but essentially and fundamentally there can be no altering of ground values. On the other hand, it should be remembered that, as the history of behaviorism shows, it sometimes takes a long time within pure psychology before the discussion becomes fundamental.

3.4.2. Systematic versus Field Psychology

Generally speaking, the method of the empirical cycle is adequate for pure systematic psychology. Pure systematic psychology is best suited for investigation in laboratories, where psychological phenomena are studied under controlled conditions. Pure field psychology does not differ essentially from pure systematic psychology with respect to the problem of verification. The fact that in the latter case the topic of investigation arises out of the system of the discipline, whereas in the first case it arises from life situations, is not of fundamental importance for the problem of testing. If we want to speak of differences, we may find them in the fact that field psychology is mostly not very well suited for experimental investigation, and insofar as it is possible, there is the additional problem of the ecological validity (Brunswik, 1955; Neisser, 1976). Therefore research following the empirical cycle is less emphasized in field psychology than in pure psychology. As in every empirical investigation, we try to justify our conclusions with empirical facts, but in doing so we are not wholly dependent on the testing of hypotheses. In field psychology several forms of research play a part. In the first place there are certain forms of descriptive investigation in which it is not a question of "testing general theories or hypotheses, but of systematically charting one specific phenomenon, process, set or system" (Drenth, 1980). Aside from that, we also find *theory-oriented* research, wherein empirical facts are collected "to explore or test certain theoretical relations" (p. 7). This kind of investigation is characterized by a gradual transition, a continuum from "generating, raising possible theoretical explanations, to testing them. Whether a certain place on this continuum is nearer the one or the other end depends upon the degree of explicitness with which a theory or (more specifically) a set of hypotheses may be formulated" (Drenth, 1980, p. 7).

From the foregoing it is clear that the value of investigation following the empirical cycle should be regarded with some reservation. For we are here dealing with *field* investigation in which we cannot control the variables. Even so the importance of verificational research in field psychology should not be underestimated. There can only be talk of testing hypotheses if a theory has been formulated that is sufficiently explicit. A selection of variables has taken place and these variables are mutually interrelated in an often extremely complex way because the phenomena in field psychology may be of a very complex nature. The investigation in which hypotheses concerning these interrelations are tested may use a "quasi-experimental design" (Campbell & Stanley, 1963). That is to say, the investigator introduces within "natural social settings" a procedure for collecting data that resembles the procedure of experimental research "even though he lacks the full control over the scheduling of experimental stimuli" (Campbell & Stanley, 1963, p. 34). The authors make it clear that we should not condemn this kind of verificational investigation; on the contrary, it is recommended when other experimental designs are not realizable. At the same time it should be remembered that just because complete experimental control is wanting it is necessary that "the researcher be thoroughly aware of which specific variables his design fails to control." That is why a "checklist of sources of invalidity" is offered on the basis of which the investigator gets an eye for "competing interpretations of his data."

This checklist may be regarded as a list of alternative hypotheses. In other words, if the investigator is aware of these sources of invalidity, he can systematically weigh to what extent alternative behavioral determinants are acceptable. The kind of phenomena that are investigated in field psychology largely determine to what extent a "quasi-experimental design" can be applied successfully. Even though this method of testing hypotheses is useful in this field, we should still conclude that its effectiveness is not as high as in laboratory research. The scientific *discussion* in which the results of the investigation are interpreted and reinterpreted, plays a much more important role here. In this discussion the "checklist" itself may be put up for question as well but crucial experiments are not possible here.

3.5. The Investigator and the Problem of Testing

3.5.1. "Theory-laden" Data of Observation

Although in functional structuring the object as well as the subject (the investigator) plays a part, in this section the emphasis will be on the contribution of the subject. During their scientific education investigators undergo perceptual training. As a consequence considerable differences arise in the way in which investigators, claiming to study the same subject matter, perceive their data. The higher we climb up the structural ladder, the larger and more numerous those differences become.

In many quarters (Hanson, 1958, Hesse, 1970, Van Peursen, 1965) it has been pointed out that there does not exist an observational language which is independent of theory. As one moves further away from physics, a discipline in which the separation between theoretical and observation language can be carried through approximately—in other words, as one arrives in parts of reality which are a product of historical and cultural development—it is more difficult to find an observation language which is independent of theory. A theoretical framework seems to be inherent in perceptual training, and the higher the structural level the more arbitrary such a framework will be. Does this have implications for testing? In answering this question, we shall start testing at the higher structural levels, as has been described in §3.3.4. We could say that the distinctive feature of testing is that at these higher levels it is theory-immanent. It is almost circular: the new specific facts which confront us during the test are perceived according to an interpretation influenced by the theory. One could say, then, that at the higher levels the test has more of a logical than an empirical character.

The emphasis is on the internal consistency of the theory. But the closer we approach basic reality (that is, reality at the level of Strawson's basic particulars),[1] the more we are justified in speaking of an independent empirical test. This brings us back to the point in §3.3.5. where the use of the empirical cycle was discussed for organismic and humanistic theories. In order to apply the empirical cycle there, it was argued, we have to leave our own structuring and descend to a lower structural level at which hypothetical relations can be investigated in a clear-cut way.

But this is not to say that in so doing we should go so far in our reduction as to arrive at the level of physical thing language (Carnap, 1931). Even at a somewhat higher level the empirical cycle may be used. But in that case it is not so easy to reach agreement about the criteria, the basic propositions that are going to be used in testing. And as these propositions are closer to daily life the results of the test carry more weight. In connection with this difficulty of reaching an agreement about the criteria, we should like to point out once more that many operationalizations which have been carried out with a view to testing have an arbitrary and *ad hoc* character. Concepts in comparable psychological theories are often operationalized quite differently, and that is why comparison of the results of investigation is difficult.

[1]The concept of basic reality is discussed more fully in our reply to commentators.

3.5.2. Incommensurability

Modern philosophers of science argue that even in the case of established, generally accepted physical theories, *compelling* falsification is out of the question. And then we are not even taking into account the fact that falsification may also be blamed on *stořende Bedingungen* (Holzkamp, 1973), that is to say, on assumptions external to the theory, which were necessary in order to deduce the (falsified) prediction. In other words, one can never be compelled on empirical grounds to reject a theory: no theory is empirically falsifiable, let alone completely confirmable.

This state of affairs has brought Lakatos (1970) to his *sophisticated falsificationism.* According to him empirical testing is not a matter of the confrontation of one theory with empirical criteria, but it is a threefold or manifold "battle" in which it will be decided which of two or more rival theories is least contradicted (logical testing, of course, does not need a manifold battle). If such a *comparative* test is to take place, one should be able to make contradictory predictions with the help of the rival theories about what will occur under certain conditions. This implies at the same time that one is able to find perceptual facts which are relevant as criteria for the test of each of the rival theories and which are commonly accepted.

The condition which we just formulated for sophisticated falsification will offer the fewest problems if theories are variants of the same basic type, if, in Lakatos's words, they belong to the same research program or, formulated in our framework, if they are based on practically the same structuring. In those cases testing is *research program-immanent.*

It becomes a different matter if we have to choose between rival theories that structure reality in a completely different way and may—as happens quite often in psychology—have a different structural level. Does the test by way of the empirical cycle still have any use in that case? It is doubtful whether the conditions for sophisticated testing can be satisfied. In a case like that one speaks of *incommensurability.*

For an example of incommensurability in psychology, let us take the language theory of Skinner, which is based on his radical behaviorism, and psycholinguistics as it was inspired by Chomsky's structuralistic thinking. The difficulty in deciding by means of a comparative test which of two rival (incommensurable) theories should be rejected and which further developed appears to relate to finding a communal *universe of discourse* in the context of which testable predictions can be formulated. Such a *universe of discourse* can always be found. The reason for this may be found in the basic reality which is presupposed in every science and hence also in rival theories of different structural levels. On the basis of such theories predictions can be made concerning human behavior at a lower communal level which, if they are contradictory, can be tested on the basis of facts. In comparing the language theory of Skinner with the psycholinguistics of Chomsky one can, following an example given by Miller, Galanter, and Pribram (1960, p. 146), calculate the *time* it takes to learn a language. That calculation appears to offer a counterexample to Skinner's ideas. Still, we can hardly speak here of a crucial test (see MacCorquodale, 1970).

We could say that in the case just mentioned a complete comparison does not take place, as it does in the case of two variants of one prototype. At most we can speak of a marginal comparative test. Think of the logical distance between theory and prediction and the dubious representativeness, partly due to that distance, of that prediction.

Should the conclusion be that indeed we are dealing with several different possible structurings of one part of reality, while the knowledge acquired in one part is untranslatable into the language of the other part without essential aspects being lost? Furthermore, since the different structurings cannot be weighed one against another on the basis of arguments, is it not a question of personal preference or fashion (Feyerabend, 1975) which of the structurings is chosen?

The first question we shall answer in the affirmative, the second in the negative.

Scheffler (1967, p. 82) has pointed out that "lack of commensurability . . . does not imply lack of comparability." Even incommensurable works of art can be discussed in a reasonable way, can be criticized and compared. It is not true that "such a critical discussion must consist of empty rhetoric alone." Critical comparison of incommensurable theories "is itself not formulated within, nor bound by, the paradigms which constitute its objects. It belongs rather to a *second-order reflective and critical level of discourse*" (p. 83, italics ours). According to idealistic thinkers, "differing first-order persuasions inevitably step back to differing evaluative positions" (p. 83). But Scheffler opposes this position. In his view, theoreticians of science who maintain that common standards for a critical comparison are not available find themselves in a paradoxical position because they attempt to convince theoreticians who adhere to another meta-paradigm by means of arguments.

Another argument regards the problem of incommensurability in the light of whether reality bears a normative character or not (Van Peursen, 1965). In taking the view that knowledge is a product of the interaction between man and nature, we acknowledge the normativity of reality and reject the idea of incommensurability. We assume, therefore, a reality that may function as a touchstone that will sooner or later protest against structurings that are foreign to its nature. It is tempting, while discussing Van Peursen's ideas on normativity, to point to the discussion in the psychology of perception where the *object* of perception is again attracting attention. Thus, according to James Gibson, we may recognize in visual perception invariant characteristics of a higher order in the light reflected by an object. So his point of departure is not the retinal image as determining perception, but that of the "picking up" of these invariants by the organism. Neisser (1976) summarizes the implications of Gibson's (1979) theoretical position as follows:

> The most important thrust of the theory is to suggest that students of perception should develop new and richer descriptions of the stimulus information rather than ever-subtler hypotheses about mental mechanisms. (p. 19)

The test function of reality (see 3.4) arises within the context of the discussion about the developments in science and their implications for culture and society. It is typical for this discussion that concrete counterexamples are never decisive, but that it is a question of a totality of developments that become manifest in the historical process of scientifically disclosing reality. This kind of testing does not proceed according to sharply defined methodological instructions, but still it is a *way* (method) for the further development of science. This unintentional, historical, and natural testing is the basis of what has been called "rationality in retrospect" (Lakatos, Toulmin).

For those who stand in the midst of the conflicts between schools of thought and current trends it is not easy to make a choice. In this connection T. S. Kuhn speaks, and we think rightly so, of *belief* in theories and points of departure, and of *conversion* to new theoretical frameworks. Not wholly without justice the comparison is made with religious behavior. To note personal or collective irrationality, however, is not to say that this is so crucial a problem that it could not be overcome in the course of history and that irrationality is doomed to remain the dubious foundation of our scientific store of knowledge. Accordingly, irrational impulses may be important driving forces behind certain developments in science, but when the complicated process of disclosing reality has taken place and we look back upon the questions that were under discussion, then it is without doubt possible to discern rational lines of development. The concept *rational* is used here in the sense of being capable of formulation in an explicit, logically consistent train of thought. This rationality has come about by the correcting role of reality that compels us to explicate our position. Reality has functioned as a sieve in which the products of an all too fanciful and nonadequate scientific activity have been left behind and forgotten.

It is undoubtedly the case that theories of a quite different nature may be weighed one against another with reasonable and convincing arguments and may be related to each other, even though contemporaries may not be able to do so. We derive the basis for this view from the fact that although in the acquisition of knowledge we are dealing with a *human* disclosure it is still a disclosure *of something,* of something that exists independently of man, something that, as Strawson argued, is assumed in all human language.

When there are paradigm switches in science there is a process going on in which the investigator becomes conscious of the outer skin of the assumptions in which the usual formalisms are rooted. Accordingly he will perceive reality differently and his perceptual scheme changes and becomes more clearly differentiated. But what may mean a radical change in outlook on the level of semantics and have far-reaching consequences in the sphere of pragmatics appears—as witnessed in the fact that the equations from Newtonian dynamics can be deduced from Einstein's relativity theory—not necessarily to lead to incommensurability on the level of syntax. Here the newly acquired insight is detached as much as possible from human experience and introduced into the formalism, and thus the formalism becomes increasingly subtle and differentiated. When we consider the succession of formalisms, we find a cumulative, rational progression.

Just now in passing we spoke of the fact that Newton's dynamics can be deduced from Einstein's relativity theory. This thesis is disputed by some. Thus Stegmüller (1979, p. 444) states:

> Immer wird uns das Märchen erzählt, dass die Newtonse Dynamik als Grenzfall aus der relativistischen Dynamik ableitbar sei. Doch die beiden Wissenschaften verwenden verschiedene Sprachen, worüber nur eine teilweise Wortgleichheit hinweg täuscht. *Masse* und *Energie* sind in beiden Paradigmen theoretische Grössen, aber gänzlich verschiedene. Nur in der Relativitätstheorie z.B. gilt die Masse–Energie Aequivalenz.[2]

But this view is only partly correct. At the level of the formalism this borderline case is no fairy tale at all. From formalism, therefore, we derive an argument against a radical idealistic view of incommensurability.

Aside from this, at the time of paradigm switches, in the heat of scientific discussion, we do not live at the highly "dehumanized" level

[2]Time and again one is told the fairy tale that Newtonian dynamics is a borderline case deducible from relativistic dynamics. However, both sciences use different languages, which is only blurred by a partial similarity of concepts. Although mass and energy can be found in both paradigms as theoretical concepts, they are of a totally different nature. For example, it is only in relativity theory that the mass–energy equivalence holds. (translation by authors)

of formalism. Caught in our own perspective and driven by unacknowledged, or at least not sufficiently acknowledged, irrational motives, a fruitful discussion often appears impossible because only a marginal "interconceptual zone" is available to which we could refer collectively. But it does not have to remain this way. As we go on we will start to reflect on our own point of view and will attempt to place ourselves in the other parties' perspectives or structurings. Often this development takes place with a new generation of scientists. A new *metalanguage* arises, a language in which the practice of science can be justified and interpreted. This language is the fruition of an opening up of old structures and an integration into new, more-encompassing structures. Contradictions (incommensurabilities) lose their relevance and are transcended. But with this we do not come to the end of the evolution of science. New antitheses arise and will have to be transcended in their turn. The adventure of the scientific acquisition of knowledge follows this process. It is a process that, even with all the randomness and capriciousness involved, is still held within bounds and guided by reality.

4. Conclusion

In the philosophy of science there is a clearly visible tendency to replace the optimism about the possibilities of the experimental method (especially for sciences at a higher structural level) with an attitude of scepticism. Many have lost their belief in a universal science evolving in the course of all cultures and times. When we look back at the functional point of view which we have developed in this paper, we conclude that we share neither the positivistic optimism nor the idealistic pessimism.

From our point of view we should first of all stress that psychology is a science of a much more complex nature than classical physics. So we should not compare the situation in contemporary psychology with that of natural science a few centuries ago. Therefore psychology should not try to mirror the historical development of the natural sciences. In psychology, at least insofar as it does not abstract from the subjectivity of the human being, the experimental method, having as it does its roots in the sphere of "mediacy" (Wundt), is not sufficient but must be supplemented with other methods. As the intentional character of human behavior is more dominant in the subject matter under investigation, psychology needs more methods which are rooted in the sphere of "immediacy" (Wundt; see Rappard, 1979).

Although, with many methodologists, we want to take adequate account of the contribution of the investigator (context of discovery), we do not share the opinion that because of this context scientific psychology remains purely paradigm-immanent. We maintain that the development of psychology quite clearly shows rationality and cumulativity (Lakatos, 1970). From our functional point of view, in the scientific acquisition of knowledge it is a matter of interaction between the investigator and the subject matter under investigation. The contribution of the subject matter implies that the knowledge acquired cannot be merely (intra)subjective. That is to say that scientific psychology cannot be viewed as purely the product of an arbitrary group of investigators with certain interests and tastes (Feyerabend). *The subject matter of psychology has its own objectifying influence.* It has the character of giving norms (Van Peursen) and of being, when necessary, a falsifying criterion that will ensure progress—although at a slow pace.

Inherent in the subject matter of psychology there will always be a large diversity of psychologies working with different aspects of human behavior. The problem is that we as human beings make ourselves into the subject matter for investigation. And in this process we cannot do without a certain selective perspective. The difference between physics and psychology is, for one thing, that physics, even though it has undergone relatively slight historical shifts and refinements, still makes use almost entirely of just one perspective that has proved fruitful and possesses a certain universality. It is true that even in physics we are dealing with human knowledge, but the reduction of reality that has to take place there is largely undisputed. In psychology things are not that simple. Many forms of reduction appear to be possible. None of them is simply right or wrong. They all have a certain legitimacy as long as it is realized that no one perspective has a sole right to existence. *Knowledge is valid only within certain limits.*

Psychology is such a difficult science because of the fact that being a subject, as all humans are by definition, implies that every form of objectifying human reality falls short of that reality. In psychology no perspective is in itself sufficient. Only if psychology is prepared to take every *carefully thought-through perspective* seriously and is also willing to leave each one of them behind will we get a glimpse of who and what a human being is. But at that point we will have gone beyond the sphere of science.

5. References

Boer, Th. de (1980). *Grondslagen van een kritische psychologie.* Baarn, Holland: Ambo.
Brennan, R. E. (1946). *Thomistic psychology.* New York: Macmillan.

Brunswik, E. (1955). Representative design and probabilistic theory in functional psychology. *Psychological Review, 62,* 193–217.

Buytendijk, F. J. J. (1961). Het kennen van de innerlijkheid. *Academische redevoeringen* (pp. 75–98). Utrecht, Holland: Dekker & Van de Vegt.

Campbell, D. T., & Stanley, J. C. (1963). *Experimental and quasi-experimental designs for research.* Chicago: Rand McNally.

Carnap, R. (1931). Die physikalische Sprache als Universalsprache der Wissenschaft. *Erkenntnis, 2,* 432–465.

Carnap, R. (1966). *Philosophical foundations of physics.* New York: Basic Books.

De Groot, A. D.(1961). *Methodologie: Grondslagen van onderzoek en denken in de gedragswetenschappen.* The Hague: Mouton.

Drenth, P. J. D. (1980). Onderzoek in de arbeids- en organisatiepsychologie: principes en methoden. In P. J. D. Drenth, H. Thierry, P. J. Willems, & C. J. de Wolff (Eds.), *Handboek arbeids- en organisatiepsychologie* (pp. 1–42). Deventer, Holland: Van Loghum Slaterus.

Duijker, H. C. J. (1972). *De meervoudige gedetermineerdheid van menselijk gedrag.* Amsterdam: Mededelingen der Koninklijke Academie van Wetenschappen. Nieuwe reeks, deel 35, no. 3.

Duijker, H. C. J. (1980). *Psychopolis—Een essay over de beoefening der psychologie.* Deventer, Holland: Van Loghum Slaterus.

Edwards, P. (1972). *The encyclopedia of philosophy.* New York, London: Macmillan and Free Press.

Feyerabend, P. K. (1975). *Against method.* London: New Left Books.

Fortmann, H. (1971). *Inleiding tot de cultuurpsychologie.* Bilthoven, Holland: Ambo.

Gal'perin, P. J. (1980). *Zu Grundfragen der Psychologie.* Köln: Pahl-Rugenstein.

Gibson, J. J. (1979). *The ecological approach to visual perception.* Boston: Houghton Mifflin.

Hahn, H. (1933). Logik, Mathematik und Naturerkennen. *Einheitswissenschaft, Schriften herausgegeben von Otto Neurath in Verbindung mit Rudolf Carnap, Hans Hahn. Heft 2.* Wien: Verlag Gerold & Co.

Hampshire, S. (1965). *Thought and action.* London: Chatto and Windus.

Hanson, N. R. (1958). *Patterns of discovery.* London: Cambridge University Press.

Harré, R., & Secord, P. F. (1972). *The explanation of social behaviour.* Oxford: Blackwell.

Hesse, M. (1970). Is there an independent observation language? In R. G. Colodny (Ed.), *The nature and function of scientific theories* (pp. 35–77). Pittsburgh: University of Pittsburgh Press.

Hintikka, J. (1969). *Models for modalities.* Dordrecht: Reidel.

Hofstee, W. K. B. (1980). *De empirische discussie—Theorie van het sociaal-wetenschappelijk onderzoek.* Meppel, Holland: Boom.

Holzkamp, K. (1973). *Sinnliche Erkenntnis.* Frankfurt: Fischer Taschenbuch Verlag.

Koch, S. (1964). Psychology and emerging conceptions of knowledge as unitary. In T. W. Wann (Ed.), *Behaviorism and phenomenology* (pp. 1–42). Chicago: University of Chicago Press.

Lakatos, I. (1970). Falsification and the methodology of scientific research-programs. In I. Lakatos & A. Musgrave (Eds.), *Criticism and the growth of knowledge* (pp. 91–196). London: Cambridge University Press.

Leont'ev, A. N. (1979). *Tätigkeit, Bewusstsein, Persönlichkeit.* Berlin: Volk und Wissen.

MacCorquodale, K. (1970). On Chomsky's review of Skinner's verbal behavior. *Journal of Experimental Analysis of Behavior, 13,* 83–99.

Masterman, N. M. (1970). The structure of a paradigm. In I. Lakatos & A. Musgrave (Eds.), *Criticism and the growth of knowledge* (pp. 59–89). London: Cambridge University Press.

McLeish, J. (1975). *Soviet psychology, history, theory, content.* London: Methuen.

Miller, G. A., Galanter, E., & Pribram, K. H. (1960). *Plans and the structure of behavior.* New York: Holt, Rinehart, & Winston.
Morris, C. (1955). *Signs, language and behavior.* New York: Braziller.
Neisser, U.(1976). *Cognition and reality—Principles and implications of cognitive psychology.* San Francisco: Freeman.
Olst, E. H. van (1971). *The orienting reflex.* The Hague: Mouton.
Osgood, C., Suci, G. J., & Tannenbaum, P. H. (1957). *The measurement of meaning.* Urbana: University of Illinois Press.
Overton, W. F., & Reese, H. W. (1973). Models of development. Methodological implications. In J. R. Nesselroade & H. Reese (Eds.), *Life-span developmental psychology: Methodological issues* (pp. 65–86). New York: Academic Press.
Payne, T. R. (1968). *S. L. Rubinstein and the philosophical foundations of Soviet psychology.* Dordrecht: Reidel.
Peursen, C. A. van (1958). *Filosofische orientatie.* Kampen, Holland: Kok.
Peursen, C. A. van (1965). *Feiten, waarden en gebeurtenissen.* Hilversum, Amsterdam, The Netherlands: De Haan/Meulenhoff.
Piaget, J. (1970). *Genetic epistemology.* New York, London: Columbia University Press.
Polanyi, M. (1958). *Personal knowledge: Towards a post-critical philosophy.* London: Routledge & Kegan Paul.
Polanyi, M., & Prosek, H. (1975). *Meaning.* Chicago: University of Chicago Press.
Popper, K. R. (1934, 1971). *Logik der Forschung.* Tübingen: Mohr.
Quine, W. V. O. (1960). *Word and object.* New York: Wiley.
Rappard, H. V. (1979). *Psychology as self-knowledge.* Assen, Holland: Van Gorcum.
Razran, G. (1972). *Mind in evolution.* Boston: Houghton Mifflin.
Reese, H. W., & Overton, W. F. (1970). Models of development and theories of development. In L. R. Goulet & P. B. Baltes (Eds.), *Life-span developmental psychology* (pp. 115–145). New York: Academic Press.
Rogers, C. R. (1965). Toward a science of the person. In T. W. Wann (Ed.), *Behaviorism and phenomenology* (pp. 109–133). Chicago: University of Chicago Press.
Royce, J. R. (1976). Psychology is multi: methodological, variate, epistemic, world-view, systemic, paradigmatic, theoretic, and disciplinary. In W. J. Arnold (Ed.), *Nebraska Symposium on the Conceptual Foundations of Theory and Methods in Psychology* (pp. 1–63). Lincoln: University of Nebraska Press.
Russelman, G. H. E. (1982). *Van James Watt tot Sigmund Freud. De opkomst van het stuwmodel van de zelfexpressie.* Deventer, Holland: Van Loghum Slaterus.
Sanders, C., Eisenga, L. K. A., & Rappard, J. F. H. van (1976). *Inleiding in de grondslagen van de psychologie.* Deventer, Holland: Van Loghum Slaterus.
Scheffler, I. (1967). *Science and subjectivity.* Indianapolis and New York: Bobbs-Merrill.
Spranger, E. (1924). *Lebensformen.* Halle, E. Germany: Niemeyer.
Stegmüller, W. (1979). Moderne Wissenschaftstheorie: Ein Überblick. *Die Naturwissenschaften, 66,* 337–345.
Strawson, P. F. (1958). Persons. In H. Feigl, M. Scriven, & G. Maxwell (Eds.), *Minnesota studies in the philosophy of science* (vol. 2, pp. 330–353). Minneapolis: University of Minnesota Press.
Strawson, P. F. (1959). *Individuals: An essay in descriptive metaphysics.* London: Methuen.
Von Bertalanffy, L. (1967). *Robots, men and minds.* New York: Braziller.
Waterink, J. (1941). *Ons zieleleven.* Wageningen, Holland: Zomer en Keuning.
Wit, H. F. de (1977). *Onderwijspsychologische taalsystemen—Een betekenisanalyse.* The Hague: Staatsuitgeverij.
Wolman, R. B., & Nagel, E. (Eds.). (1965). *Scientific psychology.* New York and London: Basic Books.

5

Psychology and Philosophy of Science

A Commentary

Hubert C. J. Duijker

In their paper Sanders and Rappard raise so many issues, touch upon so many problems, that it is impossible in the few pages allotted to me to provide a discussion commensurate to such a wide range of topics. For this reason I shall restrict myself to some brief and rather superficial remarks on three topics only: 1. reality, 2. (psychological) knowledge, and 3. communication and conceptualization.

1. Reality

Every *empirical* science addresses itself to (and therefore presupposes) a *reality* independent of the investigator. Furthermore, empirical science is based upon the assumption that this reality is knowable.

Hubert C. J. Duijker (1912–1983) studied philosophy with H. J. Pos and psychology with G. Révész at the University of Amsterdam. In 1948 he was appointed professor of psychology at the University of Amsterdam, and he retired in 1981. During the last decades of his life, Duijker was considered to be the authoritative Nestor of psychology in the Netherlands. His numerous publications, primarily in the Dutch language, focussed on theoretical and methodological issues and he became well known for his ideas on the systematic grouping of both basic and field psychological disciplines and his views on human behavior as pluralistically determined. He held that the normative determinants (see the following commentary) were the most important in the case of human behavior. Duijker held many positions at the national and international levels in psychological and other scientific organizations. A scholar, Hubert Duijker was widely read and respected. [Editor (LPM) from notes provided by Professor C. Sanders]

Hubert C. J. Duijker • Late of the Department of Psychology, University of Amsterdam, Amsterdam, The Netherlands. Reprints may be obtained from the Editor (LPM).

Empirical knowledge is seldom easy to come by; different investigators often hold different opinions. However, in *empirical* science this is taken into account and its philosophy (or rationale) is based on the postulate that when investigators disagree reality functions as an arbiter. To use a Kantian expression: an experiment is a question put to nature. The answer obviously depends on the question and will presumably be more unambiguous as the question is more precise. (This is what Sanders and Rappard refer to as the *normative* function of reality. I prefer the term *arbitral function.*) Knowledge increases as scientists learn to formulate better questions.

This line of reasoning seems rather clear-cut and convincing, as long as one pretends to know what *knowledge* means. Serious difficulties arise when one asks the question: what is (empirical) knowledge? If one considers that form of knowledge which is considered as scientific, one may observe that according to present-day opinion it should fulfill certain requirements and moreover that these requirements have been gradually specified and refined. According to modern criteria, the equation $E = mc^2$ is scientific; the proposition "Spare the rod, spoil the child" is not.

In the natural sciences there is often a considerable distance between what is perceptionally given and the hypotheses or theorems supported or invalidated by it, for instance, between a scratch on a photographic plate and the identification of a new particle in nuclear physics or between a fossil jaw and the reconstruction of a prehistoric animal—in other words, between the final interpretation and the original datum. If, in the opinion of the scientific community, this distance is bridged adequately, the interpretation will be accepted as an addition to the store of scientific knowledge. This "legitimization" is often a laborious and time-consuming affair and may in principle always be revoked. A new and better interpretation may be found. In this respect scientific knowledge is never immune to the worm of doubt.

But there are other forms of knowledge, and they are of fundamental importance, particularly in psychology. As an example, I mention a toothache. Guided by Descartes, I may feel doubtful about even the existence of the world and all its trappings; but I cannot doubt my toothache. It is a reality that is indubitable and independent of any form of reasoning.

Still another form (or mode) of knowledge may be distinguished. If I am asked, "Do you know Dutch?" I might answer, "Yes." I am a native speaker of Dutch. This knowledge is partly explicit and partly implicit. It is explicit, for example, insofar as I am able to indicate the meaning of many words belonging to that language; it is implicit insofar as, not being a grammarian (or linguist), I cannot formulate the syntactic rules that somehow determine the way I construct (Dutch) sentences. A "language system" also seems to be "real," even in two different senses. On the one hand it determines lingual behavior; on the other, it is being studied as such by empirical scientists, that is by (some) linguists.

There are, in a psychological perspective, still other modes of reality, each to be "disclosed" (Sanders and Rappard) in its own way. I hesitate, however, to speak of different levels of reality, or of basic reality. Not only has this term an ontological flavor which I consider inappropriate in a psychological context but, and this is perhaps a more serious objection, its meaning is unclear. Psychologically speaking, *real* is what harms, hurts, hinders, and so on. A wall is real because it is an obstacle; my body is real because I cannot escape from it or its afflictions; a language is real because speakers must conform to its rules in order to be understood. In other words, in my opinion, reality is a *context-bound* concept (for this reason it is also culture-bound).

Physics is often referred to as the science of ultimate reality (Sanders and Rappard refer to reality as a whole). I do not think this is true. Physics has, as far as I know, nothing to tell us about toothaches or, for that matter, about anxiety or guilt. Nevertheless, a toothache is, psychologically speaking, far more real than a quark.

Although further elaboration would be desirable, I conclude that what is called *reality* may, or should, be considered as a class of *limitative* determinants of behavior.

2. (Psychological) Knowledge

Sanders and Rappard state that in their opinion "every form of knowledge (is) ultimately derived from *action*, from *interaction* between the knowing organism and reality" (p. 219, their italics). I must confess that I find this a rather cryptic statement (its elucidation in section 1.1. does not help much). First of all, it seems to me that we should avoid, in psychology, pronouncements about "ultimate" derivations, which we had better leave to philosophers and theologians. Second, it is dangerous to speak about "every form of knowledge," since the different meanings of the word *knowledge* are far from clear. Third, it is doubtful whether all knowledge stems from action since there is undoubtedly (unwelcome) knowledge about personal states (e.g., toothache, dizziness, etc.) which is forced upon us rather than sought. Fourth, the notion that all knowledge is derived from interaction is, to me at least, very confusing. Interaction means *reciprocal* action, that is, action on both sides. An astronomer who studies the Andromeda nebula may obtain valid knowledge. I do not believe that in any sense the actions of that nebula are influenced by those of the astronomer. Or, to give another example, that the observation of an infant's movements through a one-way screen should be called "interaction" between observer and observed. As I see it, reality may function as an arbiter precisely because it is independent of our notions, wishes, and so on.

However, in another section (1.3.1.) our authors offer a different description of (the acquisition of) knowledge. There they mention "functional structuring," and state that a functional structure is expected to "harmonize with reality," where reality is organized on the basis of an at least partially explicit point of view.

I should like to add that in this harmonization it is reality that sets the tune. If the organization distorts reality (and if unfettered scientific exploration continues to be allowed, which nowadays seems doubtful), in the course of time it will be corrected. Eventually, science will have to accept the verdict of reality. In other words, reality sets *limits* to speculation. Once again: reality does not set norms, but the scientific behavior of investigators is normatively determined by their obligation to account for all the data. Reality is the given (cf. section 2.2.).

One cannot but agree with Sanders and Rappard that there is knowledge about reality independent of and preceding scientific enquiry. They refer to Strawson's "descriptive metaphysics" and to his attempts to "lay bare the most general features of our conceptual structure" (p. 231). There are, in my opinion, serious objections to Strawson's analysis; but they are not relevant to the present discussion. What I want to point out is that Strawson's *answers* are psychology's *questions*. For instance: it is not only the *identification* of particulars that provides us with many (e.g., semantic and psycholinguistic) problems, it is also the *perception* of particulars (e.g., "object constancy") and its many ramifications.

As becomes manifest in section 3., Sanders and Rappard are to a large extent relying upon Strawson's ideas in their approach to methodology. For instance, they distinguish between "basic psychology" and "psychology in the vein of the humanities." As to the former, they say that there "the basically given could be described as things that *possess* material bodies" (italics Sanders and Rappard). Undoubtedly, such a description *could* be given; there is, at least in our part of the world, freedom of speech. But *is* it in fact given? I believe that in what Sanders and Rappard call basic psychology humans (and animals) are considered as *organisms*, that is, as very complex systems, which can (in principle) be *adequately* described in physiological (anatomical, biochemical, etc.) terms; not as *things* that *have* bodies, but as *things* that *are* bodies. I will not repeat my objections to the term *basic* but a purely organismic psychology (if it is psychology at all), in my opinion, is an incomplete, not to say deficient psychology, since it is unable to give an account of what is primarily given, such as sensory qualities, emotional states, and so on. In other words, it studies some of the (necessary) conditions for behavior (a class of limitative determinants) but neglects the other categories of determinants. It is a legitimate enterprise; but in my opinion not more basic, for example, than the study of meaning. Here again we must be careful not to confuse psychological and ontological categorizations.

I fully agree with the authors that in a very important sense the human world is a world of possibilities. This is the reason why *normative* determinants play such a dominant role in human behavior. For in everyday life explicit norms are found only where alternatives are discerned. In many countries drivers are admonished to keep to the right *because* keeping to the left is also feasible. Many other commands (or prohibitions) are being inculcated because they are plausible, or desirable, alternatives (e.g., the ten biblical commands). However, there are a great many implicit or *tacit* norms as well: among them are the rules of grammar, the rules governing many forms of nonverbal communication (eye contact, "floor-sharing," etc.), rules concerning the acceptability of foodstuffs, and so on. In such cases, what is from the point of view of psychology normatively determined is from the point of view of the layman self-evident and natural or spontaneous. The reality (i.e., the existence and effectiveness) of such normative determinants is usually discovered only by the confrontation with other (sub)cultures, where other possibilities are actualized.

All norms can be violated; this follows from the fact that they presuppose alternative possibilities. Violation, particularly of explicit norms, often is punished as the violator is supposed to have "chosen" the wrong action. In other words, in everyday life people assume that deviants *might* have behaved differently and therefore are responsible for or guilty of their "misdeeds." In popular psychology people are capable of "free" choices. It is a moot and unresolved question, whether in scientific psychology complex notions such as *free, choice, responsible,* and *guilty* are adequate. Ideas that are prevalent, perhaps even indispensable in everyday life do not necessarily provide the most fruitful "functional structure" in science. We *see* the sun "rise," but we *know* that the earth turns.

I cannot resist the temptation to add a few words on the "logic of perception" (section 2.5.3.). I am second to none in my admiration of the logician's art, but I cannot forget that logic and psychology are not identical. To be quite honest, I do not believe that a logician is competent to prescribe how statements about perception should be formulated while in psychology the problems of perception are far from solved. As there are many forms of knowledge, so there are many modes of perception; thus the perception of hunger is different from the perception of a running man. In Hintikka's example, cited and commented upon by Sanders and Rappard, it is suggested that there is an observer on the one hand and a "state of affairs" on the other. In my opinion, however, both the observer and the observed are included in the "state of affairs." Stronger yet, the distinction between observer and observed, between self and environment (world) is a phenomenal one and is therefore not absolute but determined by the situation, which includes *both* person and environment. Consequently, it depends on the situation whether, in psychology, we should speak about "propositional attitudes" or "observational propositions."

Sanders and Rappard state (section 2.5.4.): "A description of human behavior is meaningful only in the context of the realization of *possibilities,* which can only be called possibilities in the context of alternatives" (italics Sanders and Rappard). Here, again, I feel doubtful. Let us compare three descriptions: (a) he jumped from the roof, (b) he fell from the roof, (c) he was pushed from the roof. These statements are *meaningful.* One might say that only (a) refers to behavior, the others only to *movements.* (Such a distinction is not always easy to make: consider the stuntman in the movies.) However, to restrict the meaning of *behavior* to *intentional* or *purposive* behavior is dangerous as well as difficult. A few examples: "The newborn baby is crying." "Mr. A. sometimes walks in his sleep." "She was startled by the explosion." "He is stuttering." Such statements are, in my opinion, *meaningful* (understandable) descriptions of behavior. Of course, in a very trivial sense, they refer to possibilities: *everything* factual is *eo ipso* possible. But what are the *psychological alternatives* (for the persons concerned)? Surely psychology cannot deny the existence of or refuse to pay attention to unintentional behavior. This would imply as unwarranted a restriction of its scope as (in quite another fashion) orthodox behaviorism attempted. The difference between intentional and unintentional is in psychological description unclear and ambiguous and in psychological explanation one of the most difficult problems.

3. Communication and Conceptualization

No form of knowledge should be called scientific unless it is communicable. A person may have found the solution to a very difficult problem, but as long as he keeps silent his solution—even if it is correct—

does not become available to the scientific community (or the "forum," as De Groot calls it) and therefore cannot contribute to the common store of scientific knowledge.

In psychology, however, this simple requirement leads to very complex problems. For example, human behavior is to a large extent guided (or determined) by sensory impressions. But strictly speaking these impressions are uncommunicable. To a totally color-blind person I cannot *explain* what I refer to if I use words like *red* or *green*. The same applies to communication about subjective states, such as wishes, desires, expectations, and fears. It is well known that in the past attempts have been made to bypass (or eliminate) these problems by declaring them to be irrelevant, that is, by considering them as mere *epiphenomena*. Present-day methodology, on the contrary, is developing a variety of techniques and models in order to improve the possibility of communication about subjective states. Apparently, they are nowadays taken seriously, as potential determinants of human behavior. Perhaps one may venture the statement that the fundamental (and undefined) concepts at the base of these recent developments are comparison (which of course implies discrimination) and preference.

Promising (and even necessary) as the "new" methodology may be, its impact on research and theory is still limited. This is illustrated by the still popular use of *ad hoc* operationalizations. There is in psychology a plethora, a deluge of operationalizations, of measurements of all kinds, and therefore of so-called theories that are not only mutually independent, but even incommensurable. Operationalism has, in a sense, been "imported" from physics. But there is a very considerable difference between physical and psychological measurement. I give just one example. A thermometer may be used to provide information about the temperature of water. A multiple-choice question may be used to test a person's knowledge. The difference is that the structure and the function of the thermometer are part of the *same* conceptual system that is used in describing the temperature of the water. But even so simple a response as indicating (e.g., by a pencilled cross) the correct item involves a number of psychological processes that are to a large extent still unknown, such as seeing, reading, the performance of intentional movements, the willingness to follow instructions, and the like.

There is yet another aspect of the problem. Measurement operations may contribute to, and may facilitate, communication, which is one of the necessary conditions for extending scientific knowledge. A precise description of the way an event (a glimpse of a section of reality) is measured is helpful, to say the least, in assessing the scope and the reliability of the data obtained. In other words, a description of the data combined with a description of the way in which they were obtained is useful (not to say indispensable) to know, in order to know as exactly as possible what we are (in science) talking about. I believe that these considerations are relevant to the questions about "types of theory and structural level" raised by Sanders and Rappard (section 3.3.1.).

Effective communication is made possible by unambiguous conceptualization. Exact concepts (or conceptual systems) are the result of formalization (not identical with quantification). Generally speaking, the natural sciences show a higher degree of formalization than the humanities. There are, in science, *levels of formalization.*

It is, in my opinion, far from evident that the level of formalization achieved in a field of scientific enquiry corresponds to a level of reality. In other words, the most formalized empirical sciences (such as physics) are not necessarily investigating the most basic reality. We should consider the possibility that, if there are levels of reality, the most basic reality might be the most resistant to formalization. Seen in this perspective, psychology might well turn out to be far more difficult than even the most abstruse areas of physics.

Perhaps the humanistic approaches to psychology are aimed at the fundamental level of (psychological) reality. But they lack clear concepts and their level of formalization is not noticeably different from that of common language (this is probably one of the reasons for their popularity). But, presenting themselves as scientific, they promise more than they can give (that some of its adherents reject cold and abstract science is a demonstration of weakness rather than of strength). Humanistic theories underrate greatly the fundamental problem of communication.

A similar underestimation is manifest in everyday language. (This is, by the way, the reason why poets, who in their fashion strive after exactitude, are continually involved in a struggle against conventional language.) But everyday language is adequate insofar as it serves the purposes of social (inter)action; effectiveness rather than scientific understanding is its aim. There are thousands of "natural" languages. Each language embodies certain preconceptions about the structure of reality. This is why it is dangerous to draw general psychological conclusions about human cognitive functioning from the analysis of just one language, to wit, English.

In psychology there are at least two language barriers, the first with respect to its internal communication. The various approaches and schools each have their own vocabularies and therefore are to a large extent unable to communicate with one another, even though they (presumably) are dealing with the same phenomena. The second is in connection with the even larger variety of natural languages.

Psychology, if it is to fulfill its mission to understand human behavior, will have to transcend these barriers. At the moment it is regional instead of worldwide and it is divided instead of united. Cross-cultural research is neither a fad nor a luxury; on the contrary, it is the only way to get rid of the blinkers, the tacit preconceptions that culture and language impose.

Comparisons between psychology and other empirical sciences are difficult and often misleading: psychology is not only, not even mainly, interested in the result of cognitive activities, but in these activities themselves; not so much in the theories as in the theorists; not so much in knowledge as in the *knower*. Of course, as Sanders and Rappard rightly remark, it should not neglect the products of human endeavor, scientific, artistic, and otherwise. But it uses such products in its own way: it inspects them for cues about the determinants of the behavior of their makers.

All the same, in one respect the development of other sciences may serve as a guideline for psychology. In particular the natural sciences have succeeded in transcending the barriers referred to above. They have not a regional but a mondial character and are truly international. On the one hand, this means that scientists with a large variety of personal and cultural backgrounds have been able to develop a system of concepts (a conceptual structure) that permits adequate communication; on the other hand, that they have at least one thing in common, namely, the (psychological) possibility to perform the same cognitive tasks, to function cognitively in the same way.

This can also be expressed in another fashion. The international sciences have succeeded in obtaining knowledge that is considered valid by investigators from every part of the world. The reality disclosed (Sanders and Rappard) by these sciences is *shared* on a worldwide basis. Their example demonstrates not only the presence of cognitive (including normative) determinants that transcend national and cultural barriers; it also carries the promise that in the future psychology will be able to achieve international consensus about the structures of those realities in which it is specifically interested.

4. Final Remarks

If one is invited to discuss a paper, obviously one tends to look for issues that are debatable, where a difference of opinion, or at least in emphasis, may obtain. Consequently, a discussion is focussed on areas of potential disagreement, though there may be many other points on which the discussant finds himself in complete accordance with the authors whose paper he has been commenting upon. This is the situation I find myself in. Therefore I should like to conclude with the remark that even a simple listing of the problems on which I am of the same opinion as the authors would require at least as many pages as the previous discussion.

5

Psychology, Philosophy, and Scientific Research Programs

Willis F. Overton

Sanders and Rappard have presented a most thoughtful and insightful critique and integration of the relationship between psychology and the philosophy of science. In reflecting upon this work, I find myself in agreement with many of the general and specific arguments which they make.

The major point on which I will focus in the present paper is the question of whether Sanders and Rappard are making a commitment to an *epistemology* of *empiricism* or *rationalism* in their consideration of both psychology and the philosophy of science. I believe that this question is crucial to an understanding of a number of their comments, and it will serve as a context within which to explore their perspective.

Before proceeding directly to a general consideration of the question, I will give a brief example in an attempt to demonstrate that it is a question with significant implications. Sanders and Rappard begin from the assumption that all knowledge derives from activity of the organism in interaction with reality. Certainly the emphasis on activity seems to entail a rationalist commitment. However, both here and throughout the paper there is a heavy emphasis placed upon the notion of *reality*. If by reality Sanders and Rappard mean some form of commonsense understandings, or even abstract scientific understandings, that we have of the flux which we call the world, then they stay consistent with a rationalist perspective. In this case they will also commit themselves to other implications deriving from contemporary rationalistic interpretations of the philosophy of science. That is, following philosophers of

Willis F. Overton • Department of Psychology, Temple University, Philadelphia, Pennsylvania 19122.

science such as Hanson, Hesse, Feyerabend, Kuhn, Lakatos, and Laudan, they will agree that there are no bare uninterpreted data; there is no neutral data base; all data (whether that of physics or psychology), as Hanson has said, are theory-laden. In addition, from this perspective, they will understand that the order and organization that the scientist seeks with respect to the empirical world is one which scientists construct in the sense of offering potential solutions to the puzzles or problems that confront the investigator.

On the other hand, if by reality Sanders and Rappard mean a fixed and fully structured material existence operating independently of the knower they are in fact committing themselves to an empiricist epistemology. Here, although activity may be implicated, it is not an activity that is constructive in nature. Rather, it is an activity that simply serves to arrange and order an external really real. In this case they will also commit themselves to an empiricist philosophy of science which claims that there are fixed and theory-free veridical data bases. They will, in addition, view the scientist's job as one of attempting to disclose or uncover the realistic order of a fixed reality (see Sanders and Rappard, section 3.2.2.).

There are other implications that could be drawn from this example, but we hope these are sufficient to suggest that the question is an important one. In order to address the question more directly I will now turn to a brief but more systematic elaboration of empiricist and rationalist proposals concerning the philosophy of science.

As a first point, it should be noted that the main problem for the philosophy of science, regardless of its epistemological commitment, is to establish normative criteria that demarcate science from nonscience or pseudoscience. As Lakatos (1978b) has argued, unless this demarcationist strategy is accepted science becomes a nonrational or irrational activity. That is, unless a demarcationist strategy is accepted science slides between skepticism wherein "anything goes" and elitism wherein only individual scientists can establish standards. Only with universal criteria does it become possible to decide rationally on the scientific acceptability of various propositions and to judge the progress or degeneration of any evolving scientific field.

Lakatos (1978a) describes four rival demarcationist sets of criteria or rival methodologies that have served as standards for the acceptance of theories or theoretical propositions in science. These are the methodologies of positivism, conventionalism, conventionalism with falsification, and Lakatos's own methodology of "scientific research programmes." Recently Laudan (1977) has extended Lakatos's methodology into what he refers to as scientific "research traditions."

The strategies of positivism, conventionalism, and conventionalism with falsification were the direct outgrowth of an empiricist epistemology which asserted that *ultimately* all knowledge derives from an external fixed reality. These strategies, articulated by philosophers such as Ayer, Carnap, Hemple, Popper, and Shapere, held a dominant position in the evaluation of psychology and other sciences until the early 1960s. The common core of each of these methodologies is that in the final analysis science and its progress are judged on the basis of neutral observations and the primary aim of science is to discover universal truths about the real world.

The strategy of scientific research programs or research traditions is the outgrowth of a Kantian rationalist epistemology which asserts that knowledge is the product of reason and observation. This strategy owes much to the contribution of philosophers such as Hanson, Hesse, Feyerabend, Toulmin, and Wartofsky. But its primary path of development, which originated in the late 1950s, begins with Hanson and goes from there to Thomas Kuhn and then to Lakatos and Laudan. This position starts from the proposition that there are no neutral observations and goes from there to show how deep-seated ontological and methodological commitments exert an essential and formative influence on all scientific activity. Here science and its progress are judged not in terms of discovering truths about a real world, but in terms of solving empirical problems in a coherent, systematic preplanned fashion.

The rationalist strategy of research programs and traditions has been particularly influential in my own attempts to articulate the ways in which deep-level conceptual foundations (world views, or models, or paradigms) play an essential and formative role in the development of psychological theories and day-to-day research approaches (see Overton, 1976, 1982, 1984; Overton & Reese, 1973, 1981; Reese & Overton, 1970). I believe that this strategy would also best serve as an interpretative matrix for many of the arguments made by Sanders and Rappard. However, at present, it is unclear whether they would be willing to accept this strategy or whether they would be more prone to interpret their work as more closely aligned with one of the variants of conventionalism. In order to elaborate on this issue I will next briefly describe the empiricist demarcationist strategies of positivism and conventionalism.

Positivism established two universal criteria as standards of scientific legitimacy. The first was that ultimately all general propositions in science (that is, theoretical statements) must be reducible to statements describing hard data, that is, observations. Second, general propositions in science must be formulated on the basis of *inductive inference* and only inductive inference from observables.

Ultimately, positivism failed as a demarcationist position. The reason for this is that it came to be recognized that it is impossible to reduce general theories to observational propositions, and even more importantly it was recognized that the laws of science can seldom be adequately described as inductive generalizations.

Conventionalism was built directly upon positivism and it incorporates many of its features, including a heavy reliance on inductive inference (Lakatos, 1978a, p. 106). The major change in conventionalism is the recognition that not all scientific propositions can be reduced to observational statements. Conventionalism, therefore, allows for the introduction into science of general nonreducible propositions. However, the primary feature of such statements is that they serve only as convenient and conventional ways of ordering and arranging hard data, that is, observations. They do not influence the data base itself. Rather, they operate like pigeonholes to classify, arrange, and organize hard data.

As conventions, theoretical entities or models are lightly held and readily given up when simpler ways are found of organizing hard data. As Lakatos pointed out, "For the conventionalist . . . (theoretical) discoveries are primarily inventions of the new and simpler pigeonhole systems" (1978, p. 107). However, the "genuine progress of science . . . takes place (still, as with the positivists) on the ground level of proven facts (i.e., hard data) and changes on the theoretical level are merely instrumental" (p. 106).

Conventionalism, then, yields two distinct levels of scientific activity. These have been referred to as the context of justification and the context of discovery (Nickles, 1980; Overton, 1976; Reichenbach, 1938). The context of justification forms a lower level of unchanging objective data which are describable in a pure observational language upon which all can agree. This level also includes experimental manipulations and inductive generalizations, and it is at this level that genuine scientific explanation and progress take place. The context of discovery includes theoretical terms, nonobserved entities, and models. These propositions may themselves be the products of the scientist's hunches, guesses, creative imagination, or metaphysical presuppositions. It matters little because this level exerts no real influence on the essential features of science, that is, those included in the context of justification.

This outline of positivism and conventionalism facilitates a closer analysis of some core features of Sanders and Rappard's work. It appears that the key to their general position is entailed by their concept of "functional structuring." By this they mean that, depending upon the investigator's intention, a particular domain of reality will be cognitively ordered in a particular manner (section 1.3.1.). Note here that there is the clear implication that the reality is a fixed neutral base and that the ordering is simply a convention (determined by intention, i.e., function) for arranging this.

Further, Sanders and Rappard propose that given the plurality of possible arrangements these may themselves be ordered in terms of their individual distances from reality (i.e., the "level of spatio-temporal materiality"— section 1.3.2). Thus different forms of knowledge (particularly among the empirical sciences, but I presume Sanders and Rappard would also include philosophical knowledge) constitute a ladder of increasing abstractness and ambiguity ("meaning variance") ranging from physics through "psychology, history, sociology, and so forth" (section 1.3.2.).

If one takes these two proposals together, a number of other features of their work fall into place. For example, explanation is limited to establishing causal laws (i.e., antecedent–consequent relationships) and it is reserved primarily for those at the lower levels of the ladder or structural hierarchy, those closer to reality. Understanding, or *Verstehen,* based on internal coherence, applies to the upper levels. Similarly, empirical testing is reserved for the lower levels. Also, the extent to which observations are theory-laden increases as one moves to higher levels. Finally, within any given level (e.g., psychology) a neutral observational language can be found that will permit comparative tests among rival theories.

Despite their explicit rejection of a radical split between the context of discovery and the context of justification (section 3.3.5.), the sum of Sanders and Rappard's work suggests that it is highly compatible with a conventionalist strategy. If this, in fact, is the claim they would wish to make I suggest several weaknesses inherent in it. The primary problem with the concept of functional structuring is that it yields no universal standards by which one can judge the totality of scientific activity. At any given level structuring itself offers no *rational* explanation for why a particular cognitive order and a particular intention are selected. The effect of this is that any given science can be judged only on psychological or sociological grounds. But if science is judged by psychological or sociological standards then it is predominately irrational, and as Laudan (1977) pointed out, "then there is no reason to take its claims any more (or less) seriously than we take those of the seer, the religious prophet, the guru, or the local fortuneteller" (p. 2).

A related set of problems exists with respect to Sanders and Rappard's "levels" of science. To order the sciences in terms of their distance from reality is at best a vague, controversial, and potentially divisive strategy. First, it is highly questionable whether, for example, physics as it focuses on issues of the Schrodinger wave function, quarks, and antimatter is less distant from any level of spatio-temporal materiality than is either psychology or sociology. Second, ordering science in terms of abstractness closely resembles a failed strategy employed by the early positivists. The positivists suggested that a unity could be established among the sciences by ordering the sciences in such a way that ultimately the propositions of the more abstract sciences could be reduced to those of the more concrete sciences and from there to observations. Both historical examples and philosophical analysis demonstrate that this positivist strategy failed.

Despite their disclaimer, there is also in Sanders and Rappard's concept of levels a distinct suggestion that the natural sciences are primarily concerned with issues relating to the context of justification whereas other sciences are more involved with issues involving the context of discovery. Thus, explanation and experimentation are reserved for the one group and understanding for the other. This suggestion is particularly divisive because it promotes different scientific standards for different sciences and such fractionation ultimately leads to the question of which are the *real* sciences.

As a final point, it also should be noted that the concept of levels of science effectively displaces philosophical thought, for example, ontological commitments, either outside the body of science or to a level that is so abstract as to be largely irrelevant to science. This strategy is compatible with a conventionalist interpretation, but I believe that such an interpretation errs in failing to recognize the essential and formative influence that deep-seated philosophical commitments have upon all sciences.

In summary, then, if Sanders and Rappard's proposal is interpreted as deriving from a commitment to an empiricist epistemology and consequently a scientific demarcationist strategy of conventionalism, the proposal exhibits significant problems. I believe that, in contrast, if the proposal is interpreted within the framework of a rationalist epistemology, and specifically within the demarcationist strategy offered by Lakatos and Laudan, these problems are resolved and a number of Sanders and Rappard's arguments maintain a viability with respect to the relationship between philosophy and psychology.

In the following section, I will describe Lakatos's and Laudan's strategy in an attempt to show the way this strategy might serve to provide more rational criteria for the nature of science and for the relationship between psychology and philosophy. Prior to describing this rationalist position I will mention briefly a final conventionalist strategy to which Sanders and Rappard make reference, Popper's conventionalism with falsification (Popper, 1959).

The main point with respect to Popper's strategy is that it remained steadfastly within an empiricist epistemology. Popper's proposal that falsifiability be accepted as the standard of scientific acceptability led to the admission of theories into the essential body of science. However, for Popper, other more general propositions such as metaphysical propositions, could only provide the external stimulus that leads the scientist to falsifiable theories. Metaphysical propositions, since they are inherently nonfalsifiable, could not themselves enter the essential body of science.

For both Lakatos and Laudan, the demarcation of science from nonscience begins with a unit of analysis that is broader than any observational data base and broader than any isolated theory or conjunction of theories. Lakatos referred to this unit as a scientific research program and it is possible to have several rival programs operating in a field at any given time. A research program may be thought of as being composed of three levels arranged in a hierarchy. The most general level is called the *hard core* of the program. At a somewhat less general level there is the *positive heuristic* of the program. These first two levels constitute a set of ontological and methodological do's and don't's. They define problems and outline the construction of the third level, called the *belt of auxiliary hypotheses* that are embodied in a family of specific theories.

The hard core of a program may contain various types of propositions including metaphysical propositions. The importance of this fact is that, in contrast to rival strategies, research programs admit into the essential body of science propositions that may have no potential falsifiers. Metaphysical propositions, as they may enter the hard core of a scientific research program, are not then idle psychological or sociological curiosities. Metaphysical propositions are, in this case, essential components of scientific activity. They exert a formative influence on lower levels and give meaning to the theoretical concepts of specific theories. For example, Lakatos describes how Cartesian metaphysics, that is, the mechanistic view of the universe, operated as a hard core of a scientific research program which discouraged work on scientific theories that were inconsistent with it, for example, Newton's theory of action at a distance, while encouraging work on auxiliary hypotheses which might have saved it from counterevidence (Lakatos, 1978a, p. 48).

The hard core of a scientific research program or "research tradition" (Laudan, 1977) is the vehicle by which philosophical commitment influences scientific choice. The hard core cannot be refuted, and to reject features of the hard core is to repudiate the particular program or tradition in question.

The positive heuristic of a research program is influenced by the hard core but describes the long-term research policy of the program. The hard core provides the ontological prescriptions and proscriptions, the positive heuristic the methodological ones. To quote Lakatos, the positive heuristic consists of "a partially articulated set of suggestions or hints on how to change, and develop the 'refutable variants' of the research programme, how to modify, and sophisticate the 'refutable' protective belt of auxiliary hypotheses" (Lakatos, 1978a, p. 50).

Taken together, the hard core and the positive heuristic of a given program constitute a *conceptual framework* which generates specific theories that within any given program constitute a family of theories. These theories in turn embody the belt of auxiliary hypotheses, the sets of observational hypotheses that constitute the falsifiable or refutable component of the scientific research program.

For Lakatos and Laudan the research program or tradition serves as the context within which any science is evaluated, be it physics, biology, psychology, sociology, or whatever. The aim of science within this strategy is not that of discovering universal truths about a fixed reality. The aim of science generally, regardless of the particular domain of inquiry, "is to maximize the scope of solved empirical problems, while minimizing the scope of anomalous, (i.e., empirical falsifications of observational hypotheses) and conceptual problems" (Laudan, 1977, p. 66). Any given program or tradition can be judged in terms of its progressiveness (or regressiveness) as it meets (or fails to meet) this aim of science across the course of time. Further, rival research programs or traditions can be rated and hence compared according to this scientific standard of progressiveness.

The research program strategy would accept Sanders and Rappard's concept of functional structuring. However, the domain in question would not be a domain of reality. The domain would be a set of empirical problems. Empirical problems would in turn be recognized, not as veridical bits of unambiguous data but as issues that are always determined and colored by the features of some conceptual network or other. Further, the functional structuring that occurs both across different sciences and within any given science is determined by the normative structures of the particular research program. It is not a product of simply cultural or psychological factors.

A research program strategy, in contrast, would not admit a concept of levels of science. From a research program perspective a concept of levels pertains to the arrangement of propositions (hard-core, positive heuristic, theories, observational hypotheses) as they form a stratified organization (Overton, 1976) both across the different sciences and within any given science. From this perspective each science must ultimately submit to empirical evaluation, and one science can never be reduced to another. One science does not exist in the realm of the context of discovery and another in the context of justification. Each science fully participates in discovery (hard core, positive heuristic, theories) and justification (observation and hypothesis testing).

To this point I have suggested a possible relationship between the philosophy of science, general philosophical propositions, and science by describing the nature of the scientific research program or traditions. I will now attempt to describe these components of scientific activity in relationship to the specific science of psychology. For a number of years my colleague Hayne Reese and I have maintained that two rival world views, models, or paradigms have strongly influenced concepts, theories, and research in several of the sciences including psychology (Overton, 1976; Overton & Reese, 1973, 1981; Reese & Overton, 1970). Following others (e.g., Pepper, 1942), we referred to these general sets of propositions as the mechanistic and organismic world views or world models. Recently, however, I have come to believe that these models and the corollaries that derive from them are best understood as rival scientific research programs or traditions (Overton, 1982, 1984).

The hard core of the mechanistic program is expressed in an ontological commitment to a Locke–Hume philosophy of being wherein uniformity, stability, and fixity are considered basic while change and organization are understood as the result of contingent or accidental factors only. The positive heuristic or research policy of this program—or what Overton and Reese earlier referred to as corollary model issues—encourages the practitioners of the program to work within a framework of elementaristic or reductionistic analysis, to consider all change and organization as the product of contingent antecedent factors, and to represent all change as strictly additive or continuous in nature. The positive heuristic also establishes that all explanations will be contingent explanations based on efficient or material factors (see Overton & Reese, 1981).

The hard core and positive heuristic form the conceptual basis for a family of theories. The theories constitute a family because, although they may demonstrate mutual inconsistencies, each adheres to, or at a minimum does not violate, the ontological and methodological commitments of the hard core and positive heuristic. In psychology, the mechanistic family includes but is not limited to: behavioristic and neobehavioristic theories, operant and classical conditioning theories, observational learning theories, mediational learning theories, and some variants of information-processing theories. Finally, the theories themselves lead to testable auxiliary hypotheses with respect to various issues in the particular empirical domains under investigation, for example, cognition, language, perception, social-emotional.

In contrast to the mechanistic program, the hard core of the organismic program is expressed in an ontological commitment to a Kant–Hegel philosophy of becoming, wherein organization, activity, and change are understood as natural and necessary features of the cosmos and not simply the product of contingent accidental factors. Accidental factors can here affect activity, organization, and change, but they cannot explain them. The *organismic positive heuristic* encourages its practitioners to work within a holistic-analytic framework, to consider change and organization as necessary and consequently open to a structure–function analysis, and to represent change as both continuous and discontinuous. The positive heuristic also establishes that both formal and contingent explanations are legitimate and that each serves a different role in providing general explanations (see Overton & Reese, 1981; Rychlak, 1977).

The family of theories generated by this program includes, but is not limited to: contemporary structuralist theories such as those of Chomsky, George Kelly, Kohlberg, Piaget, and Werner; humanistic theories such as those described by Rychlak; gestalt theories; and ego development theories such as those of Erikson and Bowlby. Finally, this family too, like the mechanistic family, leads to testable auxiliary hypotheses in the empirical domains under investigation.

To describe the various ways in which these two programs function as rivals in solving empirical and conceptual problems is beyond the scope of this paper (see Overton, 1982, 1984; Overton & Reese, 1981). However, several points should be mentioned briefly as they relate to the Sanders and Rappard proposal. First it should be apparent that the two positions are general research programs and not specific theories as suggested by Sanders and Rappard. Thus, the division between the organismic and mechanistic programs is not analogous to Sanders and Rappard's structural levels (section 3.3.3.).

Further, the organismic program forms the conceptual base for the generation of humanistic theories among others. Humanistic theories do not stand as a separate category. The importance of this lies in the fact that it must be maintained that all theories as they are generated by either program necessarily lead to testable observational hypotheses designed to solve empirical problems. If any theory does not lead to testable observational hypotheses then it simply cannot count as a *scientific* theory. The fact that some humanists have avoided or even derided the process of testing does not form an adequate justification for including them as a separate category of scientific theories (see Rychlak, 1977).

Both the organismic and mechanistic scientific research programs lead to theories that are designed to *explain* empirical events, and both adhere to the position that the explanations offered must be tested by means of empirical methods. However, these programs are competitive rivals in that each defines the context of inquiry differently. For example, the organismic program asserts that both structure–function explanations (formal explanations) and contingent explanations are necessary in order to solve the empirical problems that comprise the domain of psychology. In contrast, the mechanistic program asserts that only contingent antecedent–consequent causal explanation is necessary. This does not mean that the mechanistic program is "closer" to the natural sciences, because the natural sciences are themselves replete with noncausal laws (see Overton & Reese, 1981). What it does mean is that each program vies with the other to be judged as the most scientifically progressive.

In conclusion, in my judgment, a rationalist interpretation in the form of scientific research programs or scientific research traditions provides the most adequate contemporary perspective within which to consider the relationship of the philosophy of science, general philosophical propositions, science and psychology. Sanders and Rappard's proposal offers many interesting insights and important arguments about this relationship. However, I believe that their position would be significantly enhanced were they to explicitly repudiate an empiricist epistemology and instead formulate their proposals in the context of scientific research programs and research traditions.

1. References

Lakatos, I. (1978a). *The methodology of scientific research programmes: Philosophical papers* (vol. 1). Cambridge: Cambridge University Press.

Lakatos, I. (1978b). *Mathematics, science and epistemology: Philosophical papers* (vol. 2). Cambridge: Cambridge University Press.

Laudan, L. (1977). *Progress and its problems.* Berkeley: University of California Press.

Nickles, T. (1980). Introductory essay: Scientific discovery and the future of philosophy of science. In T. Nickles (Ed.), *Scientific discovery, logic and rationality.* Boston: D. Reidel.

Overton, W. F. (1976). The active organism in structuralism. *Human Development, 19,* 71–86.

Overton, W. F. (1982). Scientific methodologies and the competence–moderator–performance issue. Invited address presented at the annual symposium of the Jean Piaget Society, Philadelphia.

Overton, W. F. (1984). World views and their influence on psychological theory and research: Kuhn–Lakatos–Laudan. In H. W. Reese (Ed.), *Advances in child development and behavior* (vol. 18, pp. 119–226). New York: Academic Press.

Overton, W. F., & Reese, H. W. (1973). Models of development: Methodological implications. In J. R. Nesselroade and H. W. Reese (Eds.), *Life-span developmental psychology: Methodological issues* (pp. 65–86). New York: Academic Press.

Overton, W. F., & Reese, H. W. (1981). Conceptual prerequisites for an understanding of stability–change and continuity–discontinuity. *International Journal of Behavior Development, 4,* 99–123.

Pepper, S. (1942). *World hypotheses.* Berkeley: University of California Press.

Popper, K. R. (1959). *The logic of scientific discovery.* London: Hutchinson.

Reese, H. W., & Overton, W. F. (1970). Models of development and theories of development. In L. Goulet & P. Baltes (Eds.), *Life-span developmental psychology: Research and theory* (pp. 115–145). New York: Academic Press.

Reichenbach, H. (1938). *Experience and prediction.* Chicago: University of Chicago Press.

Rychlak, J. (1977). *The psychology of rigorous humanism.* New York: Wiley.

5

Psychological or Philosophical Issues?

Reply to Commentators

C. Sanders and H. V. Rappard

In his commentary, Duijker addresses our conception of *reality*, which he finds to be "unclear" and to have an "ontological flavor." Furthermore, he objects to the notion of *basic reality* and *levels of reality*.

To begin with, we reject every ontological assumption. We have expressed no opinion about reality-as-such. What we termed basic reality is in no sense intended to be ontological. In accord with Strawson, we began from the way people *speak* and *think* about the world. Strawson maintains that they do so "as containing particular things, some of which are independent of ourselves." This is a point of view which is also shared by Duijker. Again, following Strawson, we concluded that in our thinking about particular things, the existence of things which are or possess material bodies is always assumed. In other words, material bodies and persons "are what primarily exist" (Strawson, 1959, p. 39). Finally, we discussed the difference between material bodies and persons and noted that the word *person* is used in a specific sense which is not identical with the conception of a person in personalistic thinking.

When speaking of basic reality, the point at issue is what *primarily exists*. We emphasized that what primarily exists does not denote reality as such, but a conceptual structure that underlies and is a condition for our way of speaking about reality. Similarly, the concept of levels of reality is not used in the sense of "layers" of reality as such, but rather as levels of interpretation–structuring–of reality. Therefore, the levels of

C. Sanders and H. V. Rappard • Department of Theoretical Psychology, Free University, De Boelelaan 1081, 1007 MC Amsterdam, The Netherlands.

reality are products of human activity, of a human disclosing or structuring of the given.

Another point that Duijker raises concerns (psychological) knowledge. First of all, we should make clear that our approach was primarily epistemological whereas Duijker emphasized the acquisition of knowledge as a psychological process. The difference between our positions, it would seem, is that Duijker is less convinced than we are of the necessity of choosing an epistemological position prior to conducting research. This becomes clear from his statement: "As I see it, reality may function as an arbiter precisely because it is independent of our notions, wishes, and so on" (p. 272). Thus the observer behind a one-way screen studying the movements of a child is apparently merely an onlooker and does not interact with the object of his research.

However, the behavior of the child is not an independent reality which acts as an arbiter; rather, the behavior of the child and our structuring of it (reality) form one system which is constituted from a specific human perspective which is not an absolute one. Thus we again find scientific research rooted in the interaction between observer and observed.

Duijker attaches as important a role to reality in the process of acquiring knowledge as we do ourselves. However, whereas Duijker speaks of a *limitative* role, we speak of a *normative* role. This is more than a terminological problem, for *normative* refers to a guiding and restricting principle operating in time, that is, a historical principle, whereas *limitative* denotes an inner-paradigmatic role.

When discussing the logic of perception, the problem of balancing the philosophical (that is, the logical) and the psychological is again at issue. Once more Duijker maintains the priority of the psychological: "I do not believe that a logician is competent to prescribe how statements about perception should be formulated while in psychology the problems are far from solved" (pp. 273–274). In psychology many modes of perception are known; some of them justify the use of propositional attitudes; others, however, are better served for the description of the observed by means of "observational propositions." So from the psychological point of view the general validity of the logic of perception described by us is at least open to question. And yet this conclusion is not justified. From an epistemological point of view, one can very well maintain the modal character of the logic of perception. For example, in some situations the role of the observer is minimized because of perceptual training which results in a uniform description of observations within a particular scientific community. Within that community it is no longer necessary to mention that the reported observations have the character of propositional attitudes in disguise. In practice it is merely a matter of observational statements. However psychologically different the various modes of perception may be, from our epistemological point of view they are always reports by an observer, by a human subject who concretizes his perception in an act of knowledge.

Duijker's remark on the meaningfulness of behavioral descriptions is also understood by us as originating from a different emphasis on the need for a philosophical starting point. From our perspective, meaning was connected with the structuring activity of the knowing subject, and in that light *meaningfulness* is synonymous with *intention*. We do not understand why this should lead to "fuzziness." Indeed, some forms of behavior cannot in this sense be termed meaningful. When Duijker claims that these forms of behavior are just as meaningful as intentional actions, he understands *meaningful* as "intelligible," which is rather different from the sense in which we used it.

Finally, when Duijker states that it is far from evident that the level of formalization achieved in a field of scientific inquiry corresponds to the level of reality, we should point out that we spoke rather of levels of *structuring* and not of levels of *reality*. And we affirm a correspondence between levels of formalization and levels of structuring: lower levels of structuring correspond to relatively simpler formalizations.

The major point of Overton's commentary is whether we are committed to an epistemology of empiricism or rationalism. However, we reject the alternatives and opt for a third possibility, namely, a functional epistemology in which the separation between subject and object, so characteristic of both rationalism and empiricism, is abandoned and in which knowledge is seen as relational.

To elucidate our position we must again address, as we did in our reply to Duijker, our conception of reality. Reality is in our opinion not an *objectively* recognizable "fixed and fully structured material existence operating independently of the knower" (p. 280); rather, all data are theory-laden, including the data of physics. Neither, however, is reality a mental construction. It is a datum that transcends human knowledge or, as the Dutch philosopher Van Peursen puts it, "reality is, in a way, the very resistance against which the knowing subject can develop his knowledge." In the acquisition of knowledge a transcendental principle can be recognized (that is, reality) which should be distinguished from knowledge since it is the *condition* of knowledge. Reality is what regulates the acquisition of knowledge in a *normative* manner; it constitutes an appeal for progress and maintains the continuing search for knowledge (see Van Peursen, 1965, p. 45).

We refer to this transcendental principle in a concrete way when qualifying knowledge as correct or incorrect, by asking ourselves time and again whether or not we are mistaken (that is, by testing our views).

The concept of reality with which science works, however, does not directly refer to this transcendental reality-as-such, but to a reality which has been disclosed in the interaction between subject and object; in short, our position is not that of rationalism, as Overton describes it, because in our opinion there is a place for an independent concept of reality as a touchstone for knowledge. And neither is empiricism at stake because in our view reality, the "real world," is never available for science as a pure uninterpreted fact, as neutral observation.

Overton observes that "the main problem for the philosophy of science, regardless of its epistemological commitment, is to establish normative criteria that demarcate science from pseudoscience" (p. 280). Starting from four rival methodological positions, that is, positivism, conventionalism, conventionalism with falsification, and Lakatos's scientific research programs, Overton states that "the common core of each of the first three methodologies is that in the final analysis, science and its progress are judged on the basis of neutral observations and the primary aim of science is to discover universal truth about the real world" (p. 281). In the case of the last methodology, it is a question of an assumption whereby "there are no neutral observations. . . . Here science and its progress are judged not in terms of discovering truth about a real world, but in terms of solving empirical problems in a coherent, systematic, preplanned fashion" (p. 281). Overton concludes that it is difficult to tell how our position should be evaluated. According to him there are indications of a conventionalism. But in view of our emphasis on the role of knowledge already at hand and the significance of norms and values for functional structuring and also on the roots of knowledge in the concrete reality (Strawson), this judgment is, in our opinion, difficult to maintain. However, perhaps the problems indicated by Overton do not originate in a lack of clarity about our starting point but rather are the result of his classification of methodological positions which leave no room for other starting points.

An already familiar problem that Overton raises is his objection to the concept of levels of reality. First of all, we must point out again that we are *not* concerned with ontologically conceived levels of reality, but with levels of *reality-structuring*, that is, with levels of disclosed reality. Reality can be structured in many ways. If we compare those structurings, they appear to differ to the extent to which they reduce meaning (*Sinn*)—not, as Overton states, abstractness—as developed in the course of history. The most extreme reduction ends up with the spatial-temporal structuring when man looks at reality solely from the perspective of physics.

It appears that knowledge acquired on the basis of a certain way of structuring is not independent of the knowledge gained by using another structuring. Often one may come across a hierarchy in which knowledge of a certain higher level cannot be reduced to knowledge of a lower level without losing the essentials, for example, intentional behavior reduced to "behavioristic" behavior. But this does not mean that the laws found at the lower level would cease to hold true for structuring at a higher level, only that they are insufficient at that level. Therefore, we do not classify theories according to their level of structuring on the grounds of their distance from reality, but on the grounds of their level of reduction or meaningfulness.

A last issue concerns our distinction, from a descriptive perspective, between mechanistic, organismic, and humanistic theories. We stated that these theories reflected differences in levels of structuring. In science we may expect mechanistic theories at lower levels of structuring, whereas organismic and humanistic theories are at higher levels. We did *not* claim, as Overton seems to assume, *that an intrinsic connection exists between natural sciences, biosciences and human sciences on the one hand and mechanistic, organismic, and humanistic theories on the other hand*. We did, however, make the point that it is apparently easier and often also more successful to start from a mechanistic paradigm in those sciences which study a strongly reduced aspect of reality than in less reductive sciences. We do not hold the view that spacial-temporal reality is mechanistic in nature. In principle it is possible to use an organismic frame of reference but, of course, such a science would lie at a higher structuring level than its mechanistic counterpart. Just as for Overton, so for us the time-honored mechanistic-organismic cleft is unbridgeable in the sense that we do not conceive of a continuous transition. Once more, however, we would like to point out that this chasm is a result of man-made structurings. The mechanistic and organismic may not be taken as isolated and unchangeable positions between which one has to choose. *Unbridgeable* does not mean a radical incommensurability between mechanistic and organismic starting points. From our metatheoretical framework the mechanistic approach is no less correct than the organismic, but it should be seen as a legitimate and in certain cases appropriate simplification of organismic structuring.

1. Reference

Peursen, C. A. van (1958). *Filosofische orientatie*. Kampen, Holland: Kok.

Peursen, C. A. van (1965). *Feiten, waarden en gebeurtenissen*. Hilversum, Amsterdam, The Netherlands: De Haan/Meulenhoff.

Strawson, P. F. (1959). *Individuals: An essay in descriptive metaphysics*. London: Methuen.

6

The Problem of Theoretical Pluralism in Psychology

Joseph R. Royce

Abstract. The contemporary morass of facts and theories has resulted in a state of intellectual paralysis in the discipline of psychology. In short, we need a pragmatic basis for dealing with theoretical pluralism. Because extant theories are refuted by better theories rather than by direct refutation I end up with the paradoxical conclusion that the best way to deal with theoretical pluralism is to produce more theory—but with the qualification that the new theory be more theoretically powerful than its predecessors and/or its competitors. The paper concludes with an elaboration of the role of dialectic analysis in the evaluation of complementary and competitive theories.

1. The Problem

In other papers on theoretical psychology (e.g., Royce, 1978) I presented a program of metatheoretic analysis based on a philosophy of constructive dialectics (Royce, 1977). A major conclusion of these earlier analyses is that a metatheory for psychology will be viable only if it confronts the major problems of the discipline. One such problem is the fact of theoretical pluralism—that is, the existence of a maze of theories in each of the major domains of psychology (Royce, 1970).[1]

Why is this a problem? It is a problem because of the conceptual confusion engendered by such diversity, because of the demands it places on the research investigator, and because such multiplicity is usually an indication that not much is known about the subject in question. All of these points about psychology's theoretical pluralism are

[1]The primary purpose of this paper is the modest one of bringing the problem of theoretical pluralism to the attention of fellow psychologists in the hope that this will lead to further metatheoretic analysis. The issues this problem entails are complex, and their eventual resolution will require extended and penetrating critical analysis.

Joseph R. Royce • Center for Advanced Study in Theoretical Psychology, University of Alberta, Edmonton, Alberta T6G 2E9, Canada.

true. However, I do not agree with the conclusion implicitly held by many and explicitly expressed by Koch (1974), that this means psychology has failed as a scientific endeavor. The major reason for my disagreement with this conclusion is the fact that psychology has not yet been put to the test as a theoretical science. Furthermore, it is important to point out that theoretical pluralism is a characteristic of all sciences, both mature and immature, but that it is particularly characteristic of immature sciences such as psychology. A major reason for this state of affairs has to do with the corrigibility of the entire theory construction enterprise. This situation is elegantly summarized by the philosopher Mario Bunge:

> Every scientific theory is built, from the start, as an *idealization* of real systems or situations. That is, the very building of a scientific theory involves *simplifications* both in the selection of relevant variables and in the hypothesizing of relations (e.g., law statements) among them. . . . Such simplifications are, of course, deliberate departures from the truth. . . . When such theorems are subjected to empirical tests a discrepancy with the outcomes of such tests will sooner or later emerge—the finer the empirical techniques the sooner. And such a discrepancy will force the theoretician to gradually complicate the theoretical picture. . . . But the improved theory will *no less* refer immediately to an ideal model than the previous theory. This cannot be helped, because *theories are conceptual systems rather than bundles of experience.* (Bunge, 1967, vol. I, p. 388; italics mine)

2. The Growth of Science and Theoretical Pluralism

When viewed from the perspective of evolutionary epistemology (e.g., see Campbell, 1960; Popper, 1974) it becomes clear that theoretical pluralism is a necessary condition for progress. That is, theoretical variation is a precondition for the subsequent adoption of a new theory more adequate than the current one.

The recent research on the question of how science advances comes primarily from both the historiography and the philosophy of science. The major difference between these two approaches is that the philosophic approach is focused on the logic of the scientific enterprise whereas historical analysis is focused on "what actually occurred." Although the issue of theoretical pluralism emerges from both of these sources, most of the analyses are focused on pluralism as a historical sequence of paradigm shifts. I will refer to this form of pluralism as sequential theoretical pluralism, a form of pluralism which manifests itself primarily in the mature sciences.

But I will also defend the view that theoretical pluralism is a characteristic of immature sciences as well—but that it manifests itself simultaneously rather than sequentially. The major reason for this difference is that the immature sciences have not been able to evolve a single paradigm capable of assimilating a discipline's full range of empirical findings and problem-solving techniques. Although Kuhn (1970) focused on mature science (especially physics), his latest version of a paradigm is also relevant to theoretical pluralism in all scientific disciplines. He now describes a paradigm as "what the members of a scientific community, and they alone, share," where a scientific community is defined as "the members of a scientific specialty" (Kuhn, 1970, p. 460). He also points out that the size of a scientific community varies widely and that it can often be less than 100. Thus, he clearly means to include specialists within any one scientific discipline as well as specialists between scientific disciplines (i.e., *any* group of scientific specialists). A variety of philosophers of science, such as Lakatos (1968) and Feyerabend (1970), also support the view that theoretical pluralism is characteristic of all science; however, it is the philosophers Popper (1963, 1974) and Naess (1972) and the psychologist Campbell (1960) who strongly support this view of science. Campbell's view arises from his conception of epistemology as evolutionary. Popper's (e.g., see Schilpp, 1974) view is very similar. It is based on the idea that normal science consists of many "small revolutions or trials, most of which will be errors" (his motto is "Science in permanent revolution")—that is, there is no such thing as normal science in Kuhn's sense. However, Naess comes the closest to the view I am espousing—namely, that the production of many theories is characteristic of the scientific enterprise regardless of degree of maturity. He argues that it cannot be any other way because of the metatheory which undergirds the production of theories. For example:

> If a description of scientific theorizing tries to do justice to very different ways of doing biology, chemistry, physics, astronomy, this must affect the very essence of how "scientific theorizing" is conceived in its complete generality. A very broad metatheoretical concept is needed, within the frame of which a plurality of kinds of theories and their methodologies can be placed. A pluralist approach is needed, otherwise one way of doing things is explained as an unsuccessful attempt at doing it another way. (Naess, 1972, p. 97)

Although psychology is not as broad as all of science, it does range from the biological to the social sciences (Royce, 1983). Thus, the sheer complexity of the field is a major reason why it contains so many theories (i.e., no one theory can assimilate the full range of psychological phenomena). This point is brought out quite clearly by Rosenzweig (1937) in his analysis of the complementary roles of psychology's major schools (see Table 1).

This view is also buttressed by Brunswik's (1939, 1952, 1956) and Royce's (1970) more recent analyses of the conceptual and epistemic foundations of psychological theories. But if we accept the conclusion that the prolific production of alternative theories is a more accurate reflection of the typical situation, then we need a method for appraising theories in order to determine their degree of theoretical power (Royce, 1978).

3. Theoretical Pluralism and Complementary Theories

Determining the adequacy of a theory is a difficult problem. For example, the historiographers have pointed out that theories are not refuted or accepted in an all-or-none manner. Rather, there is cumulative evidence for and against. This means that there are instances of confirmation or disconfirmation for a specifiable segment of a theory. Furthermore,

> these instances do not determine the truth or falsity of a theory. The question is rather: What appears to be the best research strategy right now. Saving [theory] *x* most often means something like 'finding out how to continue to work with *x*', whereas 'accepting *x*' usually means 'working with *x*'. It is a question of 'accepting as tenable', not of 'accepting as true'. (Tennessen, 1981, pp. 39–40)

Thus, theories can range from low to high tenability or plausibility. However, no theory, even those that are minimally plausible, is regarded as "refuted." Instead, metatheoretic analyses suggest that they should be placed on "hold," meaning "suspend action until further notice." Why place an implausible theory on hold? Because we have no convincing way of deciding when a theory will continue to be viable and when it will not. For example:

> By postulating an almost massless and chargeless particle, the existence of which to physicists "seemed outstandingly impossible to verify," an important hypothesis about the conservation of momentum and energy could be saved. So-called "data" or "facts" which we considered strongly disconfirmatory could be seen as miraculously changed into wonderfully confirmatory instances by the change in hypotheses elsewhere. Later on, physicists were able to give independent, empirical confirmation of the existence of such a particle. (Tennessen citing Robert Cohen in Tennessen, 1981, p. 39)

On the other hand,

Table 1. A Complementary Pattern of Psychological Schools[a]

			Functionalism		
Name of school	Structuralism	Gestalt psychology	Behaviorism	Psychoanalysis	*Geisteswissenschaftliche* or *verstehende Psychologie*
Characteristic problem field of research	Sensation	Perception	Learning	Motivation	Character
Temporal span of observation in typical methods	●	○	→	→	→∞○
Conceptually allied sciences	Physical e.g., physics		Biological e.g., physiology	Social e.g., sociology	

[a]From "Schools of psychology: A complementary pattern" by S. Rosenzweig, 1937, *Philosophy of Science, 4*, p. 97. Copyright 1937 by The Williams & Wilkins Co. Reprinted by permission.

> by the end of the century Priestley was almost alone in defending the doctrine of phlogiston. It is often possible by adding a number of new special auxiliary postulates to a conceptual scheme to save the theory—at least temporarily. Sometimes, so modified, the conceptual scheme has a long life and is very fruitful: sometimes, as in the case of the phlogiston theory after 1785, so many new assumptions had to be added year by year, that the structure collapsed. (Conant & Nash, 1957)

The point the proponents (e.g., Lakatos, Feyerabend, Popper, and Kuhn) of the pluralist view are making is that science progresses because of the acceptance of increasingly better theories rather than "refutation" of inadequate theories. Thus, there is no clear way to determine the plausibility of a theory. There is no algorithm, for example; and therefore, the passage of time is needed in order to allow for a retroductive assessment of a theory's value. However, it must be pointed out that these conclusions are primarily concerned with the mature sciences and the case of sequential theoretical pluralism. But Kuhn indicates that theoretical pluralism is a symptom of a science in crisis. And he alludes to the idea that pluralism is also a characteristic of immature science: "In the early 1770s there were almost as many versions of the phlogiston theory as there were pneumatic chemists" (Kuhn, 1970, p. 132). The point is that paradigmatic pluralism and theoretical pluralism are symptoms of any science in crisis, whether the science is mature or immature. As I see it, the major difference in the two situations is that pluralism is typically sequential in the case of the mature sciences whereas it is typically simultaneous in the case of the immature sciences. But the critical point is the similarity in the two cases—namely, that in either case the science is in a state of crisis. On this Kuhn says: "The significance of crises is the indication they provide that an occasion for retooling has arrived" (Kuhn, 1970, p. 138). And for Kuhn, such retooling occurs only at a time of revolutionary or extraordinary science. But my point is that an immature science such as psychology is in a perpetual state of retooling. It follows, therefore, that an immature science is typically multiparadigmatic and multitheoretic. This means that simultaneous theoretical pluralism constitutes "normal science" for immature sciences because that is the only procedure they have for covering the full terrain of their domain. That is, since no single paradigm is universal (i.e., covers the entire discipline), normal science in Kuhn's sense is impossible. In short, the only other way for an immature science to cover the full scope of the discipline is to produce a large number of complementary minitheories. Failure to do so would mean simply not investigating large segments of the discipline in question. Furthermore, I will argue that simultaneous theoretical pluralism should aim to keep all the options open until one paradigm, or a small number, emerge(s) as dominant. But how can this be accomplished?

First of all, Kuhn says: "It *waits* [because historical analysis shows that] each of the currently established sciences has emerged from a previously more speculative branch of natural philosophy" (Kuhn, 1970, p. 245). But the mere passage of time will not result in progress. During this time we must analyze and test our theories. However, Laudan (1977) says that the attempt to test theories in terms of whether they are true or false has failed. Furthermore, says Stegmüller (1979, p. 168), logical and historical analysis reveals that

> an old theory cannot disappear due to refutation or foundering on experience; the community of its holders simply dies out. . . . Similarly, revolutionary changes are not primarily based on theoretical insights but on value-judgments and on decisions.

And Weimer (1979, p. 33) carries this view to its logical extreme when he says that all theories are false, and

> one must abandon the conception of testing as a problem of 'how to replace a theory that is refuted by the facts' and replace it with the appraisal of which alternative among many competing theories to accept, when it is known in advance that they are all false.

But if all theories are false how does science progress? Science progresses by developing better and better theories, a better theory being defined as one capable of solving the most significant problems of a given scientific domain (Laudan, 1977). Lakatos has been particularly eloquent on this point, as when he says:

> the crucial element in falsification is whether the *new theory* offers any novel, excess information compared with its predecessor and the extent to which this excess information is corroborated. . . . We are no longer interested in the thousands of trivial verifying instances nor in the hundreds of readily available anomalies; *a few crucial excess-verifying instances are decisive*. (Lakatos, 1970, pp. 120–121, italics mine)

What we learn from this line of thought is that our major goal in dealing with the problem of theoretical pluralism is *not the direct elimination of refuted theories*, but rather, the identification of a better theory to which we can now commit ourselves (e.g., see Kuhn's description of the conversion experience).

This means that theory appraisal is particularly critical. This involves both a method of analysis and a set of criteria. For the time being I shall assume the availability of a set of criteria (this will be elaborated in the next section) and I shall merely allude to a recommended method. The recommended method is that of dialectic analysis, a procedure which involves a detailed analysis of each theory, both separately and through dialectic confrontation (this will be elaborated in the last section). At this juncture the point I wish to make is that the major goal of conceptual analysis (including dialectic analysis) should not be the elimination of theoretical alternatives. Rather, our goal is to identify the most promising alternatives. This means that we will be extremely reluctant to recommend the elimination of a theory, even though it appears to be strongly disconfirmed. Furthermore, since it is so difficult either to confirm or refute a theory, the best we can do in these circumstances is to develop a pragmatic basis for action. We begin with the fact that, despite the corrigibility of the procedure, we can make an assessment of each theory in terms of degree of plausibility or tenability. However, such judgments are not scaled—rather, they represent crude guesses the meaning of which is unclear. I propose, therefore, the following three-category taxonomy as the basis for action: (1) that we regard all theories assessed as highly plausible or tenable as *tentatively accepted,* (2) that we regard all theories assessed as minimally plausible as *tentatively refuted,* and (3) that the remaining theories (of average plausibility) be placed on hold—meaning that these theories should be regarded as *suspended until further notice.* Why the reluctance to eliminate theories? Primarily because immature science is typically unable to produce any one theory as clearly the most powerful. Thus, the basic strategy is to stall for time because the development of theories with a high degree of explanatory power is a slow and difficult process which usually requires several generations. Furthermore, "In the absence of a [universal] paradigm, all of the facts . . . are likely to seem equally relevant. As a result early fact-gathering is a far more merely random activity [and, as a result] . . . it produces a morass" (Kuhn, 1970, pp. 77–78).

Kuhn then indicates that this kind of factual and theoretical chaos eventually disappears because various scientific subcommunities become increasingly specialized. This results in a restriction of the scope of a given scientific subcommunity. Although such subcommunities do not adopt a discipline-wide (universal) paradigm, they typically involve an area-wide (regional) paradigm. This means it is likely that the many theories of an immature science are covering different aspects of the domain in question—that is, that most of the theories are actually complementary rather than competitive. However, this state of affairs cannot be assumed. Rather, extensive and continual critical analysis must be conducted in order to determine the actual state of affairs in a given domain of investigation. Thus, an early objective of conceptual analysis is to determine the actual (in contrast to the intended) scope of each of the theories. In general this means determining that several apparent theories of x (intended scope) are actually shown to be theories of x_1, x_2, or x_3 respectively. An example might help clarify this point. It is taken from the individual differences domain.

The most potent extant factor theories of individual differences of personality are those of Cattell, Eysenck, and Guilford. However, close analysis reveals that in each case the actual scope of the theory in question is more limited than the intended scope. In all three cases the intended scope is all of personality. However, in actuality they cover only two subdomains of personality—affect and cognition. All three theories deal with affect, but only two (Cattell and Guilford) make claims about cognition. Furthermore, close analysis reveals that in the cognitive domain Guilford's theory is focused on first-order factors whereas Cattell's theory is focused on higher-order factors. We have a similar situation in the domain of affect. But in this case Eysenck is focused on a small number of third-order factors, Guilford is focused on a large number of first-order factors, and Cattell is concerned with a different set of primaries and higher-order factors. Thus, the conclusion is that, although these three theories appear to be competitive, they are in fact complementary—that is, the theories of Cattell and Guilford are complementary in the cognitive domain, and Eysenck's theory of personality is not relevant to cognition; and all three theories are complementary in the affective domain.[2] There are three conclusions to be drawn from this example: (1) None of the three theories deals with the broadly intended scope of personality; (2) all three theories are complementary when analyzed in terms of their actual (i.e., more limited) scope; (3) however, at this stage of the analysis, since none of the three has been clearly refuted (i.e., since they are assessed to be of at least average plausibility), they should all be retained.

Since all three theories constitute the best available theoretical structures in the subdomains they cover, they will all be retained until a better alternative comes along. A better alternative could appear in either of the following forms: (1) a better version of the subdomain in question

[2]I have been involved in the development of a general theory of personality and individual differences for the past 17 years. It involves building on the contributions of the major contributors in this domain (Cattell, Eysenck, and Guilford). For an overview of the total theory see Powel & Royce, 1981a,b and Royce & Powell, 1981; for an analysis which includes the complementary roles of Guilford and Cattell in the domain of cognition see Diamond & Royce (1979) and Powell & Royce (1982); and for an analysis in the domain of affect see Royce & McDermott (1977) and Royce & Diamond (1980).

Although this research has involved an intimate awareness of the extant theories in this domain, it has not included a penetrating metatheoretic analysis of these theories. The latter would be required for a complete dialectic analysis.

or (2) a version of the entire domain (i.e., personality) which is at least as good as each of the subdomain theories and which is supported by a total theory adequately incorporating each subdomain as a viable segment of the new theory. My point, then, is that theoretical pluralism in an immature science usually indicates that different theories are covering different segments of the total terrain. The challenge to the theorist in such a situation is to (1) determine the actual scope of each theory, (2) show the complementary role of each theory, and (3) evolve a general theory which makes special cases of each of the partial theories.

4. The Principle of Incommensurability and the Appraisal of Competitive Theories

Although the principle of complementarity will account for most of the theoretical pluralism in immature sciences, it will not take care of all of it. The problem is that the scope of some of the theories will cover either all or a segment of the *same terrain,* and that these theories are competitive rather than complementary.

Such a situation will call for more penetrating metatheoretic analysis and the eventual determination of the most powerful theory or theories. The task of deciding on the most powerful theory will be somewhat easier if the competing theories are of the same genre (e.g., psychoanalytic or behavioristic). However, to the extent that competing theories are from highly disparate genres the task of deciding on the most powerful theory will become increasingly difficult. The major reason for this is that highly divergent theories will also be based on divergent criteria as to what constitutes a good theory. Furthermore, there will also be differences in the extent to which different methodological principles (i.e., how to proceed) are assumed to be valid. According to the philosopher Naess:

> If you adopt methodology M_1 theory T_1 is confirmed and T_2 is disconfirmed: if you adopt M_2, T_2 is confirmed and T_1 disconfirmed. Or T_2 appears untestable from the point of view of M_1 and T_1 if M_2 is adopted. (Naess, 1972, p. 97)

Such theories are what Kuhn (1970) and Feyerabend (1970) refer to as incommensurable. Kuhn (1970) refers to this state of affairs as a flip-flop or a gestalt switch. He refers to the figure of the duck–rabbit as a perceptual demonstration of his point concerning incompatibility. The point is that there are no loci where the two percepts overlap. One sees a figure of either a duck or a rabbit for varying periods of time, but never some combination of the two.

The duck–rabbit or pelican–antelope figures are two of many such figures, including the reversible staircase and the Necker cube. They are all examples of an illusion known as perceptual reversals or perceptual alternations (Gregory, 1974). Although the psychology of perceptual illusions is not totally understood, it is believed that the reversal phenomenon is due to the fact that the brain transforms the information from the sensorium, which is ambiguous, in different ways. That is, sensory information is not transmitted through the mind/brain as pure sensation; it is modified en route by such things as mental set, motivation, and values. But the most important transformations of sensory inputs are provided by the cognitive system. And it is the perceptual subsystem which is the major link between the sensorium and cognitive processing (Diamond & Royce, 1979; Kearsley & Royce, 1977; Powell & Royce, 1982). Thus, when confronted with an ambiguous stimulus, which is the case with illusions, the perceptual system (unconsciously) tries different hypotheses—namely, two in the case of the alternation phenomenon. Furthermore, such psychological constructions as concepts, symbols, styles, and values constitute a contextual backdrop in the processing of sensory inputs. This means that one's world view and philosophy of life are also part of perceptual processing (Royce, 1983; Royce & Powell, 1983). The point is that the perceptual alternation phenomenon strongly favors Kuhn's claim that a different image of reality is induced by paradigm shifts. And the incompatibility of perspectives is just as complete in the case of paradigm shifts as it is in the case of perceptual alternations. The major implication of the principle of incommensurability in the context of theoretical pluralism is that there is no way to translate fully between theories that are generated from different paradigms, thereby involving different presuppositions, methods, and exemplars. The most striking case in point is the concept of waveicle, a compromise concept which attempts to provide a bridge (as in pantelope or antelican) between the particle theory of light and the wave theory of quantum mechanics.

Although the concept of waveicle appears to be basically unworkable, it does maintain the tension between the most viable theoretical alternatives. And the idea behind retaining both theories is to allow time for an eventual resolution. However, the view that light sometimes behaves as a corpuscular phenomenon and sometimes as a wave phenomenon continues to be the dominant view among contemporary theoretical physicists.[3] It is difficult to know what the outcome will be if both theories are kept on hold indefinitely. One possibility is that incompatible theories will be forced into acceptance if it is decided that the facts continue to demand both theories. In this case this would mean,

[3]I am indebted to Burt Voorhees, a theoretical physicist colleague in the center, for this information.

for example, that the long-standing criterion of logical consistency would be sacrificed in the metatheoretic evaluation of theories. It is also possible that the two theories will eventually be seen as complementary rather than competitive. In fact, Radnitzky (1970) takes this view when he claims that "polarity is changed into . . . complementarity . . . as science grows." However, there is no broader theoretical synthesis available at present which makes special cases of the quantum and particle theories. In this connection Tennessen may be right when he says that the current overload of knowledge means that "no road leads back towards a unification of physics, astronomy, metallurgy, biology, psychology, history, or any other discipline" (Tennessen, 1970, p. 64).

Thus, when we are confronted with competitive rather than complementary theories the stakes are high. This is why Kuhn referred to this kind of change as revolutionary science. The revolution involves a change in all aspects—all the way from the underlying philosophical assumptions to the observed facts. Although the principle of incommensurability was originally developed in the context of sequential theoretical pluralism, the waveicle example suggests that it applies to the context of simultaneous theoretical pluralism as well. What it means in this case is that two or more highly plausible paradigms exist at the same time, and, in terms of the perceptual alternation example, we have to shift from one paradigm to the other if we want to see the different views of reality each provides. For example, if we have viewed the findings of a given domain through theory A and we now go on to theory B we will not be able to comprehend what theory B has to say unless we switch from paradigm A to paradigm B. (N.B. I am told by Voorhees that this is exactly what theoretical physicists do in the case of waveicle theory.) It should be noted, however, that a similar shift would be inappropriate if we are *evaluating* different theories. The point is that the critical evaluation of competitive theories can be fruitful only if all relevant theories are evaluated on the basis of a standard set of criteria. The fact that competitive theories involve different paradigms is, of course, an important reason why it has been so difficult to evolve a universally applicable set of criteria for the appraisal of theories.

5. The Role of Dialectic Analysis in the Appraisal of Theories

The primary goal of dialectic analysis is to maintain the tension between viable alternatives. To the extent that the term *dialectic* refers to only two alternatives it is misleading and should be replaced by a term such as *multilectic*. The point is that the analysis will be carried out on as many alternatives as there are—two, five, or twenty. Thus, the first job of the metatheoretician is to winnow the alternatives in a given subdomain to a short list of the most plausible or tenable theories. Let us continue with my example of extant factor theories of individual differences in order to exemplify this process. We start with the fact that the number of candidates is actually larger than three in the cognitive subdomain. Thus the analysis would also have to include Thurstone's multiple-factor theory, Spearman's *g* theory, Thomson's sampling theory and Thorndike's theory of S–R bonds. It is my guess, however, that all four of these theories would be found wanting, but that two of them, namely those of Spearman and Thurstone, would probably be placed in the "hold" category. However, Spearman's theory would subsequently be viewed as highly implausible because its most fundamental claim has been shown to be false. The Spearman claim is that *g* is a factor which is common to *all* the measurements there are in the cognitive domain. In spite of 50 years of research, this claim has never been empirically demonstrated. However, what has been demonstrated is that there are several general factors that are common to a large number of cognitive measures. Thus only a modification of the Spearman model is now acceptable. And Cattell's model, which involves two general factors, a general factor of crystallized intelligence and a general factor of fluid intelligence, is the best available alternative in this genre.

The problem with the Thurstone model is that he identified only seven factors at the primary level. Guilford has since postulated some 120 primary factors, over 100 of which have been empirically identified. And, although nobody knows how many cognitive factors there are, it is now clear that there are more than seven. Although Guilford's original version of his theory did not include any higher-order factors, he has since postulated 101 higher-order factors. To date, all 16 of the third-order factors and about half of the 85 second-order factors have been empirically confirmed (Guilford, 1981). Thus, in this example, I would say that a more thorough conceptual analysis would indicate that the theories of Spearman and Thurstone would be the only additional contenders for the short list in the cognitive domain but that they would subsequently be eliminated upon more penetrating analysis.

The situation is simpler in the affective domain because there are no other extant contenders. That is, the theories of Cattell, Eysenck, and Guilford constitute the short list in the area of factor theories of personality. Thus, our conclusion is that there are two strong contenders in the cognitive domain and three strong alternatives in the affective domain. They would now be subjected to closer scrutiny.

Before we employ the cutting edge of dialectic analysis, however, let us review the nature of the conceptual analysis up to this point. Each theory (including those of Thomson, Spearman, Thurstone, and Thorndike, in addition to those of Guilford, Eysenck, and Cattell) would be metatheoretically analyzed. This involves the application of a standard set of evaluative criteria to each theory. The criteria in question have to do with the appraisal of a theory. Elsewhere (Royce, 1978) I have addressed this issue under the rubric of theoretical power, theoretical power referring to a multidimensional set of characteristics such as empirical testability, degree of empirical-formal fit, comprehensiveness or scope, parsimony, depth of penetration (i.e., degree of abstractness in the nomological network), degree of formalization, degree of cohesiveness, and explicitness of conceptualization. Philosophers of science have been wrangling over the issue of how to evaluate theories for decades without achieving a convergence of views. However, there has been a gradual shift from the demand for certainty and justificatory criteria to a more corrigible view of the scientific enterprise and the realization that theory preference can best be answered by giving good reasons (Weimer, 1979). The rational view of Popper (1974) and the historiographic analysis of Kuhn (1970) have been influential in this shift. And, although there are differences in views, Wisdom (1974) points out that their views are similar on the following metatheoretic features: "the need for refutability, the significance of unanticipated novelty necessary for a new theory to reject an older one, the need to risk being wrong, the role of observation and experiment, [and] the dominance of theory" (footnote 20, p. 842).

Another line of convergence lies in Radnitzky's (1980) agreement with Popper that theory preference should be in terms of identifying the theory which most surely advances knowledge, which means identifying the theory that is able to penetrate most successfully to the underlying reality or the "deep" problems of the field. Andersson (1979, p. 12) provides a highly condensed but accurate summary of the Popper–Radnitzky strategy:

> By choosing between two theories we should consider their falsifiability (before testing) and (after testing) how severely they have been tested (degree of corroboration), their predictive and explanatory successes and failures (as a fallible indicator for verisimilitude) and the scientific importance of the problems they try to solve.

Thus, if their analyses are right, Radnitzky and Popper retain the traditional concern for establishing a rational basis for scientific progress without inheriting the problems of justificationism. Furthermore, Radnitzky is aware that it is also necessary to devise a method of theory appraisal in the form of an "achievement profile." But, although he refers to achievement "items," he never presents them. We are, therefore, left with a promising but programmatic scheme for appraising theory.

However, there are three analyses available by Bunge (1967), Naess (1972), and Madsen (1974) which can serve as a starting point for developing a standardized method of appraising theories. These three authors provide conceptual analyses of theories at three levels of abstraction: the descriptive stratum (i.e., the observables), the explanatory level (i.e., the hypothetical construct level), and the most abstract level, the level of philosophic and methodological assumptions (the metatheoretic level).

After this kind of analysis is applied to each viable candidate we then decide on the current status of each theory. Those that survive this kind of independent analysis are then put through further analysis by means of dialectic comparison. This involves using the method of paired comparisons in order to provide a scaled score for each theory on each of the items in the theory appraisal inventory. A variety of indices, such as degree of scope and theoretical power, could then be obtained through appropriate weighting and combining of items.

To continue with the individual differences example of theory appraisal in the case of the affective domain we would compare Eysenck with Cattell, Eysenck with Guilford, and Cattell with Guilford in each item of a (not presently available) standardized theory appraisal inventory. My guess is that we would selectively retain all three theories primarily on the grounds that they are both complementary and equal in their theoretic power.

We would carry out a similar comparative analysis in the cognitive domain. It is my guess that such an analysis would reveal that the original version of Guilford's theory has about the same degree of scope and theoretical power as Cattell's theory, but that the recent incorporation of higher-order factors into the structure of intellect model has significantly increased the scope and theoretical power of this theory. Thus, as a result of the dialectic analysis in the cognitive domain, Thurstone's theory would be viewed as refuted or replaced by Guilford's theory, Spearman's theory would be regarded as refuted or replaced by Cattell's theory, and Guilford's modified structure of intellect model would be assessed as the most potent of the extant factor theories of cognition.

Thus the affective domain exemplifies a complementary analytic strategy, and the cognitive domain exemplifies a competitive analytic strategy. In the complementary case (affect), the major task for the future is to evolve a more general theory which makes special cases of the Cattell, Guilford, and Eysenck theories. This new theory can either build directly on one or more of the three theories, thereby demonstrating how it fits into the subsequent theory of greater scope; or the new theory can be an entirely new alternative which includes a demonstration of how it subsumes the terrain currently covered by Cattell, Guilford, and Eysenck. In either case the new alternative must demonstrate that it adequately covers the terrains of the partial theories (or better yet, that it covers all three partial terrains better than its predecessors).[4]

The situation is more complicated in the case of simultaneous, competitive theories. That is, we cannot have recourse to a theory of broader scope which makes special cases of the previous major contenders. What we are up against in this case is an all-out war for an indefinite period of time. That is, we must continue the dialectic analysis until one of the existing contenders or an entirely new alternative emerges as the most powerful theory. The situation is more difficult in the case of competitive theories because of the principle of incommensurability. However, it should be kept in mind that the existence of competitive theories and incommensurability are also signs of scientific progress—that is, they are characteristic of the more mature sciences. And, if a scientific discipline reaches the point at which its competitive theories are sequential rather than simultaneous that is an indication of even greater maturity because it means that the entire scientific community of a specifiable domain is governed by a single (i.e., a universal) paradigm rather than several.

6. Recapitulation and Conclusion

The current state of affairs in theoretical psychology—that is, holding most of the theoretical alternatives in suspension— comes close to putting us in a position of permanent suspended animation (i.e., mental paralysis). Furthermore, the insights available from the philosophy of science are not adequate to the task. In short, despite the existence of a vast literature, there is no hard metatheoretic knowledge or wisdom available to which we can turn for the answers. Since definitive answers are not available, I recommend a pragmatic course of action as the only available alternative. This means that we must evolve a basis for action, even at the risk of erring in one direction or another. And, of course

[4]See the Royce and Powell theory (Royce & Powell, 1981, 1983; Powell & Royce, 1981a, b), a broad-scoped theory which builds on the contributions of Cattell, Eysenck, and Guilford (and others) as special cases. But time is needed in order to allow for more severe empirical testing of this theory so that its degree of corroboration or verisimilitude can be assessed.

the risks are great. In fact, Radnitzky (1980) says that making decisions as to which theory to adopt on the basis of currently available methods of theory appraisal is comparable to playing the stock market. That is, one makes a decision on which stock to buy on the basis of the investment counselor's evaluation of the plus-and-minus features of a short list of "best buys." By analogy, the researcher in a given discipline must also decide on which theory he should invest in. In this case, however, the investment involves the much more serious one of committing one's professional life (the very limited resource of a researcher's talent, time, and energy).

In short, we need a pragmatic basis for dealing with theoretical pluralism. My analysis of the situation indicates that this is most effectively achieved by a combination of constructive dialectics and continuous theory building. A major conclusion arising from this analysis is that extant theories are refuted when they are replaced by better theories, not by direct refutation of theory *x*. This means that the primary purpose of metatheoretic analysis is to clarify the characteristics of extant theories and that we must also produce new theoretical alternatives which have more theoretical power than their predecessors. Thus we end up with the paradoxical conclusion that the best way to deal with theoretical pluralism is to produce still more theories. But there is an important qualification—the only way theories can be refuted and thereby eliminated is to produce new theories that have more theoretical power than either their complementary or their competitive alternatives.

This analysis also indicates that proliferation of theory is a characteristic of all the sciences, particularly those that are immature or in a state of crisis. Since psychology is a relatively immature science, it follows that we should carry as much theoretical baggage as possible. This includes the retention of theories at all levels of theoretical power, weak as well as strong. Such a policy implies that the load of theoretical baggage can only be minimally reduced—that is, by eliminating only a few theories which theory appraisal overwhelmingly demonstrates to have little or no potential for improvement. This strategy is required, however, if each proposed theory is to have the opportunity to develop to its full potential. Only those claims that survive the tests of severe and extensive critical analysis (this includes both conceptual and empirical tests) are admitted to a given body of scientific knowledge. And the testing of a wide range of theories is necessary in order to cover the full terrain of a scientific domain. It appears, therefore, that the proliferation of theory is necessary for the advance of science. However, as a science matures, typically a relatively small number of theories will achieve paradigmatic status.

Thus, my analysis indicates that the production and critical evaluation of a large number of theories is both necessary and critical to the advancement of scientific knowledge. The recommended pragmatic action, therefore, is to promote the proliferation of potentially viable theory and simultaneously to develop a small number of theories that show the greatest potential for eventually becoming conceptually powerful.

7. References

Andersson, G. (1979). Presuppositions, problems, progress. In G. Radnitzky & G. Andersson (Eds.), *The structure and development of science* (pp. 3–15). Dordrecht, Holland: Reidel.

Brunswik, E. (1939). The conceptual focus of some psychological systems. *Journal of Unified Science, 8*, 38–49.

Brunswik, E. (1952). The conceptual framework of psychology. *International encyclopedia of unified science* (vol. 1, 10). Chicago: University of Chicago Press.

Brunswik, E. (1956). Historical and thematic relations of psychology to other sciences. *Scientific Monthly, 83*, 151–161.

Bunge, M. (1967). *Scientific research* (vols. 1 & 2). New York: Springer-Verlag.

Campbell, D. T. (1960). Blind variation and selective retention in creative thought as in other knowledge processes. *Psychological Review, 67*, 380–400.

Conant, J. B., & Nash, L. K. (Eds.). (1957). *Harvard case histories in experimental science* (Vol. 1). Cambridge: Harvard University Press.

Diamond, S., & Royce, J. R. (1979). Cognitive abilities as expressions of three "ways of knowing". *Multivariate Behavioral Research, 15*, 31–56.

Feyerabend, P. (1970). Against method. In M. Radner & S. Winokur (Eds.), *Analyses of theories and methods of physics and psychology* (pp. 17–130). Minneapolis: University of Minnesota Press.

Gregory, R. (1974). *Concepts and mechanisms of perception.* New York: Charles Scribner's Sons.

Guilford, J. P. (1981). Higher-order structure-of-intellect abilities. *Multivariate Behavioral Research, 16*, 411–435.

Kearsley, G. P., & Royce, J. R. (1977). Multifactor theory of sensations: Individuality in sensory structure and sensory processing. *Perceptual and Motor Skills, 44*, 1299–1316.

Koch, S. (1974). Psychology as science. In S. C. Brown (Ed.), *Philosophy of psychology* (pp. 3–40). London: Macmillan.

Kuhn, T. S. (1970). *The structure of scientific revolutions* (rev. ed). Chicago: University of Chicago Press.

Lakatos, I. (1968). Criticism and the problem of scientific research programs. *Proceedings of the Aristotelian Society, 69*, 149–168.

Lakatos, I. (1970). Falsification and the methodology of scientific research programs. In I. Lakatos & A. Musgrave (Eds.), *Criticism and the growth of knowledge* (pp. 91–195). Cambridge: Cambridge University Press.

Laudan, L. (1977). *Progress and its problems.* Berkeley: University of California Press.

Madsen, K. B. (1974). *Modern theories of motivation: A comparative metascientific study.* New York: Halsted Press.

Naess, A. (1972). *The pluralist and possibilist aspect of the scientific enterprise.* London: Allen & Unwin.

Popper, K. R. (1963). *Conjectures and refutations.* New York: Harper.

Popper, K. R. (1974). Replies to my critics. In P. A. Schilpp (Ed.), *The philosophy of Karl Popper* (pp. 961–1197). La Salle, IL: Open Court.

Powell, A., & Royce, J. R. (1981a). An overview of a multifactor-system theory of personality and individual differences: I. The factor and system models and the hierarchical factor structure of individuality. *Journal of Personality and Social Psychology, 41,* 818–829.

Powell, A., & Royce, J. R. (1981b). An overview of a multifactor-system theory of personality and individual differences: III. Life-span development and the heredity–environment issue. *Journal of Personality and Social Psychology, 41,* 1161–1173.

Powell, A., & Royce, J. R. (1982). Cognitive information processing: The role of individual differences in the search for invariants. *Bulletin of Academic Psychology, 4,* 255–287.

Radnitzky, G. (1970). *Contemporary schools of metascience.* Göteborg: Akademieforlaget.

Radnitzky, G. (1980). Progress and rationality in research. In M. D. Grmek, R. S. Cohen, & G. Cimino (Eds.), *On scientific discovery.* Dordrecht, Holland: Reidel.

Rosenzweig, S. (1937). Schools of psychology: A complementary pattern. *Philosophy of Science, 4,* 96–106.

Royce, J. R. (1970). The present situation in theoretical psychology. In J. R. Royce, (Ed.), *Toward unification in psychology* (pp. 10–52). Toronto: University of Toronto Press.

Royce, J. R. (1977). Toward an indigenous philosophy of psychology. *The Ontario Psychologist, 9,* 16–32.

Royce, J. R. (1978). How we can best advance the construction of theory in psychology. *Canadian Psychological Review, 19,* 259–276.

Royce, J. R. (1983). Personality integration: A synthesis of the parts and wholes of individuality theory. *Journal of Personality, 51,* 683–706.

Royce, J. R., & Diamond, S. R. (1980). A multifactor-system dynamics theory of emotion: Cognitive-affective interaction. *Motivation and Emotion, 4,* 263–298.

Royce, J. R., & McDermott, J. (1977). A multifactor theory of affect: A system dynamics model. *Motivation and Emotion, 1,* 192–224.

Royce, J. R., & Powell, A. (1981). An overview of a multifactor-system of personality and individual differences: II. System dynamics and person-situation interactions. *Journal of Personality and Social Psychology, 41,* 1019–1030.

Royce, J. R., & Powell, A. (1983). *A theory of personality: Factors, systems, and processes.* Englewood Cliffs, N.J.: Prentice-Hall.

Schilpp, P. A. (Ed.). (1974). *The philosophy of Karl Popper.* La Salle, Ill.: Open Court.

Stegmüller, W. (1979). A combined approach to the dynamics of theories. How to improve historical interpretations of theory change by applying set theoretical structures. In G. Radnitzky & G. Andersson (Eds.), *The structure and development of science* (pp. 151–186). Dordrecht, Holland: Reidel.

Tennessen, H. (1970). Science of history and notions of personality. In W. Yourgrau & A. D. Breck (Eds.), *Physics, logic, and history* (p. 59–77). New York: Plenum Press.

Weimer, W. B. (1979). *Notes on the methodology of scientific research.* Hillsdale, NJ: Lawrence Erlbaum Associates.

Wisdom, J. O. (1974). The nature of normal science. In P. A. Schilpp (Ed.), *The philosophy of Karl Popper* (pp. 820–842). La Salle, IL: Open Court.

6

Theoretical Pluralism and Complementarity

Michael E. Hyland

Royce develops two major ideas in his paper, both of which are consistent, in general, with his earlier work (e.g., Royce, 1978). I shall summarize these two ideas, referring to Royce's general orientation to theoretical psychology, and then comment on part of just one of those ideas.

Royce argues that psychology is at an immature stage in its development as a science. While acceding that scientific psychology has not been particularly successful to date, he believes that this is a characteristic of the immaturity of the discipline rather than any inherent defect in the notion of a scientific psychology. Royce believes that psychology can eventually become a mature science but that a mature scientific psychology will be achieved only through increased theoretical sophistication. Immature sciences become mature in stages. The present paper includes one view of how scientific development takes place, though a slightly different type of analysis is found in an earlier paper (Royce, 1978).

There are two substantive ideas in this paper. One is the idea of theoretical pluralism; the other is an analysis of theory appraisal.

Theoretical pluralism means simply that there are many theories in a discipline. Royce distinguishes two sorts of theoretical pluralism. Simultaneous pluralism occurs when there are many theories at the same time. Sequential pluralism occurs when there is a historical sequence of many theories and when the discipline develops by accepting new theories and discarding the old.

Michael E. Hyland • Department of Psychology, Plymouth Polytechnic, Plymouth, Devon PL4 8AA, England.

Simultaneous pluralism is a characteristic of immature sciences and takes two forms. First, simultaneous pluralism occurs with complementary theories when the different theories "explain different segments of the total terrain" (p. 302). Second, simultaneous pluralism is competitive when there are two theories which explain the same data.

Sequential pluralism is a characteristic of mature sciences and is characterized by competitive theories, that is, one theory is replaced by another. The difference between competitive theories in the mature and immature science is that in the immature science the competitive theories are more likely to coexist at any one moment in time.

The second major idea in Royce's paper concerns how we should evaluate the truth of a theory. Along with modern philosophers of science, Royce believes that science proceeds by the acceptance of better theories rather than by the refutation of unacceptable theories (the latter is what Chalmers, 1978, calls "naive falsification"). Royce develops a method of dialectic analysis as a way of appraising the plausibility of a theory. The main feature of this method is that it is carried out in the context of the total pool of available theories: theories should not be appraised in isolation.

Royce's approach to theory appraisal is best characterized by the word *tolerance*. One should tolerate a theory even if it is not very plausible because it may turn out that the theory, or perhaps a reformulated version, can still play a major role in the development of the discipline. Even when there is evidence against a theory, it should not be completely discarded but should be placed on "hold." Royce's dialectic analysis is consistent with his positive orientation to the discipline of psychology. He disagrees with Koch's (1974) assertion that scientific psychology should be abandoned. The general message appears to be that we are basically on the right track but should learn from earlier mistakes.

In general, I find myself in complete agreement with Royce's orientation to theoretical psychology and with many of the points he makes in the present paper. There is one area of possible difference and it is this one area which I wish to expand and comment on: it is the idea of complementarity.

1. Complementarity and the Domain of Psychology

According to Royce, complementarity is a feature of an immature science: "It is likely that the many theories of an immature science are covering different aspects of the domain in question—that is, that most of the theories are actually complementary rather than competitive"

(p. 304). Also: "My point, then, is that theoretical pluralism in an immature science usually indicates that different theories are covering different segments of the total terrain" (p. 306).

When Royce says that theories are complementary he means that the data to be explained (the *explanandum*) differs between the theories. The theories are explaining different sorts of data. At the same time, the idea of complementarity embodies the notion that the different sorts of data are relevant to each other in some way. The different *explananda* fall within the same "domain." Royce illustrates his argument with personality theories. When one is studying personality it is evident that the different theories (i.e., the *explananda* of the theories) are relevant to each other. The domain of personality theory is recognized as a domain, if only because of the structure of psychology courses. However, Royce does not explain when we should recognize that theories are complementary to each other rather than simply irrelevant. Theories can be called complementary only when the *explananda* of the two theories are relevant to each other, and one would suppose that not all theories in science are relevant to each other. As Royce says himself, "Although psychology is not as broad as all of science, it does range from the biological to the social sciences" (p. 299). Are biological theories with a psychological aspect complementary or irrelevant to sociological theories with a psychological aspect? And if they are complementary, are they complementary in the same or a different way to theories in cognitive psychology? These questions can be answered only if we have a good understanding of what the domain of psychology actually consists of.

It seems evident that complementarity does exist. What is slightly puzzling is that the idea of the domain of psychology has received so little conceptual analysis in the past. What exactly is the domain of psychology? The assertion that psychology is the study of behavior does not get us very far in that the meanings attached to the word *behavior* are quite varied. Any attempt to answer these questions is clearly beyond the scope of this present comment. For the moment I wish to show that our understanding of complementarity is slight. The possibility that conceptual analysis of behavior might be weak was noted by Tolman in 1959, who said, "I still feel that 'response' is one of the most slippery and unanalyzed of our present concepts" (Tolman, 1959, p. 95).

Behavior, like other observables, is a theory-laden concept (Chalmers, 1978; Hanson, 1958). An author's description of what behavior consists of usually reflects what he thinks is responsible for behavior. Descriptions of behavior reflect an author's theoretical assumptions. When an author describes psychology he provides either a list of behaviors to be explained or a list of theories which do the explaining; to an extent these two sorts of lists are interchangeable. Here are some examples of such lists.

Tolman (1936) suggests that behavior has three different aspects:

> The final dependent behavior has three component aspects. Sometimes it is one and sometimes it is another of these aspects which the given experiment is interested in. No one of these aspects can, of course, be missing, but a given experiment can vary one of them somewhat independently of the other two. They are (a) direction, (b) quantity or persistence, and (c) efficiency or skill. . . . Thus, for example, in a given discrimination problem, using Lashley's technique, the rats (a) jump to one door rather than the other; (b) they do this a certain percentage of times; and (c) they exhibit a certain degree of skill in their actual jumping technique. (p. 119)

Tolman's account of behavior is similar to another analysis prepared at about the same time by Duffy (1941, 1951). Duffy (1951) maintains that "all behaviour shows variation in goal direction and intensity and apparently *only these two dimensions*" (p. 30, italics in original). Duffy argues that two concepts are needed, one to explain behavioral intensity and the other to explain directionality, as the two behavioral dimensions vary independently. "Dual concepts, or concepts which incorporate two aspects of behaviour which may vary independently have retarded the progress of psychological investigation" (p. 39).

Not everyone in those early days of theory construction would have agreed with Tolman or Duffy as to the subject matter of psychology. For instance Stevens (1935) asserts boldly, "It is the sole business of psychology to test and measure the discriminatory capacities of the organism" (p. 325).

The more recent attempts to systematize psychology—or parts of psychology—tend to be more complex. For instance, Mischel (1973) proposes a list of five person variables which are inferred from an individual's "cognitive activities and behaviour patterns" (p. 265). Mischel describes each of his five person variables in turn, and although he admits that "these variables obviously overlap and interact" (p. 265), he does not say how they do so.

Powell and Royce (1981) propose a "multifactor-system theory" which they derive from factor analysis. According to Powell and Royce, there are six major subsystems (some of which are similar to Mischel's person variables): "each of these six systems is organised as a multilevel hierarchical system . . . in turn each of the individual systems is also conceptualized as a multilevel, hierarchical system" (p. 281).

Evidently, there are many approaches to understanding the domain of psychology. It is worth noting that none of the examples cited above refers to any of the others. There seems to be no accepted account of the theory-laden concept behavior. Nor is there any consensus over the domain of psychology beyond the trivial assertion that it is the study of behavior.

2. Different Forms of Complementarity

When Royce uses the word *complementary,* he uses it in a specific sense. However, there is another sense in which the word could be used in psychology. The aim of this section is to describe another sort of complementarity and compare it with that so labeled by Royce.

A theory is sometimes divided into two separate though related components. The *explanandum* is the sentence which describes the data to be explained. The *explanans* is the sentence which does the describing. In psychology, the *explanandum* is the sentence which describes some aspect of behavior; the *explanans* is what is normally meant by labels such as theory, model, or process.

In an account of simultaneous theoretical pluralism there are two ways in which theories could be said to be complementary: they could be complementary in terms of the *explanans–explanans* complementarity or they could be complementary in terms of their *explananda–explanandum* complementarity. In practice these two forms of complementarity may well be mixed but for the moment I shall discuss them separately.

What is *explanans* complementarity? Suppose that there are two theories T_1 and T_2, both of which explain behavior B. Conceptual analysis of the *explanans* of T_1 and T_2 shows that the two *explanans* are related in some way. For instance, T_1 and T_2 may each deal with a different part of a causal sequence leading to B, that is, they are related sequentially in time. Alternatively, some terms in T_1 and T_2 may be common so that they could more properly be seen as forming part of some more general theory; yet again they may be hierarchically arranged. In whatever way they are related, it is evident that one discovers the relationship between T_1 and T_2 by examining the relation between the *explanans* of the theories.

Explanandum complementarity is the sort of complementarity called complementarity by Royce. Suppose two theories T_1 and T_2 explain two different behaviors B_1 and B_2 respectively. And suppose that there is some relation between the two behaviors B_1 and B_2 so that they could be said to represent part of a more general field B'. To understand the relation between T_1 and T_2 we must first find how B_1 and B_2 fit into the more general field B'. There may be no conceptual overlap between T_1 and T_2—as there must be in the case of *explanans* complementarity. In the case of *explanandum* complementarity the relation between T_1 and T_2

is only incidental. What is important is the relation between B_1, B_2, and B'.

Although Royce is concerned principally with *explanandum* complementarity, the idea of *explanans* complementarity does appear, at least by implication in his work. One example of *explanans* complementarity is discussed in the present paper when Royce compares first- and second-order factor theories of personality. However, a better example is found in the hierarchical organization of Powell and Royce's (1981) multifactor-system theory. In the present paper, Royce suggests that an aim of psychology is to "evolve a general theory which makes special cases of each of the partial theories." That is, we should try to incorporate the *explanans* of the individual theories within some more general, all-encompassing *explanans*.

In the previous section I described some examples in which authors describe the domain of psychology. Are these examples of *explanandum* complementarity, *explanans* complementarity, or some kind of combination of the two? There really does appear to be a need for more work to be done on the way in which psychological theories are complementary. An understanding of *explanadum* and *explanans* complementarity and the way in which the two are related is a very important requirement for theoretical psychology, for it is only through such an understanding that we will gain a unitary view of psychology. It is only through such an understanding that we will gain a picture of the discipline as a whole.

3. Complementarity and the Development of a Science

Royce suggests that complementarity—in the sense of *explanandum* complementarity—is a characteristic of immature sciences and hence by implication is something which we should aim eventually to discard. I wish to argue that complementarity is sometimes found in mature sciences, and moreover, it may be a sensible goal to which psychology should aim.

Royce uses the instance of the wave–particle theory of light in physics as an example of two competitive theories, but he notes that "it is possible that the two theories will eventually be seen as complementary rather than competitive" (p. 308). The possibility that the wave–particle theory should be understood within the context of complementarity—rather than in competition—is one that has been argued for many years (see reviews in Feyerabend, 1958; MacKay, 1958). Royce's quotation from Radnitzky (1970) that "polarity is changed into . . . complementarity . . . as science grows" is consistent with that view. There is, naturally, a body of opinion which supports the view that complementarity is a stagnating course for a science to take (e.g., Feyerabend, 1958). However, there are also those who suggest that complementarity is a positive way for scientific development. In particular, many feel that complementarity as employed in microphysics (it is, for example, a feature of the uncertainty principle) provides a more true view of the world than the opposing view in which complementarity would eventually be eliminated. Niels Bohr, who was one of the earliest advocates of the complementary viewpoint in physics, says:

> The evidence obtained under different experimental conditions cannot be comprehended within a single picture, but must be regarded as complementary in the sense that only the totality of the phenomena exhausts the possible information about the objects. (Bohr, 1949, p. 210)

MacKay (1958) extended the idea of complementarity as a logical property of all sciences, and although there is no space here to review the many points he makes, it is worth repeating some of his conclusions. MacKay defines complementarity in a way which is slightly different from Royce's and my own descriptions above, but the general message remains the same. MacKay concludes:

> 1. Complementary statements are essential when one chosen language-system is logically precluded from representing some of the distinctions or relationships which are discernible in the subject-matter. 2. The current tendency to decry complementarity as an optional notion borrowed from ephemeral physical theory is based on a mistake. . . . 4. Complementarity may arise between statements at the same logical level . . . or at different logical levels. (pp. 121–122)

MacKay distinguishes different sorts of complementarity. One of his distinctions is between complementarity at the same logical level and complementarity at different logical levels. In my earlier examples of ways in which psychologists have defined the domain of psychology, those examples were in fact limited to instances wherein complementarity, if it exists, is at the same logical level. However, other authors do argue for the notion of complementarity at different logical levels in psychology. One such example is found in Rose's account (1973) of different levels of explanation. Another example is my own analysis of different kinds of hypothetical construct uses in psychology (Hyland, 1981). Both these views are consistent with MacKay's point that complementarity is a logical property of scientific theories rather than a specific property of microphysics; and both views are also consistent with MacKay's suggestion as to how complementarity might feature in psychology.

Whatever one's views on the eventual role of complementary theories, it is evident that complementarity is a complex idea; it is an idea which has had important theoretical consequences in the physical sciences; and it is an idea which may be equally, if not more important to the development of psychology. In my view, it is a crucial area for development in theoretical psychology.

4. References

Bohr, N. (1949). Discussions with Einstein on epistemological problems in atomic physics. In P. A. Schilpp (Ed.), *Albert Einstein: Philosopher–physicist* (pp. 199–241). Evanston, IL: Library of Living Philosophers.

Chalmers, A. F. (1978). What is this thing called science? Milton Keynes, England: The Open University Press.

Duffy, E. (1941). The conceptual categories of psychology: A suggestion for revision. *Psychological Review, 48,* 177–203.

Duffy, E. (1951). The concept of energy mobilization. *Psychological Review, 58,* 30–40.

Feyerabend, P. K. (1958). Complementarity I. *Proceedings of the Aristotelian Society, 32,* 75–104.

Hanson, N. R. (1958). *Patterns of discovery: An inquiry into the conceptual foundations of science.* Cambridge: Cambridge University Press.

Hyland, M. (1981). *Introduction to theoretical psychology.* London: Macmillan.

Koch, S. (1974). Psychology as science. In S. C. Brown (Ed.), *Philosophy of psychology* (pp. 3–40). London: Macmillan.

MacKay, D. M. (1958). Complementarity II. *Proceedings of the Aristotelian Society, 32,* 105–122.

Mischel, W. (1973). Toward a cognitive social learning reconceptualization of personality. *Psychological Review, 80,* 252–283.

Powell, A., & Royce, J. R. (1981). An overview of a multifactor-system of personality and individual differences: I. The factor and system models and the hierarchical factor structure of individuality. *Journal of Personality and Social Psychology, 41,* 818–829.

Radnitzky, G. (1970). *Contemporary schools of metascience.* Göteborg: Akademieforlaget.

Rose, S. P. R. (1973). *The conscious brain.* London: Weidenfeld and Nicholson.

Royce, J. R. (1978). How we can best advance the construction of theory in psychology. *Canadian Psychological Review, 19,* 259–276.

Stevens, S. S. (1935). The operational basis of psychology. *American Journal of Psychology, 47,* 323–330.

Tolman, E. C. (1936). *Operational behaviorism and current trends in psychology.* Paper presented at the 25th anniversary celebration of the inauguration of graduate studies at the University of Southern California, Los Angeles, CA. Reprinted in E. C. Tolman (1966). *Behavior and psychological man* (pp. 115–129). Berkeley, CA: University of California Press.

Tolman, E. C. (1951). Operational behaviourism and current trendsin psychology. In E. C. Tolman (Ed.), *Collected papers in psychology* (pp. 115–129). Berkeley, CA: University of California Press.

Tolman, E. C. (1959). Principles of purposive behaviour. In S. Koch (Ed.), *Psychology: a study of a science* (vol. 2, pp. 92–157). New York: McGraw-Hill.

6

The Problem of Singularism

Stig Lindholm

I have read Royce's chapter with keen interest and a good deal of empathy and agreement. On many points I can only support and further emphasize his views. On other points, however, I am a bit doubtful, and still other points have been even more thought-provoking, serving as points of departure for taking the enterprise a little bit further. I shall use my space here mainly to address this third category.

Royce often seems to subscribe to a fairly broad concept of pluralism. Nevertheless I trace some ambivalence in his discussion, as, for instance, in the distinction he makes between mature and immature sciences and what sort of pluralism we can accept where. I think that, deep down, many of us share an ambivalent attitude to "retooling" in a radical sense our metatheoretical conceptual system. We are all parts of our contexts, offsprings of our *Zeit-* and *Ortsgeister*. That is also part of the story, and the paradigm of metatheoretical debate would profit from our appropriating such an insight. Maybe we would then find less of patronizing *von-oben* attitudes, accusations of irrationality, and things like that and could achieve more collective creativity.[1] In order to be able to see ourselves from the outside we need to pull ourselves up by our shoestrings, or do we? Maybe a better analogy would be that of trying to walk up a loose sand dune (cf. Jantsch, 1975, p. 120). Alone it is difficult. Two persons linking hands can help each other keep momentum.

As to the general problem of pluralism, I agree with Royce that it can be interpreted as a sign of crisis, be the sciences mature or immature.

[1]To avoid misunderstandings I should make clear that this thought has in no way been provoked by Dr. Royce's article. It has grown out of extensive reading in the philosophy of science elsewhere and can of course be seen as moralizing and patronizing on a metalevel (cf. Lindholm, 1981).

Stig Lindholm • Department of Education, Stockholm University, S-106 91 Stockholm, Sweden.

Only I would add that the crisis is part of not only our existential but also our ontological and epistemological conditions. I shall return to this.

In his conclusions Royce describes the state of affairs in theoretical psychology with words like "mental paralysis." To me that sounds adequate, and I have chosen the title of my article in order to hint at one of the factors behind that paralysis. I think we have learned a lot from "the insights available from the philosophy of science" (p. 312). I agree that they are not enough. Especially, I think, *wisdom* is lacking. What is wisdom? It is, among other things, cognition plus valuation (facts plus values). It has also to do with the art of living, which is not completely logical and cannot be fully described in a consistent and linear language. Do we, then, have no recourse but to sacrifice "the long standing criterion of logical consistency . . . in the metatheoretical evaluation of theories" (p. 308)? Let us see where we end up! Royce's final recommendation is the pragmatic action of promoting the proliferation of theories. I believe in such a proliferation,[2] and I will now add another kind of proliferation: that of making explicit underlying assumptions.

Pluralism with respect to what? Theories! Can we come further if we take a look at the underlying *concepts of truth*? Conventionally we talk about the theory of correspondence (truth is that which corresponds with reality—in whatever way you assess *that*), the theory of coherence (the truth of a statement is determined by its coherence with a system of other statements), and pragmatism (truth is what is fruitful and useful). As I read Royce, he starts by tacitly assuming the correspondence theory and ends up by espousing the pragmatic one. That is all right with me as long as we are conscious of such a shift of perspectives.

To introduce another element of confusion: Can we discuss psychological theories without making clear what sort of *knowledge interests* we are after? I am, of course, referring to Habermas's (1972) famous triad: the technical, the hermeneutic, and the emancipatory knowledge interests. We can add other ones.

As for the concepts of truth and the knowledge interests, my stand is that the above-mentioned ones are special cases of a much larger number of possible—and necessary!—truth concepts and knowledge interests. It seems to me that we are living in a period of history where more and more thinkers are finding it legitimate to take the difficult step from one to two—and maybe three and further. The first would be the step from the received view of only one truth (that of correspondence) and only one legitimate knowledge interest (the technical one).

[2]Not necessarily in the retention of all theories "weak as well as strong." I doubt that all of them deserve retention. Here wisdom should help.

How do we judge the fruitfulness of different concepts and interests? They are useful and necessary in different walks of life! Not a very precise conceptualization I admit, but the "pragmatic basis for dealing with theoretical pluralism" requires discourses not only in epistemology, but also in ontology and axiology. Here we are really confronted with what Royce in another context terms "ambiguous stimulus," and our "world view and philosophy of life" do become part of the perceptual processing. Formal logic and the assessing of facts are necessary in our ontological and axiological orientation. They are not enough. As any psychoanalyst will tell us there are logics other than formal logic, and what is a fact today may be fiction tomorrow. I think that an important factor behind mental paralysis is the denial of these circumstances. If we do not accept that the sand of the dune is loose but try to walk up it as if we were on firm ground, then we will certainly not reach very high. Such a lack of acceptance of looseness is part of the certainty-ridden scientific culture in which we have been living for quite some time. Since I have an evolutionary view of the development of knowledge and science, I guess that this epoch of craving certainty has been necessary. Again, we have learned a lot here. But this should be only a stepping-stone from which to proceed. This being said, I have exposed a Western belief in progress. Such an attitude can be questioned, but this is where I now stand, and my interpretation of progress is that what once appeared to be the Truth, the general case, later on is seen as, at best, a *special case* of a broader context.[3] Compare the physics of Newton with that of Einstein! To put it a bit roughly: The received view has taught us that the Truth and the only truth is that which corresponds with reality and that the only knowledge interest worth pursuing is the technical one. Such a view leads to, among other things, "methodological monism" (cf. von Wright, 1971, chapter 1). It also leads to seeing pluralism as a *problem*. As my title indicates, the real problem is singularism, the refusal to accept pluralism as part of our ontological and epistemological condition.

Among the strongest and often tacit assumptions of the received view is the assumption of an underlying ultimate reality which can be described *relevantly* and *unambiguously* in a scientific language. Furthermore, this reality is assumed to be essentially *static* (if there are changes, look for the permanent pattern of the changes). Why else these dreams of finding *invariances*, that is, relations which are independent of time, space, the consciousness of the subject (the researcher) and of the object (cf. Galtung, 1977, p. 47). On the level of knowledge and values important

[3]For a more thoroughgoing discussion of this see Lindholm, 1981.

assumptions are that the *objective* (not influenced by the knowledge process) world is the only real and relevant one and that formal logic, measuring, quantification, and the like constitute the epitome in the development of man's knowledge tools. The list of assumptions can be extended, but I think this one will do for my purposes. The underlying paradigm or pet model is that of a machine, nowadays updated to a computer.

If these are the assumptions, then theoretical pluralism does become a problem. My point is that the reality defined by such assumptions is *one* reality (a "special case") and that there are other ones, *relevant and knowable*—if not exactly with the same approaches and methods. With such a view, pluralism becomes the natural and general case, and singularism becomes a temporary narrowing down of vision. A narrowing down which is sometimes necessary and fruitful but which becomes a catastrophe if totalized into the only way of seeing. How, then, do we discuss possible contexts for our special case and for other possible cases?

Let us turn to psychology. Which parts of the realm of the psyche can fruitfully be described unambiguously, objectively, as invariant patterns? There are certainly such parts. Man is not only a conscious being, he also behaves at times in a machine-like way and must then be approached with reifying methods. The question is where, when, and how do we decide upon the approaches and the methods. Man is not only a machine; he is also an ever-changing, conscious, and intentional being (some even say spiritual). For me this is the essence of man, and this essence cannot be captured by approaches emanating from the received view—which is not to say that it cannot be approached fruitfully and coherently, maybe even scientifically, depending upon our concepts of science.

The reality called man can be described in different ways. There are hierarchic approaches. So, for instance, we have the old idea of "levels of being" (see Schumacher, 1978, Chapter 2): mineral, plant, animal, and human, where man consists of material, life, consciousness *and* self-consciousness. Here follows what Schumacher (p. 50) calls the requirement of *adequatio*: "The understanding of the knower must be *adequate* to the thing known." Lack of *adequatio* is, I think, another part of the mental paralysis problem.

Another approach is that of system thinking. What kind of systems do we assume men to be? "Rigidly controlled" or "deterministic" systems, being able to pursue "prescribed operational targets"? Maybe systems that can "pursue prescribed goals or multigoal patterns but select the corresponding operational targets"? Or do we dare to assume men to be "purposeful systems," formulating and selecting policies "in the light of the long-range outcomes . . . of their own and their environment's dynamics"? The concepts and the quotations are from Jantsch

(1975, p. 69). I have chosen to quote people outside psychology, persons like Schumacher and Jantsch who have a firm grounding in praxis, natural science, and general system thinking. In order to walk up the steep sand dune we need guides from other clans. We need to desocialize ourselves to some extent from the culture of our own tribe in order to be able gradually to resocialize, or rather to create new scientific cultures more open to intercultural intellectual exchange. That would be pluralism on a level beyond the one we are discussing here.

My main thesis here is that *the problem of pluralism arises as soon as we try to totalize assumptions on any of the levels indicated above*: if man is assumed to be only an animal ("nothing-but-ness") or only self-conscious, only a deterministic or only a purposeful system, and so on. The phenomenon of totalizing perspectives I have dealt with extensively elsewhere (Lindholm, 1981). I think this phenomenon is based on what Jantsch (1975, p. 203) calls "man's mythical need for security, which itself goes back to fear." It is also based on our overall world view as described by David Bohm (1980), professor of theoretical physics at London University. Our way of thinking is "*thinking about things*" he states, and continues:

> That is convenient and useful mainly in the domain of practical, technical and functional activities. . . . However, when this mode of thought is applied more broadly to man's notion of himself and the whole world in which he lives (i.e., to his self-world view), then man ceases to regard the resulting divisions as merely useful or convenient and begins to see and experience himself and his world as actually constituted of separately existing fragments. (p. 2)

Here we hit upon another effect of current singularism: the fragmentation of knowledge. Assuming man to be essentially thing-like leads to fragmentation. Fragmentation as ideology is *denial of context*. Making context invisible reinforces singularism.

Bohm is relevant for our purposes also in other respects. The main idea in his book is that the ultimate reality is an "undivided wholeness in flowing movement." Everything hangs together and is continuously changing. The ripples and vortices of a stream are stable only temporarily. Similarly, our concepts and theories are temporary valid abstractions, "frozen" patterns that exist for a while and then disappear. The ripples and vortices are not the reality. The reality is the stream.

I have not quoted Bohm to make us psychologists turn again to physics for the answer.[4] I have chosen him for tactical and persuasive reasons. He is a man of science reenacting the forgotten knowledge of

[4]What about changing the pattern for a while and seeing the mind as a model for the universe instead of vice versa?

what is sometimes called *perennial philosophy*. The ultimate reality is deep and multifaceted, the merging of contradictions and the resolution of paradoxes. This reality can never be fully known by us; it can not be caught by the loosely woven net constituted by our languages and our sciences. This is not a paralyzing assumption, it is a challenge! Our languages and sciences are able to capture temporarily stable patterns, parts of this ever-changing fantastic reality. There is plenty of room for progress! On the way I hope we shall be able to retool many times, also to reconceptualize science and the relations of science to other forms of knowledge. A last quotation from Bohm (1980, p. xiv):

> My suggestion is that at each stage the proper order of operation of the mind requires an overall grasp of what is generally known, not only in formal, logical, mathematical terms, but also intuitively, in images, feelings, poetic usage of language, etc.

How can contemporary psychology contribute to the survival and evolution of man and mankind? Pluralism, retooling, and reconceptualizations are *sine qua nons*, not primarily for internal scientific reasons (however legitimate these are) but for very worldly reasons. We can use our special insights—practical and theoretical ones—to shed light upon the processes of developing knowledge and world views. A first step might be to focus the light on ourselves, on our paradigmatic singularism and other hang-ups—and on our mythical need for security. Then we will be free to discover that reality is a "pantelope" and simultaneously an "antelican," the one exceeding the velocity of light and the other conducting Beethoven's tenth symphony.

1. References

Bohm, D. (1980). *Wholeness and the implicate order.* London: Routledge & Kegan Paul.

Galtung, J. (1977). *Methodology and ideology. Essays in methodology* (vol. 1). Copenhagen: Christian Ejlers.

Habermas, J. (1972). *Knowledge and human interest.* London: Heinemann Educational Books.

Jantsch, E. (1975). *Design for evolution: Self-organization and planning in the life of human systems.* New York: George Braziller.

Lindholm, S. (1981). *Paradigms, science and reality: On dialectics, hermeneutics and positivism in the social sciences.* Research Bulletin Vol. IX:1, Institute of Education, University of Stockholm, S-106 91, Stockholm.

Schumacher, E. F. (1978). *A guide for the perplexed* (Abacus ed.). London: Sphere Books, Ltd.

von Wright, G. H. (1971). *Explanation and understanding.* London: Routledge & Kegan Paul.

6

Is Theoretical Pluralism Necessary in Psychology?

Richard F. Kitchener

There is no phenomenon in contemporary psychology more striking and puzzling than its theoretical pluralism. Not only is there a plethora of quite different and apparently irreconcilable theories in different subdomains of psychology—physiological psychology, learning, cognition, motivation, personality, developmental psychology, social psychology, psychotherapy—but even within a single field such as psychotherapy, for example, there is a steady increase in the number of new theories. When I was a graduate student in psychology there were already 36 systems of psychotherapy and by now the number may well have doubled. I have lost count.

Many (perhaps most) psychologists take this as a lamentable sign of the immaturity of their science. Largely because the views of Thomas Kuhn are now so popular, psychology is often taken to be in its preparadigm, protoscience stage, a period of theoretical pluralism. Someday, it is hoped, this prescientific period will give way to a paradigm period of *real* science in which theoretical monism will replace theoretical pluralism, basic disagreements concerning theoretical, methodological, and philosophical issues will be resolved, and there will be a consensus about these fundamentals. When this occurs, psychology will have become a mature science like physics, astronomy, or biology.

Assuming there is a theoretical pluralism in contemporary psychology, several questions arise concerning this state of affairs: (1) Why is there a theoretical pluralism in psychology? (2) Is this theoretical pluralism unavoidable and necessary, or is it merely a passing phase? (3) Is this theoretical pluralism good or bad? (4) If it is bad, how can one avoid

Richard F. Kitchener • Department of Philosophy, Colorado State University, Fort Collins, Colorado 80521.

such pluralism and/or what should be done to overcome it? According to many psychologists—those adopting what I will call a classical Kuhnian interpretation—the answers to these questions are obvious: (1) There is a theoretical pluralism in psychology because psychology is immature and preparadigmatic. (2) This theoretical pluralism is not necessary (otherwise psychology would forever be "immature") and someday will be replaced by a theoretical monism. (3) Theoretical pluralism is bad. (4) It is to be avoided by somehow convincing all (or most) psychologists to adopt the same paradigm.

Anyone concerned with metatheory in psychology (as Royce is) eventually must address these and related questions concerning the problem of theoretical pluralism. It is not surprising, therefore, that Royce discusses these (and related issues) and makes several suggestions concerning them that are both provocative and sensible. His major conclusion is the "paradoxical" one that "the best way to deal with theoretical pluralism is to produce more theory—but with the qualification that the new theory be more theoretically powerful than its predecessors and/or its competitors" (p. 313). In order to decide which theory is more powerful, Royce advances a method of analysis—constructive dialectics—and offers a list of criteria of theory evaluation including comprehensiveness, parsimony, degree of formal-empirical fit, and so forth. Since I am in fundamental agreement with much of what Royce says concerning these central issues, my comments will focus on what I believe to be aspects of Royce's views that are not sufficiently clear (and in need of more discussion and clarification) or are questionable (and in need of reevaluation).

According to Royce's paradoxical conclusion, we ought to advance theories that are more powerful than the ones we have today. Although such advice can hardly be questionable, there is, I think, an important ambiguity underlying this claim. Although Royce's Abstract claims that we ought to produce more theory "but with the qualification that the new theory be more theoretically powerful than its predecessors and/or its competitors," (p. 297) this point is put somewhat differently in the closing section of Royce's paper. There we read that we ought "to promote the proliferation of potentially viable theory and simultaneously to develop a small number of theories that show the greatest potential for eventually becoming conceptually powerful" (p. 314).

The difference here is not a trivial semantic point but one concerning a crucial issue in the philosophy of science. For, as Royce's discussion of Lakatos's views clearly brings out, the judgment that a theory is (or has been) powerful or more powerful than its competitors can often be made only with historical hindsight. At the time a theory is initially advanced, it may be altogether unclear (to many scientists) how much potential it does have or how much theoretical power is really there. A particular individual may believe it has great power and potential, but (as in other areas) potential requires developing. An individual may thus commit himself or herself to developing that potential and may turn the theory into a universally recognized powerful one. The point, however, is that the majority of scientists may not share the individual's initial assessment of its potential power, but with time the particular individual may prove to be right and the majority of scientists wrong.

It is unclear to me, therefore, whether Royce is claiming that we ought to produce more theories *tout court* (some of which may later prove to be powerful) or rather that we ought to produce more theories only if they are judged by most scientists at their inception to be more powerful than their competitors. As I have suggested, the latter view is too narrow and restrictive, but the former ("let a 1,000 theories bloom") seems to be at odds with Royce's attempt to reduce theoretical pluralism in psychology. If it were followed, there would be so many new theories that there would indeed be a staggering increase in the "conceptual confusion" faced by the average psychologist, the demands placed upon him or her, and the feeling that psychologists do not know very much.

The underlying issue here is whether there really is a problem when one has theoretical pluralism and if so, what the problem is. On the one hand, there is the standard Kuhnian answer to these questions, a view that Royce often seems to adopt. Psychology (Royce often seems to be saying) is immature (and in its preparadigm period) because of its "simultaneous theoretical pluralism." This is to be overcome, however, by producing ever more comprehensive, powerful theories, some of which will become dominant in the field. Simultaneous theoretical pluralism (which is bad), therefore, will give way (someday) to theoretical monism (which is good), which in turn will give way to successive theoretical pluralism (which is also good).

On the other hand, I also detect another, somewhat different motif running throughout the paper (although it tends to be muted). In this view, simultaneous theoretical pluralism is both unavoidable (necessary) and valuable. Popper, Lakatos, and Feyerabend (*contra* Kuhn) are usually taken to be adopting this point of view and hence encouraging perpetual theoretical pluralism, whereas the standard (philosophical) interpretation of Kuhn (1962) is that simultaneous theoretical pluralism is only characteristic of non-normal science. Perpetual theoretical pluralism is necessary (so it is claimed) in order to avoid dogmatism and to foster the critical evaluation of currently accepted theory. This in turn arises from the fact that truly critical evaluation of another theory (or paradigm) can be made only from a theoretical perspective and not on the basis of pure data alone.

Is theoretical pluralism necessary, therefore, and is it inherently valuable (or is it merely an instrumental means to employ toward the goal of theoretical monism)? It is unclear how Royce would answer these questions. In another paper (1976), for example, he argues that contemporary psychology is multitheoretic, multiparadigmatic, and so on. But it is unclear whether he is really claiming that "psychology is *necessarily* pluralistic" (1976, p. 2) and that such theoretical pluralism is inevitable (1976, p. 37), or just that current psychology *is* pluralistic but that that situation can (and should) change in the future toward a theoretical monism. I think the same ambiguity is present in the current presentation and reflects an uncertainty on Royce's part about the standard Kuhnian interpretation. Such a problem of reconciling theoretical pluralism with "the paradigmatic maturity of a science" is explicitly recognized by Royce (1976, p. 38), but he devotes very little attention to it there—merely a footnote—and none in the present discussion.

For my part, I would suggest that the standard Kuhnian interpretation be seriously reevaluated by psychologists. Kuhn's (1962) work, clearly among the two or three most significant works in twentieth-century philosophy of science, has had some very unfortunate effects in psychology and contains some seriously misleading notions. Several of these can be found in Royce's discussion.

The notorious thesis of incommensurability is one such questionable idea. Royce employs such a notion in his paper, but I think it is not clear what precise sense he is giving it. He says, for example, that two paradigms are incommensurable with respect to their views of reality and that in order to understand what such a paradigm has to say we must make a gestalt switch from one to another. On the other hand, two paradigms are not incommensurable with respect to their criteria of evaluation, which is standard. This seems to me to be at odds with what Royce says earlier, as well as with what Kuhn and Feyerabend originally meant by "incommensurable." Royce opts for a weaker version of incommensurability—one that Kuhn now (1977) also adopts—namely, that "there is no way to translate fully between theories that are generated from different paradigms." This takes the sting out of incommensurability but it also removes the necessity of a gestalt switch! This, in my view, is a change for the better, since many psychologists have appealed to Kuhn's notion of incommensurability to support their dubious claim that since different paradigms are incommensurable there is no objective way to evaluate their respective merits. Relativism, personal choice, and existential commitment is thus the only solution. However, if one abandons radical incommensurability for a weaker version (as Kuhn and Royce do) rational choice is once more possible.

Even if the notion of incommensurability were freed of its problematic features, there are other disturbing aspects of the Kuhnian model underlying much of Royce's paper. The image of progress and growth we are given is one in terms of which different areas of psychology are synthesized and incorporated by more and more comprehensive theories. But is there a limit to such theoretical synthesis or unification, a limit determined by the heterogeneous nature of the subject matter of psychology? True, we must (as Royce says) "strive for as much theoretical unification as possible" (1976, p. 43), but the question is: How much is possible? Royce does not think that a single theory could cover "the range, complexity, and richness of psychological phenomena" (1976, p. 38) and (presumably) that this ambitious degree of theoretical unification is not possible. But a similar point could be made about each of the subdomains of psychology. Hence a thoroughgoing conceptual pluralism would seem to be unavoidable, a conceptual pluralism which is (contra Royce) incompatible with theoretical unification. In this context, the suggestion to "strive for as much theoretical unification as possible" is an idle suggestion.

Of course, Royce might reply that it is an empirical (not an *a priori*) question as to whether we can produce theoretical unification in, say, social psychology and that "only time will tell." But then the same response could be made to Royce's suggestion that psychology as a whole cannot be expected to have a single paradigm covering all subdomains. In short, I do not think Royce has solved the problem of reconciling conceptual pluralism with theoretical unification.

I do not have any suggestions as to how one should solve this problem, but I do think that considerably more discussion must be given to (what one can call) the *scope* of a paradigm. In Kuhn's more recent views (1977), a mature science may have many simultaneous paradigms (Royce's "simultaneous theoretical pluralism"), a paradigm may have very few adherents, and even in the earliest stages of a science there are paradigms. If we are to take these views seriously, then much of the distinctiveness of Kuhn's views must be given up, including his notion of what makes a science mature. It is certainly possible, for example, that psychology can be a mature science with a great deal of theoretical (or paradigm) pluralism. This raises the question, however, of how much theoretical unity there need be in psychology as a whole or even in a single subdomain. (The question of whether this theoretical unity would be advisable is still another question.) Why could not social psychology be a mature subdomain of psychology even though it contained numerous "incommensurable paradigms"? The notion that theoretical monism is characteristic of normal or mature science (and hence necessary) appears no longer to be required, even for a Kuhnian approach.

Finally, let me say something briefly about Royce's method of theory analysis—constructive dialectics—and his criteria for theory evaluation. To begin with, it is unclear what Royce means by *dialectics*. It is said to involve a comparison of theories in which we "maintain the tension between viable alternatives" and this characterization is repeated elsewhere (1976, 1977). But this itself is not much clearer, nor is it obvious why this is called dialectics. Dialectics is a concept with a very rich (and checkered) history, and there are several distinct senses or kinds of dialectics (see, for example, Röd, 1974, and Diemer, 1976). Extremely general characterizations such as: it is the attempt to sort out the complementary roles of different theories (Royce, 1977, p. 16) are simply not very helpful as a characterization of dialectics. Of course, at one level this characterization is no doubt true, but one cannot rest content with such a general description, especially if one believes that dialectics provides an indigenous philosophy of psychology (Royce, 1977), one which can solve the problem of theoretical pluralism in psychology. At the very least, Royce owes us a more detailed and fully articulated explication of his constructive dialectics.

The motive behind Royce's dialectics is, I think, clear: it is to solve the problem of theoretical pluralism by showing how one can integrate different, contradictory theories into a larger whole and thus to unify psychology. There is, however, a danger here. Some psychologists use dialectics as a theoretical justification for a kind of global eclecticism in which "everyone is right" in what they are saying. Although the notion of dialectics contains the idea of retaining notions that are contradictory or at least contrary (e.g., behaviorism and phenomenology), it does not license retaining everything, only those contradictory aspects which are *valid*. But this, in turn, entails discerning which parts are valid and rejecting the remainder; the possibility always remains open, of course, that what is rejected is precisely what the proponent of that point of view finds most essential and characteristic. Dialectics should not be confused, therefore, with an uncritical global eclecticism. But if so, then Royce (and other dialecticians) must elucidate what the method of dialectics is, including what they mean by "contradictory" and "complementary" alternatives, how such contradictions are "dialectically resolved" or unified, how a dialectical resolution by means of *Aufhebungen* is distinctive or even possible, how it differs from other methods of resolving contradictions, and so on.

In terms of Royce's criteria of theory appraisal—empirical testability, degree of empirical-formal fit, comprehensiveness, parsimony, depth of penetration, degree of formalization, degree of cohesiveness, explicitness of conceptualization (see also Royce, 1978)—the problem is different. The nature of these criteria is not discussed, nor are we told how they differ from each other. The nature of simplicity, for example, is a crucial but controversial question. An even more pressing issue, however, is the following: How do we go about weighing and evaluating the relative importance of these particular criteria, especially when they conflict? What are we to do, for example, when one theory (e.g., Skinner's operant model) is very parsimonious but not comprehensive in scope and another theory (e.g., Chomsky's transformational grammar) scores high on degree of formalization but low on degree of empirical fit? Which of these four criteria are more important (in relation to the others), how do we determine a theory's overall score on each of these criteria and hence its theoretical power, and which theory, therefore, should we adopt? Here we are talking about a higher- (second-)order (meta) criterion for evaluating our lower-order criteria. Such a (meta) criterion (most likely) will not be an algorithm but will involve the good judgment of the individual scientist and his or her scientific common sense. What remains currently problematic about such a view, however, is how one can defend the claim that such a judgment is both rational (objective) and nonalgorithmic. Any adequate account of theory analysis and evaluation must surely address these questions.

In my very brief comments, I have only had time to address three issues that Royce's very stimulating paper raises and to suggest some possible avenues for future development. These issues appear to me to be central, however, to the major point of Royce's paper, which was to solve the problem of theoretical pluralism in psychology. It will not come as a surprise to hear that the problem is not yet adequately solved.

1. References

Diemer, A. (1976). *Elementarkurs Philosophie: Dialektik.* Düsseldorf/Wien: Econ Verlag.

Kuhn, T. (1962). *The structure of scientific revolutions.* Chicago: University of Chicago Press.

Kuhn, T. (1977). *The essential tension: Selected studies in scientific tradition and change.* Chicago: University of Chicago Press.

Röd, W. (1974). *Dialektische Philosophie der Neuzeit* (2 vols.). Munich: C. H. Beck.

Royce, J. R. (1976). Psychology is Multi: methodological, variate, epistemic, world view, systemic, paradigmatic, theoretic, and disciplinary. In W. J. Arnold (Ed.), *Nebraska Symposium on Motivation 1975: Conceptual foundations of psychology* (pp. 1–63). Lincoln: University of Nebraska Press.

Royce, J. R. (1977). Toward an indigenous philosophy of psychology. *The Ontario Psychologist, 9,* 16–32.

Royce, J. R. (1978). How we can best advance the construction of theory in psychology. *Canadian Psychological Review, 19,* 259–276.

6

The Problem of Theoretical Pluralism in Psychology

Reply to Commentators

Joseph R. Royce

In one way or another the questions raised by the commentators are all relevant to the issue of scientific progress. I will, therefore, begin with this more general issue as the context for dealing with the specific points raised by Hyland, Kitchener, and Lindholm.

1. On Scientific Progress

Although it would not be feasible to present a theory of scientific progress in this rebuttal, I will allude to three characteristics or assumptions that would be incorporated in such a theory. These characteristics are that: (1) all theory is tentative and transient; that is, no theory is regarded as permanent and ultimate; (2) regardless of degrees of theoretical power, science adopts the best available theory as the current explanation for the observables of a given domain; and (3) science advances by developing successively better theories.

Given these three points, my answer to Kitchener's query about what constraints should accompany the production of additional theories is that any successor theory $t + 1$ should be judged an improvement over any predecessor theory t.

Initially such judgments are made by the author(s) of the theory and the reviewers and editors of journals. Subsequently, such judgments are made by critics and a widening circle of the scientific community.

Joseph R. Royce • Center for Advanced Study in Theoretical Psychology, University of Alberta, Edmonton, Alberta T6G 2E9, Canada.

And, of course, since judgments are fallible, they may be wrong. But this is precisely why the dialectic process is so critical. That is, as a result of critical analysis we must inform ourselves of the conceptual characteristics of all contending theories. Furthermore, the analysis envisioned includes comparing extant theories of a given domain in detail. The implementation of such a program is, of course, a complex and difficult enterprise which is as demanding as the experimental aspect of the scientific enterprise.

In fact, I see both the data-gathering and the theorizing halves of the scientific coin as endless enterprises. That is, it is impossible to achieve scientific perfection. It is now well established, for example, that every experiment raises more questions than it answers. Similarly, we can anticipate that increments in conceptual clarity will also bring new issues to light.

In short, penetrating analysis and extended revision are needed in order to evaluate and develop the full potential of a given theory. And the task is so complex and demanding that attempts to determine an appropriate period of time for its completion are meaningless.

All of these issues highlight what is probably the most critical point raised by all three commentators—the importance of theory appraisal. I am, of course, in complete agreement with Kitchener when he says that this issue requires further elucidation. My goal in this chapter was to initiate discussion, not to close it off. Referring to this as the "tolerance" issue, Hyland appears to be in agreement with my stance on weak theory, as when he says, "One should tolerate a theory even if it is not very plausible because it may turn out that the theory, or perhaps a reformulated version, can still play a major role in the development of the discipline. Even when there is evidence against a theory, it should not be completely discarded but should be placed on 'hold' " (p. 318). But Lindholm appears to differ at this juncture when he says that he would not wish to retain "weak" theory. He wisely recommends wisdom in this situation. The bit of wisdom I recommended is that we eliminate only those theories "which theory appraisal overwhelmingly demonstrates to have little or no potential for improvement"(p. 313). The primary point is, of course, to provide fledgling theory with opportunities to manifest the directions in which it can be improved and eventually take on more theoretical power. The underlying wisdom is to nurture the potential value of any extant theory—further to develop those insights which brought the theory into prominence in the first place and to delete or modify those aspects which led to its subsequent weakness. Such insights require protection despite their inadequacies because they are so rare. Since Lindholm has not proposed alternatives, I propose to stand by the policy of tolerance until a wiser policy becomes available.

2. Theoretical Pluralism and Complementarity

Since Hyland's question about the nature of complementarity and Kitchener's guess about the desirability or necessity of theoretical pluralism are closely interrelated, I have combined them under one heading. Since advancement in science requires the production of better theory, it follows that some form of pluralism is both necessary and desirable. And, as I indicated earlier, pluralism can occur in both competitive and complementary forms. But Radnitzky (1970) argues that the polarity of theories is transformed into complementarity as a science matures. For example, he says that matter as both corpuscular and wave-like appears as a polarity when viewed within the framework of the predecessor theory of classical physics but that they are seen as complementary when placed in the context of the successor theory, quantum physics. Furthermore, in the case of highly heterogeneous disciplines such as the behavioral and social sciences, it is highly probable that several theories will be required in order to provide adequate coverage of the full range of phenomena to be explained. However, I have argued that complementary and sequential pluralism are preferable to contradictory and simultaneous pluralism. The point is that there is a tension in science between theoretical pluralism and theoretical monism, and I am of the opinion that this tension must be maintained at a high pitch and on a continuing basis because both are required for optimal progress. I have already stated the case for pluralism, and the case for monism lies in the desirability of developing theory that is both comprehensive and unifying. Thus, although the production of many theories is necessary, it is also desirable to identify a small number of viable theories that have the potential to evolve into highly explanatory theories of such broad scope and depth that they are capable of jointly unifying the entire discipline (i.e., they are complementary theories). Although there is little disagreement regarding the desirability of this goal, there is considerable difference of opinion on how to achieve it. And I have opted for the strategy of simultaneous theoretical pluralism followed by sequential pluralism where the preferred pluralism involves a relatively small number of complementary theories that span the entire discipline.

3. Other Issues

Lindholm has also raised a tangential point which is related to the tolerance issue—I am referring to his claim that I started out by assuming a correspondence theory of truth and ended with a pragmatic position. I have referred to this as a tangential point because I did not address this question directly in this paper. However, I have elaborated on it elsewhere (Royce, 1974, 1976, 1977; Royce & Mos, 1980; Royce, Coward, Egan, Kessel, & Mos, 1978) and particularly in a paper on the scientific world view (Royce, 1978). In these papers I describe a psycho-philosophical theory of knowledge which involves three ways of cognizing along with the relevant epistemic criteria. Very briefly, these are rationalism (involving the cognitive process of conceptualizing and the truth criterion of logical consistency), empiricism (involving the cognitive process of perceiving and the truth criterion of observable repeatability), and metaphorism (involving the cognitive process of symbolizing and the truth criterion of universality). In terms of these three isms, correspondence theory involves the mutual meshing of rationalism and empiricism, and coherence theory involves rationalism. However, my view, which is not subsumed by any existing standard truth claims, involves all three ways of knowing. And, finally, I wish to deny explicitly that I adopt a position of pragmatic truth.

While I do not deny the pragmatic aspect of my final position on theoretical pluralism, the nature of my pragmatism requires clarification. My pragmatism is not based on a theory of truth; rather, it constitutes an answer to an existential dilemma—namely, that some kind of action is required regardless of the adequacy of knowledge claims.

In this connection the reader is reminded of the ambiguity which inevitably accompanies the truth claims of any scientific theory, including strong theory. Bunge (1967) presents an insightful exposition of the claim that any one theory, since it is a simplified idealization, is necessarily a limited and inadequate statement about reality. And Weimer (1979) argues that, in fact, all theories are false. Thus, in this view the task at hand is not one of identifying which theory is true, but rather one of identifying and adopting theories that are the least false.

The view that all theories are false is particularly applicable to weak theory because of both empirical and rational inadequacies. That is, weak theory suffers from a poor data base and the lack of a relevant formal rational structure and is dominated by analogical metatheory. Thus its explanatory capability is minimal. What, then, is the role of weak theory? It is the pragmatic one of heuristics—of trying to identify promising leads, potentially unifying theoretical concepts, and eventually evolving more powerful theory. There is simply an insufficient epistemic basis for regarding such theory as true. The more justifiable concern is whether it carries sufficient weight to warrant further research. Thus, the major thrust of weak theory is to move beyond pure metaphor or analogy to increasing degrees of empirical and rational potency in the hope that the theory will eventually become more explanatory.

4. References

Bunge M. (1967). *Scientific research* (2 vols.) New York: Springer-Verlag.

Radnitzky, G. (1970). *Contemporary schools of metascience.* Göteborg: Akademieforlaget.

Royce, J. R. (1974). Cognition and knowledge: Psychological epistemology. In E. C. Carterette & M. P. Friedman (Eds.), *Handbook of perception. Vol. 1: Historical and philosophical roots to perception* (pp. 1–63). New York: Academic Press.

Royce, J. R. (1976). Psychology is multi: Methodological, variate, epistemic, world-view, systemic, paradigmatic, theoretic, and disciplinary. In W. J. Arnold (Ed.), *Nebraska Symposium on the conceptual foundations of theory and methods in psychology* (pp. 149–176). Lincoln: University of Nebraska Press.

Royce, J. R. (1977). Toward an indigenous philosophy for psychology. *The Ontario Psychologist, 9,* 16–32.

Royce, J. R. (1978). Three ways of knowing and the scientific world-view. *Methodology and Science, 11,* 146–164.

Royce, J. R., Coward, H., Egan, E., Kessel, F., & Mos, L. P. (1978). Psychological epistemology: A critical review of the empirical literature and the theoretical issues. *Genetic Psychology Monographs, 97,* 265–353.

Royce, J. R., & Mos, L. P. (1980). Psycho-Epistemological Profile: Manual. Edmonton, Alberta: University of Alberta Printing.

Weimer, W. B. (1979). *Notes on the methodology of scientific research.* Hillsdale, NJ: Lawrence Erlbaum.

Author Index

Italic numbers indicate pages where complete reference citations are given.

Subject Index

MIX
Papier aus verantwortungsvollen Quellen
Paper from responsible sources
FSC® C105338
FSC
www.fsc.org

If you have any concerns about our products,
you can contact us on
ProductSafety@springernature.com

In case Publisher is established outside the EU,
the EU authorized representative is:
Springer Nature Customer Service Center GmbH
Europaplatz 3, 69115 Heidelberg, Germany

Printed by Libri Plureos GmbH
in Hamburg, Germany